I Wish I'd Made You Angry Earlier

Essays on **SCIENCE, SCIENTISTS,** *and* **HUMANITY**

EXPANDED EDITION

I Wish I'd Made You Angry Earlier

Essays on **SCIENCE, SCIENTISTS,** *and* **HUMANITY**

EXPANDED EDITION

with nine new essays by the author
and an Appreciation by John Meurig Thomas

Max F. Perutz

Formerly Chairman and Member of the MRC
Laboratory of Molecular Biology
Cambridge, England

COLD SPRING HARBOR LABORATORY PRESS
Cold Spring Harbor, New York

I Wish I'd Made You Angry Earlier

Essays on Science, Scientists, and Humanity

EXPANDED EDITION

Published by Cold Spring Harbor Laboratory Press
Printed in the United States of America

Publisher and Developmental Editor	John Inglis
Project Coordinators	Elizabeth Powers Nora Rice
Production Manager	Denise Weiss
Production Editor	Patricia Barker
Desktop Editor	Danny de Bruin
Cover Designer	Ed Atkeson

Front Cover: Photo courtesy of the MRC Laboratory of Molecular Biology

Library of Congress Cataloging-in-Publication Data

Perutz, Max F.
I wish I'd made you angry earlier : essays on science, scientists, and humanity / Max F. Perutz.-- [Expanded ed.].
p. cm.
Includes bibliographical references and index.
ISBN 978-087969674-0 (alk. paper)
1. Science--Philosophy. 2. Scientists--Social aspects. 3. Creative ability in science. I. Title: I wish I'd made you angry earlier. II. Title.

Q175 .P386 2002
500--dc21

2002074056

All Cold Spring Harbor Laboratory Press publications may be ordered directly from Cold Spring Harbor Laboratory Press, 500 Sunnyside Boulevard, Woodbury, New York 11797-2924. Phone: 1-800-843-4388 in Continental U.S. and Canada. All other locations: (516) 422-4100. FAX: (516) 422-4097. E-mail: cshpress@cshl.edu. For a complete catalog of all Cold Spring Harbor Laboratory Press publications, visit our World Wide Web Site http://www.cshlpress.com/.

To my children Vivien and Robin,
for their uninhibited criticism of my manuscripts
and to Robert Silvers, Editor of the *New York Review of Books,*
for his encouragement of my writing

Contents

Preface

Every now and then I receive visits from earnest men and women armed with questionnaires and tape recorders who want to find out what made the Laboratory of Molecular Biology in Cambridge (where I work) so remarkably creative. They come from the social sciences and seek their Holy Grail in interdisciplinary organisation. I feel tempted to draw their attention to 15th century Florence with a population of less than 50,000, from which emerged Brunelleschi, Donatello, Ghiberti, Masaccio, Botticelli, Leonardo and Michelanglo, and other great artists. Had my questioners investigated whether the rulers of Florence had created an interdisciplinary organisation of painters, sculptors, architects, and poets to bring to life this flowering of great art? Or had they found out how the 19th century municipality of Paris had planned Impressionism, so as to produce Manet, Degas, Monet, Pissarro, Renoir, Cézanne and Seurat? My questions are not as absurd as they seem, because creativity in science, as in the arts, cannot be organised. It arises spontaneously from individual talent. Well-run laboratories can foster it, but hierarchical organisation, inflexible, bureaucratic rules, and mountains of futile paperwork can kill it. Discoveries cannot be planned; they pop up, like Puck, in unexpected corners.

In the past, most scientists were poorly paid; only few became famous and even fewer rich. One of the characters in Fred Hoyle's novel *The Black Cloud* remarks that scientists are always wrong, yet they always go on. What makes them continue? Often it is addiction to puzzle-solving and ambition to be recognised by their peers.

Science has changed the world, but the scientists who changed it rarely foresaw the revolutions to which their research would lead. Oswald Avery never set out to discover what genes are made of; Hahn and Meitner never intended to split the uranium nucleus; Watson and Crick were taken by surprise when their atomic model of DNA told them how the genetic information replicates itself; and when Jean Weigle and Werner Arber wondered why a bacterial virus infected one strain of coli bacteria and not another, they could not foresee that some 40 years on, their enquiry would lead to the cloning of a sheep named Dolly. Like children out on a treasure hunt, scientists don't know what they will find.

According to Paul Ehrlich, the father of immunology, scientists need the four Gs: Geschick, Geduld, Geld, und Glück (skill, patience, money, and luck). Patience may or may not reap its own reward. The astronomer Fritz Zwicky built a new kind of 18-inch telescope at Mount Palomar in California in order to obtain images over a wide field of the sky. He wanted to scan these images for exploding stars, supernovae which flare up suddenly and can be brighter than a million suns. Between September 1936 and May 1937 Zwicky took 300 photographs in which he scanned between 5000 and 10,000 nebular images for new stars. This led him to the discovery of one supernova, revealing the final dramatic moment in the death of a star. Zwicky could say, like Ferdinand in *The Tempest* when he had to hew wood:

> For some sports are painful and the labour
> Delight in them sets off; some kinds of baseness
> Are nobly undergone, and most poor matters
> Point to rich ends. This my mean task
> Would be as heavy to me as odious; but
> The mistress which I serve quickens what's dead
> And makes my labours pleasures.

The heavens were Zwicky's mistress, and mine was hemoglobin, the protein of the red blood cells. As part of my attempt to solve its structure, I took several hundred X-ray diffraction pictures of hemoglobin crystals, each taking two hours' exposure. I took some of the pictures during World War II, when I had to spend nights in the laboratory in order to extinguish incendiary bombs in the event of a German air raid. I used these nights to get up every two hours, turn my crystal by a few degrees, develop the exposed films and insert a new pack of films into the cassette. When all the photographs had been taken, the real labour only began. Each of them contained several hundred little black spots whose degree of blackness I had to measure by eye, one by one. After six years of this labour, when the data were finally complete, a London firm processed them with a prehistoric, mechanical punched card computer that produced an output of thousands of numbers. These numbers outlined not a picture of the structure I was trying to solve, but a mathematical abstraction of it: the directions and lengths of all the 25 million lines between the 5000 atoms in the hemoglobin molecule radiating from a common origin. I scanned the maps eagerly for interpretable features and was elated when they seemed to tell me that the molecule consists simply of bundles of parallel chains of atoms spaced apart at equal intervals.

Shortly after my results appeared in print, a new graduate student joined me. As his first job, he performed a calculation which proved that no more than a small fraction of the hemoglobin molecule was made up of the bundles of parallel chains that I had persuaded myself to see, and that my results, the fruits of years of tedious labour, provided no other clue to its structure. It was a heartbreaking instance of patience wasted, an ever-present risk in scientific research. That graduate student was Francis Crick, later famous for his part in the solution of the structure of DNA.

The essays grouped under the headings "Ploughshares into Swords", "How to Make Discoveries", and "Rights and Wrongs" were written for the New York and London Review of Books and other journals presupposing no scientific knowledge. The first five essays grouped under the heading "More About Discoveries" were addressed to scientists, and lay people may find them harder to follow. I was led to write the one non-scientific article "By What Right Do We Invoke Human Rights?" when I found myself having to make a speech.

Science is no quiet life. This book includes detective stories, tales of conflict and battle, a woman's love affair with crystals, a man's gruesome fascination with poison gas, cancer cures as Nobel Laureates' geriatric illusions, an onslaught on social relativists, a war hero's anticlimactic homecoming that led to a Nobel Prize, phantom perils threatening to poison us, and real perils conquered by silent heroes. Peter Medawar preached that "science at all levels of endeavour is a passionate enterprise and the pursuit of natural knowledge a sortie into the unknown." If my book convinces readers of the truth of his dictum, it will have served its purpose.

When I read books, I jot down any wise sayings which appeal to me. I keep them in my Commonplace Book, a name that goes back to antiquity when Greek and Roman orators collected metaphors to be used for speeches in public places. In the seventeenth century Milton kept a Commonplace Book from his school days in a search for truths, moral, political, and economic, with which he might serve England, mankind, and God. I doubt that I collected my quotations with such a lofty purpose in mind, but many of them have become my guiding mottos. I decided to append them to the collection of my essays in the hope that some of them might also appeal to others.

The great accompanist Gerald Moore entitled his autobiography "Am I too Loud?" After rereading some of my essays, I wondered if I should call this book "Am I Too Long?" because I may be like the father who takes his small son to the zoo. When the boy asks why the giraffe has such a long neck, the father rabbits on until he notices that the boy isn't listening anymore. So the father says: "Boy, you are not attending." And the boy responds: "No father, I don't want to know all that." If you, the reader, don't want to know all that, skip to the next essay!

Acknowledgments

I would like to thank the following publications for granting permission to reproduce articles that were originally published in slightly different forms:

"Friend or Foe of Mankind?" was originally published under the title "The Cabinet of Dr. Haber" (June 20, 1996. *The New York Review of Books* [copyright Nyrev, Inc.]).

"Splitting the Atom" was originally published under the title "A Passion for Science" (February 20, 1997. *The New York Review of Books* [copyright Nyrev, Inc.]).

"The Man Who Patented the Bomb" was originally published under the title "An Intellectual Bumblebee" (October 7, 1993. *The New York Review of Books* [copyright Nyrev, Inc.]).

"Why Did the Germans Not Make the Bomb?" was originally published under the title "War on Heisenberg" (November 18, 1993. *London Review of Books*).

"Bomb Designer Turned Dissident" was originally published under the title "Patriotic Work" (September 27, 1990. *London Review of Books*).

"Liberating France" was originally published under the title "Portrait of the Scientist as a Young Man" (May 12, 1988. *The New York Review of Books* [copyright Nyrev, Inc.]).

"The Threat of Biological Weapons" (April 13, 2000. *The New York Review of Books* [copyright Nyrev, Inc.]).

"High On Science" (August 16, 1990. *The New York Review of Books* [copyright Nyrev, Inc.].)

"Deconstructing Pasteur" was originally published under the title "The Pioneer Defended" (December 21, 1995. *The New York Review of Books* [copyright Nyrev, Inc.]).

"The Battle Over Vitamin C" and "The Hormone That Makes Nerves Grow" were originally published under the title "Two Roads to Stockholm" (October 13, 1988. *The New York Review of Books* [copyright Nyrev, Inc.]).

"A Mystery of the Tropics" (January 16, 1992. *The New York Review of Books* [copyright Nyrev, Inc.]).

"The Forgotten Plague" was originally published under the title "The White Plague" (May 26, 1994. *The New York Review of Books* [copyright Nyrev, Inc.]).

"What Holds Molecules Together?" was originally published under the title "Linus Pauling 1901–1994. Obituary" (1994. *Nature Structural Biology* **1:** 667–671).

"I Wish I'd Made You Angry Earlier" (February 23, 1987. *The Scientist*).

"Big Fleas Have Little Fleas..." was originally published under the title "Founder of Phage Genetics" (1968. *Nature* **320:** 639–640 [copyright MacMillan Magazines Limited]).

"How the Secret of Life Was Discovered" was originally published as "Co-chairman's Remarks: Before the Double Helix" (1993. *Gene* **135:** 9–13 [copyright Elsevier Science]).

"Dangerous Misprints" (September 26, 1991. *London Review of Books*).

"A Deadly Inheritance" was originally published under the title "A Deadly Inheritance and a Will of Iron" (October 6, 1995. *The Times Higher Education Supplement*).

"Darwin Was Right" was originally published under the title "A New View of Darwinism" (October 2, 1986. *New Scientist* 36–38).

"A Passion for Crystals" was originally published under the title "Professor Dorothy Hodgkin" (August 1, 1994. *The Independent*, Obituaries).

"The Top Designer" (April 22, 1999. *The New York Review of Books* [copyright Nyrev, Inc.]).

"The Great Sage" was originally published under the title "Diary" (July 6, 2000. *London Review of Books*).

"It Ain't Necessarily So" was originally published under the title "Are Fluorescent Monkeys Keys to Glowing Health?" (February 23, 2001. *The Times Higher Education Supplement*).

"By What Right Do We Invoke Human Rights?" (1996. *Proceedings of the American Philosophical Society*. **140:** 135–147).

"The Right to Choose" (October 8, 1992. *The New York Review of Books* [copyright Nyrev, Inc.]).

"Swords Into Ploughshares: Does Nuclear Energy Endanger Us?" was originally published under the title "Is Britain 'Befouled'?" (November 23, 1989. *The New York Review of Books* [copyright Nyrev, Inc.]).

"What If?" (March 8, 2001. *The New York Review of Books* [copyright Nyrev, Inc.]).

"The Second Secret of Life" was originally published under the title "Hemoglobin Structure and Respiratory Transport" (1978. *Scientific American* **239:** 92–125).

"How W.L. Bragg Invented X-ray Analysis" (February 2, 1990. *International Union of Crystallography*).

"Life's Energy Cycle" was originally published under the title "A Cycle Ride to Stockholm" (1982. *Nature* **296:** 512–514 [copyright MacMillan Magazines Limited]).

"How Nerves Conduct Electricity" was originally published under the title "Of Squids and Radar" (1992. *Nature* **320:** 639–640 [copyright MacMillan Magazines Limited]).

"In Pursuit of Peace and Protein" (January 29, 1999. *The Times Higher Education Supplement.*)

"Keilin and the Molteno" (October 1987. *The Cambridge Review*).

"Growing Up Among the Elements" (November 1, 2001. *The New York Review of Books* [copyright Nyrev, Inc.]).

"Friendly Way to Science" (July 31, 1998. *The Times Higher Education Supplement*).

"The Scientific and Humane Legacy of Max Perutz" Obituary by Sir John Meurig Thomas. 2002. *Angewandte Chemie International Edition in English* (in press).

The author gratefully acknowledges the faultless and efficient typing of many of these essays by Mrs. Mary-Ann Starkey.

Ploughshares into Swords

*Friend or Foe of Mankind?**

As far as science is concerned, there is no doubt whatsoever in my mind that to look upon it as a means of increasing one's power is a sin against the Holy Ghost.

KARL POPPER, *The Moral Responsibility of the Scientist*

FRITZ HABER: *It was never, ever my intention, to engineer more deaths by my invention.*
CLARA HABER: *Your process led to death and devastation.*
FRITZ HABER: *It saved the world that hurtled to starvation.*

These lines from Tony Harrison's play *Square Rounds* epitomize the ambiguous personality and career of Fritz Haber. He was a German chemist, born in 1868, famous for being the first scientist to have synthesized ammonia from the nitrogen in the air; this opened the way to the synthesis of the nitrogen fertilizers that have dramatically increased agricultural production throughout the world. He is also infamous for having introduced poison gas in the First World War.

Haber was larger than life in every sense. Photographs show him stiffly erect and formally dressed with a pince-nez and a starched collar turned down at the corners, lording it over his assembled laboratory staff—a *Geheimrat* par excellence. After April 1933, when the Nazis had forced him, a Jew by birth, from all official positions, Haber told a friend:

*A review of the books *Fritz Haber, Chemiker, Nobelpreisträger, Deutscher, Jude: Eine Biografie* by Dietrich Stoltzenberg (VCH, Weinheim and New York) and *Der Fall Clara Immerwahr: Leben für eine humane Wissenschaft* by Gerit von Leitner (C.H. Beck, Munich).

"I have been German to a degree which I feel fully only now." To Chaim Weizmann he described himself as one of the most powerful men in Germany:

> I was more than a great leader of armies, more than a captain of industry. I was the founder of great industries. My work opened the way to the great industrial and military expansion of Germany. All doors stood open to me.

As Dietrich Stolzenberg makes clear in his detailed biography, Haber had been devoted to the glory of Bismarck's German Reich and the German Emperor with an intensity hard for present generations to comprehend. He continued to visit the Emperor during his exile in Holland after Germany became a republic. He was a man of intellectual brilliance, with wide knowledge, overriding ambition, and a certain lack of humanity. His father, a respected businessman trading in dyes and pharmaceuticals, was more observant of the Prussian virtues of hard work, sense of duty, order, and discipline than of the Jewish rites. He compelled Fritz to enter his flourishing, carefully managed business, but when one of Fritz's impulsive transactions resulted in a severe loss, he allowed him to launch himself on what was then thought to be a badly paid academic career in chemistry. He did not foresee that one day guests invited to Fritz's Berlin residence would dine off gold plates.

Chemistry had fascinated Haber as a schoolboy. As was customary in Germany, he studied at a succession of universities, and finally landed at the Technical University in Karlsruhe. Knowing that academic careers were closed to non-Christians, he had himself baptized in the Lutheran faith.

At the turn of the century, it was hard for anyone without an independent income to follow a university career because assistant and associate professor positions were honorary posts rewarded only with the fees paid by the students they were able to attract to their lectures, and only full professors received an adequate salary. His poverty drove Haber to earn money from patents, and books, and to accept assignments from private industry. He worked furiously, determined to get to the top. When he failed to be appointed to a coveted chair in physical chemistry, chemistry's elder statesman Wilhelm Ostwald counseled him: "Achievements generated at a greater than the customary rate raise instinctive opposition amongst one's colleagues."

In 1901 Haber married Clara Immerwahr, a thirty-year-old woman, daughter of another respected Jewish family in Breslau, whom he had known as a teenager. As Gerit von Leitner's biography of Clara shows, she matched Fritz in ambition and determination, having fought against prejudice and opposition to become the first female Ph.D. in science at Breslau University. She was not pleased when shortly after the birth of their first son, Hermann, Haber left for a three-month tour of America.

In 1908, when he was only forty, Haber was appointed full professor of physical chemistry at Karlsruhe, where a contemporary described him as impulsive, temperamental, and quick-thinking, an excellent lecturer and engaging talker on virtually any subject. But in a letter to a friend quoted by von Leitner, Clara complained about his treatment of her:

> What Fritz has gained in these eight years, I have lost, and what is left of me fills me with profound dissatisfaction. I have always felt that it is only worth having lived if one has developed all one's faculties to the full, and has experienced everything that life can offer. That is what made me decide to get married, since otherwise one chord of my soul would lie fallow.
>
> If my elation was short-lived... that is due mainly to Fritz's overpowering way in his home and marriage, besides which anyone perishes who doesn't assert herself more ruthlessly than he... I ask myself whether superior intelligence makes one person more precious than another and if much of me that has gone to the devil, because it has gone to the wrong man, is not more valuable than the most important electronic theory.

If you want to make your name in science, try to accomplish something that has defeated everyone else. In 1784, the French chemist C.L. Berthelot discovered that ammonia consists of one atom of nitrogen and three atoms of hydrogen. For the next 125 years many chemists tried to make ammonia from these two gases and failed, largely because the laws governing chemical reactions were not fully understood. Haber, an excellent theoretician and talented experimenter, determined to solve the problem, at first without any thought of practical applications.[1] He and his young English collaborator, Robert Le Rossignol, made a careful study of the temperatures and pressures required to combine free nitrogen and hydrogen gas so as to produce more than tiny quantities of ammonia.

They found that the formation of ammonia required a pressure on the two gases of more than two hundred times that of the atmosphere at sea level and a temperature of 200°C (390°F)—extreme conditions never reproduced in a laboratory before. Even then the ammonia was made only very slowly. To hasten the reaction, a catalyst was needed, in this case a metal on whose surface hydrogen and nitrogen would combine faster.

Haber and Le Rossignol tried one possible metal after another until a powder of the rare metal osmium accelerated the reaction spectacularly. On July 2, 1909, they triumphantly demonstrated an experiment producing about seventy drops of ammonia a minute to the directors of the Badische Anilin und Soda Fabriken, then Germany's largest chemical firm.

At the time, saltpeter mines in Chile were the main sources of natural nitrogen fertilizer, but their output was limited and expected to be exhausted by about 1940. Some nitrate was also recovered from coal gas, but not nearly enough to satisfy demand. On the other hand, the nitrogen in the air was unlimited, hydrogen was abundant in coal gas, and their compound, ammonia, could be used as a fertilizer either by combining it with sulfuric acid or by oxidizing it so as to produce nitrates.

Convinced by the promise of Haber's demonstration, the directors of the Badische firm provided two of their ablest chemists, Carl Bosch and Alwin Mittasch, with unlimited time and resources to develop the process for industrial production. Badische Anilin took an option on the entire world stock of osmium (220 pounds), but Mittasch also performed over 10,000 tests of ammonia synthesis on 4,000 other catalysts. He finally selected a mixture of iron, which is abundant and cheap, with small amounts of the oxides of aluminum, calcium, and potassium. On September 9, 1913, the first industrial unit set up by Bosch and Mittasch started to produce between three and five tons of ammonia daily, a thousand times the output of Haber and Le Rossignol's original laboratory apparatus. Current world production of ammonia for fertilizer is about a hundred thousand times greater; all of it is still made with Mittasch's original iron catalyst, whose efficiency and durability has never been surpassed.

Haber was rewarded with generous royalties and the Nobel Prize for Chemistry in 1918; Carl Bosch received the prize in 1931 for his development of an entirely new technology for the production of ammonia under high pressure, despite an explosion of the ammonia works at Oppau on the Rhine on September 21, 1921, which killed 561 people and made 7,000 homeless (Stoltzenberg fails to mention that appalling disaster).[2] Unjustly, Mittasch was left out.

Having accomplished so much, Haber might have taken life easily, but that was not in his restless nature, and in any case he became embroiled in controversy over his discovery. His patents were immediately challenged by an Austrian firm which had suggested the possibility of synthesizing ammonia from its elements and had financed his first experiments. Other firms, which recognized the patents as a gold mine, also made claims against him. Caught up in these disputes, he no longer did original scientific work, but other opportunities now beckoned.

In 1910, the German Emperor founded the Kaiser Wilhem Gesellschaft zur Förderung der Wissenschaft, a semi-independent body for the support of research which was to prove of immense benefit to German science and learning. It was supported by Leopold Koppel, a respected Jewish banker, who also offered to pay for an Institute of Physical Chemistry in Berlin under Fritz Haber's direction. Haber said he would accept if he was also appointed to a chair at Berlin University, made a member of the Prussian Academy of Sciences, and given a salary of 15,000 marks a year (equivalent to about $85,000 today). These exacting demands were duly met and Haber accepted. Along with Max Planck and Walter Nernst, Berlin's leading physicists, he persuaded Albert Einstein to leave Zürich and move to Berlin, and he attracted many excellent young scientists to his new institute. The Kaiser Wilhelm Gesellschaft also built a second large institute for Germany's greatest chemist, Emil Fischer, who had received the Nobel Prize in 1902, partly for his work on the structure and synthesis of sugars.

In October 1912, Wilhelm II was personally to open both institutes. Two of the members, Otto Hahn and his associate Lise Meitner, later famous for their discovery of atomic fission, suggested to the Emperor's adjutant that they take him to their darkroom and show him the scintillations which alpha rays from radium produced on a fluorescent screen. The adjutant objected because Wilhelm would be frightened in the dark.[3] (Meanwhile the scientists' wives had hired a gym teacher to teach them how to curtsy to His Majesty.)

Nitrates form an essential part of explosives. When war broke out in August 1914, the British blockade cut Germany off from Chilean supplies of saltpeter, the traditional source of nitrates. The Germans captured 20,000 tons of saltpeter in Antwerp harbor after their invasion of Belgium, but had it not been for Haber's synthesis of ammonia, German nitrate supplies would have been exhausted and the Germans would have had to sue for peace. Haber volunteered for the army, in which he had served in his teens, but was rejected now on account of his age. Instead, he became chief of the chemistry section in the War Department for Raw Materials. In December 1914, he attended a test of artillery shells filled with tear gas, but he found the gas was too widely dispersed to have any effect.

According to his assistant, Fritz Epstein, Haber then suggested that the discharge of chlorine gas from cylinders would be more effective.[4] Chlorine is a greenish-yellow, blinding gas, heavier than air, which immediately produces violent coughing; corrodes the eyes, nose, mouth, throat, and lungs; and finally asphyxiates the person who inhales it. If blown by the wind toward the enemy lines, Haber proposed, it would sink into the trenches and drive the soldiers out into the open, where they could easily be killed. The idea appealed to the Chief of the General Staff, Erich von Falkenhayn. However, there was the awkward matter of the Hague Conventions of 1899 and 1907, which Germany had signed and ratified. According to the first convention: "The Contracting Powers agree to abstain from all projectiles whose sole object is the diffusion of asphyxiating or deleterious gases." The second convention also prohibited all use of poisons and poisoned weapons in war.

Falkenhayn saw a fine distinction between projectiles filled with noxious gases and gases being blown by the wind from cylinders on the ground, which the conventions had not foreseen. He put Haber in charge of a project to make such cylinders and promoted him from a noncommissioned officer in the reserve to the rank of captain. In *The Poisoned Cloud*, his scholarly book on the history of poison gas, Haber's son Ludwig writes: "In Haber the [High Command] found a brilliant mind and an extremely energetic organizer, determined, and possibly also unscrupulous."[5] Stoltzenberg confirms that Haber was without any doubt the initiator of chemical warfare.

Haber threw himself into the task. He worked himself to exhaustion organizing the manufacture of hundreds of tons of chlorine gas and thousands of gas cylinders; he trained special troops to test them; and oversaw their installation in the trenches at the front—regardless of danger to his own person.

He also recruited his own young collaborators and many other chemists for the task. When Otto Hahn objected that what he was doing was contrary to international law, Haber argued that in the autumn of 1914 the French had broken it first by firing rifle shells filled with tear gas. According to Stoltzenberg and Ludwig Haber this was untrue. Stoltzenberg writes: "When one reads the reports of Haber's activity and behavior at that time, one gains the impression that he was obsessed by his self-imposed tasks." His boundless ambition seems to have made him determined to win the war single-handed. He planned to have chlorine gas blown toward the Allied lines on a front of fifteen miles, which would either have killed the enemy soldiers or put them to flight. Massed Germany infantry were to follow and break through the Allied lines. He advised the High Command to use gas only if it could ensure victory, and he also urged that the German troops should be protected with a primitive kind of gas mask.

When the High Command asked the divisional commanders in charge of different sections of the front to cooperate with such an attack, all but Duke Albrecht of Württemberg refused. His troops were engaged in some of the fiercest fighting on the front at Ypres, twenty miles from the Belgian coast. There, Haber's special troops dug into the German trenches 5,730 cylinders capable of releasing 150 tons of chlorine gas along a four-mile-long front. It was to be blown toward the enemy when the wind came from the east, but this was known to happen on average on only one day out of three, and besides, the wind so near the sea was capricious. Ludwig Haber writes:

> Here was Haber himself, an academic in uniform, paunchy, rarely without a cigar, pockets bulging, surrounded by young acolytes who managed to look respectful, busy, and unconventional in dress and bearing.

All the wartime documents refer to him deferentially as "Geheimrat Haber." He prevailed upon Otto Hahn, who commanded a machine-gun section, to become a participating "observer," and the future Nobel Laureates in Physics James Franck and Gustav Hertz also joined him. But Max Born, another young physicist at Haber's Institute and a future Nobel Laureate, refused to take part. The chemist Hugo Stoltzenberg, the father of Haber's biographer, ran chlorine filling stations near the front. By April 11, 1915, the unwieldy cylinders, each weighing nearly 200 pounds, had been installed at night; but the masks for the German troops never arrived.

In his book dealing with his father's work on poison gas, Ludwig Haber writes:

> The first gas alert order was given on 14 April at 22.30 and cancelled at 01.45 on 15 April. The second on 19 April at 15.00, but was countermanded. By then the [High Command] had become cautious and owing to the Russian threat on the Austro-Hungarian front became reluctant to commit reserves earmarked for the east to something as uncertain as the follow-up to a gas attack. The third alert was given on 21 April at 17.00, postponed first to 04.00 on 22 April, then to 09.00 and then to the afternoon.

> The troops, the Pionierkommando, and the specialists had had very little rest and were on edge. They were sure the Allies had been alerted. They had indeed. Three weeks earlier the French, then still in the south of the Salient, were told by prisoners of the installation of cylinders, and there was visual evidence of gas cylinder explosions in March. But the French ignored these warnings...
>
> The simultaneous opening of almost 6,000 cylinders which released 150 tons of chlorine along 7,000 meters within about ten minutes was spectacular.... Within minutes the Franco-Algerian soldiers in the front and support lines were engulfed and choking. Those who were not suffocating from spasms broke and ran, but the gas followed. The front collapsed.
>
> The Germans advanced cautiously. They too were taken by surprise and followed the cloud, delayed not be resistance, but by patches of gas in low ground and ruins.... German hesitations and darkness saved the French by giving them time to regroup.... The Germans' elation at their initial success soon turned to disappointment when on 23 April their Divisions, upon being ordered to advance, met with increasingly stiff resistance.[5]

The gas attack caused 15,000 Allied casualties, 5,000 of them fatal, but Haber's great victory failed to materialize. He arrived home in Berlin a few days later, disappointed and exhausted. On the evening of May 1, the Habers had guests for dinner, and that same night, while Haber was asleep, Clara shot herself with his service pistol, waking up their fourteen-year-old son, who found her in a pool of blood in their garden. The next morning Haber dutifully left for the eastern front.

Gerit von Leitner, in her biography of Clara, and Tony Harrison, in his play about Haber's career, attribute her suicide to her disgust with Haber's activities, which included tests of chlorine and other poison gases on animals at Haber's Institute, next door to their official residence. Gerit von Leitner writes that Clara was heartbroken when the young chemist Otto Sackur, a friend of hers from student days at Breslau University, was killed in an explosion at the institute. She also describes a row in which Haber blamed Clara's gossip for the army's failure to break through.[6] Stoltzenberg could find no evidence that Clara's suicide was a protest against Haber's war work. But according to Kurt Mendelsohn's book *The World of Walter Nernst,* Clara had pleaded with Haber repeatedly not to work on chemical warfare.[7] von Leitner has found a recorded statement by James Franck saying that Haber's part in the gas war certainly influenced Clara's suicide, and that Haber blamed himself bitterly for it throughout his life.

Stoltzenberg writes that their marriage was happy at first, but that this changed after their son's birth, when Clara became increasingly concerned with domestic trivia, which irritated Haber. He also writes that Clara was hospitalized more than once for depression, a crucial point which von Leitner fails to mention. He cites Clara as one of those people "whose search for self-fulfillment makes them build a wall around themselves which becomes their self-imposed prison." Clara wrote some farewell letters which have not survived, and von Leitner suspects that Haber destroyed them.

From the front Haber wrote to a friend: "For a month I doubted that I would hold out, but now the war with its gruesome pictures and its continuous demands on all my strength has calmed me." He continued to devote himself to chemical warfare, and according to Stoltzenberg this satisfied him completely. Once Haber had unleashed chlorine gas, the Allies soon matched the German effort, and the prevailing west wind blew in their favor.

Haber's actions continued to contradict Montesquieu's belief that knowledge makes men gentle. Despite his complaints of his overwhelming responsibilities at the front, he found time to conceive strategies for research on armaments when the war was over. When some of the German military were making plans to annex all of Belgium and part of northern France, and for good measure to invade England to teach the English a lesson, Haber got his benefactor Koppel to propose to the war minister that he finance a Kaiser Wilhelm Foundation for War Technology. Haber also induced Carl Duisberg, the head of Bayer, to propose that a Kaiser Wilhelm Institute for Chemical Warfare be established with himself as director. Strangely, Haber made this proposal even though he apparently was convinced as early as 1916 that Germany would lose the war. The Emperor approved the Foundation on December 17, 1916, with Fritz Haber, Emil Fischer, Walter Nernst, and three less famous chemists on the governing board. The Kaiser Wilhelm Gesellschaft hesitated to collaborate with the Foundation at first, and one of its members objected that killing people was not the Gesellshaft's job. But in September 1918 its directors accepted, and the War Ministry assigned six million marks for the project.

After Germany's collapse two months later, Haber and Nernst were branded as war criminals by the Allies, who demanded their extradition. Haber fled to Switzerland, where he was given Swiss citizenship, a privilege reserved for the very rich. After a few months the Allies dropped their demand that he be extradited, and he returned to Germany to help with reconstruction and to continue the secret manufacture of poison gas in violation of the Treaty of Versailles.

The Spanish government sought German help in manufacturing and using chemical weapons for suppressing Abd el Krim's revolt in Morocco. The Soviet government entered into a clandestine agreement with the Germans to manufacture weapons, including poison gas, and the German War Ministry set up a secret chemical warfare factory near Wittenberg. Haber directed these enterprises through his wartime collaborator Dr. Hugo Stoltzenberg, whom Ludwig Haber describes as "a plausible rogue who in other circumstances might have been believed if he had claimed to grow mushrooms in the desert." In Spain, Stoltzenberg set up a poison-gas factory near Madrid and personally advised King Alfonso XIII and his dictatorial prime minister, Primo de Rivera, on the best gas tactics to be employed against the Moroccan rebels.

Hugo Stoltzenberg apparently negotiated contracts allowing him to set up some of the factories as his own private enterprises. In 1925 the German foreign minister, Gustav Stresemann, and his French counterpart, Aristid Briand, met at Locarno and agreed on a rapprochement which induced the German government to stop the secret manufacture of poison gas and to close down Stoltzenberg's factories. He went bankrupt and to his fury Haber refused to support his claims for compensation.

Chemical warfare had failed to break the stalemate on the western front, but it had succeeded on the southern front, where the Austrian and Italian armies faced each other near the present frontier between Italy and Slovenia. In my youth in Austria I learned of "our" great victory at Caporetto (now Cobarid in Slovenia) in October 1917, where the Austrian and German armies broke through the Italian lines and advanced seventy miles westward to the River Piave. (The Italian retreat forms the background to Ernest Hemingway's moving novel *A Farewell to Arms*.)

The books by Stoltzenberg and Ludwig Haber show that the Austrians owed their breakthrough to an attack on the unprotected Italians with a mixture of chlorine and phosgene gas that had been prepared by Otto Hahn and other co-workers of Haber. Otto Hahn's autobiography suggests that he regretted this later.[8] In September 1939, after attending a meeting at the German Army Ordnance Department where the possibility of exploiting his discovery of nuclear fission for an atomic bomb was discussed, Hahn declared: "If my work leads to a nuclear weapon, I will kill myself." He sounded desperate when he heard of Hiroshima during his internment at Farm Hall in England. Fortunately, he would have found it difficult there to carry out his threat, had he still wanted to do so.[9]

Emil Fischer killed himself in 1919 in despair over the loss of his son in the war and over Germany's defeat and its postwar chaos. It seems that Haber had no regrets. He justified his invention of chemical weapons by claiming that the French had used them first, which was untrue, and that these weapons were more humane than high explosives because most soldiers survived the chemical attacks. He did not mention that many of the survivors were broken in both body and spirit for the rest of their lives. Haber continued until 1933 to advise Germany's government on its secret production of chemical weapons, but his main energies were devoted to the rebuilding of his institute as a leading center of fundamental research, to the revival of German science, and to the restoration of contacts with scientists abroad. While Haber was overbearing and dictatorial at home, he was wise enough to give his young collaborators scientific freedom. After their seminars he would say apologetically that he may not have been able to follow all their arguments, and then summarized them more lucidly than they had themselves. The discussions were animated by the search for the scientific truth, regardless of one's rank or fame. Haber's institute again became an outstanding center of chemical research, and it still bears his name.

The Treaty of Versailles made Germany's huge reparations payable in pre-war gold marks, which crippled Germany's recovery. Haber had read that a ton of sea water contains between five- and ten-thousandths of a gram of gold, which meant that the oceans might contain as much as eight million tons of it. Once again Haber set out to save Germany single-handed. He decided to devise chemical methods to extract the gold, and to use it to pay Germany's reparations. In strictest secrecy, he raised money and recruited fourteen young collaborators. Disguised as members of the crew, he and his assistants took a German passenger liner to New York and then another to Rio. Some of the initial analyses they made on board confirmed the earlier high estimates, but their variability made Haber decide to ship all samples of sea water back for analysis at his Berlin laboratory. After careful analysis of about five thousand samples from sources throughout the world, Haber's assistant Johannes Jaenicke reported a mean gold content of no more than a thousandth of the original estimates. It was a shattering blow.

In 1917 Haber married Charlotte Nathan, an attractive, independent, enterprising woman twenty-one years his junior, who lacked his Prussian sense of duty and had a passion for travel. Marriage to a man who was absent from home on important business most of the time and returned exhausted satisfied her as little as it had the very different Clara, and the marriage broke up after ten years. Ludwig Haber, the author of *The Poisonous Cloud*, is a son of that marriage.

Early in 1933, a few weeks before the Nazis seized power, Haber wrote to a friend: "I fight with ebbing strength against my four enemies, insomnia, the economic claims of my divorced wife, lack of confidence in the future and awareness of the grave mistakes I have committed in my life." He did not specify the mistakes, but feelings of guilt toward Clara may have been among them.

In April 1933, the Nazis ordered that all Jewish civil servants be dismissed, including employees of the Kaiser Wilhelm Gesellschaft. Max Planck, its president, used his official courtesy call on the newly appointed chancellor to plead that Haber and other Jewish scientists be allowed to continue their work. Hitler retorted that he had nothing against the Jews, but that they were all Communists. When Planck remonstrated and pointed out that Germany would harm itself if it expelled all of its excellent Jewish scientists, Hitler slapped his knee, talked faster and faster, and whipped himself into such a rage than Planck had to leave the room.

Haber now devoted all of his remaining energy to securing work abroad for his Jewish staff. Einstein happened to be visiting the United States, where he stated publicly that he would not return to Germany because it no longer recognized "civil liberty, tolerance and equality of citizens before the law." The Nazi press responded with a flood of abuse and the commissioner in charge of the Prussian Academy of Sciences demanded that disciplinary action be taken. Planck believed that as a German, Einstein should have stood up for Germany abroad, whatever the faults of the new regime, and decided that Einstein had made his continued membership of the Academy impossible. When Planck put this view to the assembled members, Haber concurred and only the physicist and Nobel Laureate Max von Laue had the courage to object to the shameful decision. Einstein deeply resented it. When a friend asked him later if he could take greetings from him back to Germany, Einstein replied: "Only to Laue." "Really no one else?" "No, only to Laue."

Haber himself eventually fled to Cambridge in England, where the Professor of Chemistry, William Pope, his adversary in chemical warfare, received him with honors, but the laboratory technicians who had fought in the trenches shunned him. After a short stay, Haber traveled to Switzerland, where he died of a heart attack after his arrival in Basel in January 1934, aged only sixty-five.

Had he lived, he would have had to face the most gruesome of his mistakes, to which Tony Harrison alludes in *Square Rounds*, when Clara and a veiled chorus sing:

> He'll never live to see his fellow Germans use
> his form of killing on his fellow Jews.

In 1919, when Allied inspectors of his institute prevented further research on chemical warfare against human beings, Haber turned to chemical warfare against agricultural pests. He became National Commissioner for Pest Control and founded a new firm, the German Society for Pest Control. The firm developed a preparation combining hydrocyanic acid, which is highly toxic, with a sweet-smelling, volatile, nontoxic irritant; both were absorbed in a porous powder. Another firm, Tesch and Stabonov, undertook to spread the powder in insect-contaminated fields and buildings. When it was spread on an open field, the acid evaporated, killing the insect pests, and the irritant warned people to keep away. The preparation was called Zyklon B. In 1943 Dr. Peters, the director of the pest-control firm, received a secret order from an SS officer to deliver Zyklon B *without the irritant* to Auschwitz and Oranienburg. He was told that it would be used to kill criminals, incurables, and mentally deficient persons, and he was threatened with the death penalty if he broke the secret.[10] So the pesticide which began in Haber's institute ended up as an instrument of the Holocaust, in which some of Haber's own relations perished.

In 1946 Dr. Tesch, the sole owner of the firm Tesch and Stabonov, was convicted by a British Military Court of delivering Zyklon B to Auschwitz and hanged; ironically, Hugo Stoltzenberg, the secret poison-gas manufacturer of the interwar years, was appointed by the British as trustee of the firm. In 1949 a Frankfurt court sentenced Peters to five years in prison for complicity in manslaughter; later, a Wiesbaden court sentenced him to six years for complicity in murder. In 1955 he was acquitted for lack of proof that he had known what was going on at Auschwitz.

Stoltzenberg's excellent biography is written with scholarly detachment. He confesses that he found it hard to imagine himself in the mind and role of Haber and to understand his unquestioning nationalism and sense of patriotic duty. Stoltzenberg's documents show that Haber continued to be held in the highest esteem in Germany as a great patriot, scientist, and statesman, despite the widespread public disillusionment with World War I; he also captivated people by his liveliness, charm, Old-World courtesy, and quick repartee. By a terrible irony of fate, it was his apparently most beneficent invention, the synthesis of ammonia, which has also harmed the world immeasurably. Without it, Germany would have run out of explosives once its long-planned blitzkrieg against France failed.[11] The war would have come to an early end and millions of young men would not have been slaughtered. In these circumstances, Lenin might never have got to Russia, Hitler might not have come to power, the Holocaust might not have happened, and European civilization from Gibraltar to the Urals might have been spared.

Haber's synthesis of ammonia for fertilizer was an extremely important discovery, but, unlike relativity, it did not take a scientist of unique genius to conceive it; any number of talented chemists could, and no doubt would, have done the same work before very long.

*Splitting the Atom**

Lise Meitner's career as a scientist spanned most of the heroic age of atomic physics, from the discovery of radioactivity in 1896 to the discovery of atomic fission in 1938. Born in Vienna in 1878, she spent her working life in Berlin, where Einstein called her "our Marie Curie." Her life became tragic when the Nazis drove her from all she had lived for; and the dropping of the atomic bomb on Hiroshima made her realize that her passionate devotion to atomic physics had prepared the way for a weapon of unimagined destructiveness.

She came from a liberal Jewish family in Vienna—her father was a lawyer—and she grew up in what she herself described as a remarkably stimulating intellectual atmosphere. Toward the end of the last century and until the First World War, Vienna had one of the world's leading medical schools and a renowned university. It was also a lively center of literature, music, and the arts. To illustrate the city's intellectual ferment, Meitner's biographer Ruth Sime mentions Sigmund Freud, Viktor Adler, and Theodor Herzl, but neither Adler, a socialist leader, nor Herzl, a founder of Zionism, contributed much to the cultural life of Vienna. She fails to mention Gustav Mahler or Arthur Schnitzler, or Otto Wagner and Adolf Loos, the two great pioneers of modern architecture, or Josef Hoffmann and Koloman Moser, who founded the Wiener Werkstätte, the cradle of modern design. Sime deplores the views of Karl Lueger, Vienna's anti-Semitic

*A review of *Lise Meitner: A Life in Physics* by Ruth Lewin Sime (University of California Press).

mayor, but she fails to note that Emperor Franz Josef was a philo-Semite who appointed Mahler director of the Vienna Opera at the age of only 37 and elevated many prominent Jews or men of Jewish descent to the nobility, among them the fathers of the philosopher Ludwig Wittgenstein and of the poet Hugo von Hofmannsthal.

Lise Meitner was determined from an early age to become educated like the men around her, but higher education was barred to girls, whose public schooling ended at the age of fourteen. Undeterred, she found private tutors to help her pass the entry exams for the University of Vienna, where she began the study of mathematics and physics. She had the good luck to be taught by Ludwig Boltzmann, one of the greatest physicists of all time.

At the beginning of the century the study of radioactivity was the most exciting subject in physics. Ten years after its discovery, when Meitner started her research, it was known that radium emitted three different kinds of radiation: alpha rays, which were positively charged helium nuclei shot out of the nuclei of radium atoms at a speed of more than 9000 miles a second; beta rays, which were negatively charged electrons; and gamma rays, which were electromagnetic waves like X rays, only more penetrating. Meitner began research on alpha rays in Vienna, but after Boltzmann had killed himself in a fit of depression she decided to continue her studies in Berlin. She intended to stay for a few semesters, and remained for thirty-one years.

At first she just attended lectures, but in 1907 she met the young chemist Otto Hahn, and they decided to study radioactivity together. They asked the famous chemist Emil Fischer for space in his laboratory, but he would tolerate no women there, and reluctantly allowed Meitner to install herself in the wood workshop in the basement provided she did not set foot anywhere else. To go to the toilet, she had to make her way to the nearest café.

She lived frugally with support from her father, but for her, as for Marie Curie, science was a vocation for which she was prepared to suffer penury. On return from a summer vacation in 1908, Meitner had herself christened into the Lutheran faith, inspired, apparently, by the example of the great physicist Max Planck, who personified the German Protestant ideal of "excellent, reliable, incorruptible, idealistic and generous men, devoted to the service of Church and State"; for Church and State she would have substituted Science.

Hahn and Meitner were the same age. They not only shared an interest in radioactivity, but their skills complemented each other's. Hahn was an accomplished chemist, but lacked knowledge of physics and mathematics, while Meitner was a physicist inexperienced in chemistry. They soon made their names with the discovery of two new radioactive elements and of two different mechanisms leading to the emission of beta rays.[1] In their relations with each other, Meitner and Hahn never deviated from the strict Victorian code for relations between the sexes: they addressed each other as Fräulein Meitner and Herr Hahn, and avoided eating or going out for a walk together, signs of intimacy that might have invited gossip. It took sixteen years and the post-First World War revolution before they called each other Lise and Otto and used the familiar *du*. Neither of them was paid a salary until 1912, when Hahn was made a member of the newly founded Kaiser Wilhelm Institute for Chemistry and Meitner became assistant to Planck. In 1913 she, too, became a member of the Institute, but at first at a salary below Hahn's.

When war broke out in August 1914 Meitner rushed to Vienna. She was carried away by the patriotic fervor of the crowds as they saw the eager young soldiers, including her own brothers, off to the front, and by the excitement of the early German victories. Despite her strong moral convictions, her letters show no evidence that she questioned the morality of the Austrian attack on Serbia or of the German invasion of Belgium. Hahn expected a quick German victory. But her enthusiasm evaporated when, working as an X-ray technician and nurse behind the lines on the Russian front, she came face to face with the severely wounded and dying young soldiers. In 1916 she was transferred to the Italian front and then back again to the Russian one. Feeling useless there, she returned to Berlin, where she was soon promoted to be head of the physics section of the Kaiser Wilhelm Institute with a salary equivalent to Hahn's. In 1919 she became a full professor. Her academic career does not seem to have been seriously hampered either by her sex or by her being born Jewish.

Hahn and Meitner succeeded in isolating an important new radioactive element, protactinium, which others had failed to find. She became one of the stars in Berlin's galaxy of great physicists, which included Albert Einstein, Max Planck, Max von Laue, James Franck, and later, Erwin Schrödinger. All the same, she remained diffident. She wrote to Hahn:

> Did I write to you that I recently gave a colloquium on our work, and that Planck, Einstein, and Rubens [the professor of experimental

physics] told me afterwards how good it was? From which you can see that I gave quite a decent lecture, even though I was, stupidly enough, again very self-conscious... .

In the nineteenth century each chemical element was believed to consist of only a single kind of atom, but radioactivity soon showed that certain elements consist of a mixture of atoms of slightly different weights; these were called isotopes. It was also considered impossible to turn one kind of chemical element into another; but radioactive elements were found to transform themselves spontaneously into a succession of different, slightly lighter elements. For example, the heaviest element then known, uranium, is a mixture of isotopes that are 238, 235, and 234 times the weight of a hydrogen atom, the lightest of the elements. Uranium 238 disintegrates into a succession of lighter elements, one of which is radium, whose radioactivity becomes reduced every 1690 years to half of its original value.

In Cambridge, England, Ernest Rutherford first achieved an artificial transmutation of elements by bombarding nitrogen with alpha particles from radium. This turned each atom of nitrogen into a heavier atom of oxygen, plus one lighter atom of hydrogen. Alpha particles penetrated and transformed lightweight atoms carrying nuclei with few positive charges; but being positively charged, they were repelled by the multiple positive charges of heavy nuclei like that of uranium.

In 1932 James Chadwick in Rutherford's Cambridge laboratory discovered the neutron, a particle with the same weight as a proton, the nucleus of a hydrogen atom, but without its positive charge. Enrico Fermi in Rome realized that neutrons would not be repelled by atomic nuclei, however strongly charged. He irradiated all the chemical elements with neutrons, transmuting them into other elements and creating a great many new radioactive ones.[2] When he bombarded uranium with neutrons, he produced a complex mixture of radioactive elements, some of which Fermi thought he had identified as new ones heavier than uranium, which he called transuranes.

Hahn and Meitner were skeptical of one of Fermi's results and decided to reinvestigate it, together with the gifted young chemist Fritz Strassmann. In their experiments, the irradiation of uranium with neutrons produced three separate series of radioactive elements, some of which they also believed to represent transuranes.

This work was in full swing when Hitler incorporated Austria into the German Reich in March 1938. Until then Meitner's Austrian citizenship had protected her from the Nazi laws; now one of her colleagues denounced her as a Jew whose presence endangered the Kaiser Wilhelm Institute, and her position there became untenable. Several foreign colleagues invited her to work abroad, but she hesitated until a new law forbidding technical experts to leave Germany trapped her. Her Austrian passport was invalid and she was refused a German one.

In this ominous situation the Dutch physicist Dirk Coster persuaded the chief of the Dutch border guards to issue instructions to admit her without a valid passport. He traveled to Berlin, and on July 13, 1938, smuggled Meitner into Holland through a small, lightly guarded frontier station. To avoid suspicion, she crossed the border with two small suitcases and no more than the legal currency allowance, a derisory ten marks. When Hahn said goodbye to her, he gave her, for emergencies, a diamond ring he had inherited from his mother. On the train Coster kept it for her in his pocket. To his and Meitner's disappointment, he failed to find her a job or even a small grant in Holland. But this saved her life because the Nazis would have caught her there a few years later and sent her to Auschwitz. Instead, the Swedish physicist Manne Siegbahn offered her a place in his new Stockholm laboratory, which she accepted. Neutral Sweden proved a safe haven, but Meitner, now aged fifty-nine, found herself stranded, without money, equipment, or collaborators, in a country whose language she could not speak.

Meitner never married, nor does she ever seem to have had a lover, but she had a great talent for friendship, and in Berlin the Plancks, the Hahns, and the Laues, all anti-Nazis, had treated her as part of their own families. Apart from Siegbahn, who had no use for her, the Swedes she met were neither inhospitable nor cold; yet without her friends and her work she felt forlorn, in her own words, a "wind-up doll...with no real life in her."

Before Meitner fled, she had discussed with Hahn and Strassmann a strange new radioactive element, discovered after irradiation of uranium with neutrons by Irène Curie, Marie's daughter, and Pavel Savitch in Paris. A few weeks later, Curie and Savitch reported that this element behaved chemically as if it were a radioactive isotope of lanthanum, an element of only a little more than half the weight of uranium, which could have formed only by the splitting of the irradiated uranium atoms. Hahn and

Strassmann refused to believe that this splitting had occurred and decided to repeat the work of Curie and Savitch. After irradiating a sample of uranium with neutrons, they detected traces of a radioactivity which behaved as if it came from elements chemically similar to radium, though its activity halved in hours rather than years. How could they isolate and identify the elements responsible for such behavior?

When faced with a problem of this kind, chemists used to add some compound of a known, non-radioactive element to their solution as a carrier. When it was made to precipitate, i.e., separate itself, from the solution as an insoluble salt, it would carry the unknown radioactivity along with it; and more refined methods would later separate the radioactivity from the carrier. Since the unknown activity behaved chemically like radium, they dissolved their irradiated uranium in acid and then added to the solution a salt of barium, a non-radioactive element chemically similar to, but much lighter than, radium. Precipitation of the barium as an insoluble salt did indeed carry the new radioactivity with it, leaving the uranium behind in solution. Had the radioactivity come from radium itself or an element similar to radium, Strassmann could now have separated it from barium, using a method pioneered years earlier by Marie Curie. But all his attempts at separation failed.

This implied, even though they could not yet believe it, that the uranium atom had broken into pieces, and that one of the pieces was a radioactive isotope of barium. Hahn later recalled that at this stage "the possibility of a breakdown of heavy atomic nuclei into various light ones was considered as completely excluded." On October 25 he wrote to Meitner: "A great pity that you are not here with us to clear up this exciting Curie activity." On November 13 they met in Copenhagen, but there is no record of what was said at that meeting.

Strassmann now wondered if the traces of the new radioactive elements were just too small to be separated from barium. To test that objection, he added equally small traces of known chemical elements to the barium salt, but found no difficulty in separating them from barium. Still incredulous, he wondered if the particular salt (the chloride) of the new element which he had made just happened to be inseparable from the chloride of barium. So he transformed the barium chloride into five other barium salts in turn. Each time the radioactivity was transformed with the barium and in none of the salts could it be separated from barium salt, proving that it must be caused by radioactive isotopes of barium. This unique chemical identification of minute traces of a short-lived radioactive element was a remarkable feat. Lise Meitner once told me that no one else could have done it at that time.

Despite the incontrovertible evidence that the uranium atoms had split—a momentous event in the history of science—Hahn was still torn by doubts; he asked Meitner in a letter dated December 19 if there might not be an element heavier than barium, but with the same chemical properties, and added: "We know ourselves that [uranium] can't actually burst apart into [barium]... . If there is anything you could propose that you could publish, then it would still in a way be work by the three of us"—a reference to the fact that a joint paper with a Jewish émigré was politically ruled out.

Two days later he wrote: "How beautiful and exciting it would be just now if we could have worked together as before. We cannot suppress our results, even if they are perhaps physically absurd. You see, you will do a good deed if you can find a way out of this." On the same day Meitner wrote to Hahn that she found it hard to accept a complete rupture of the uranium nucleus, "but in nuclear physics we have experienced so many surprises, that one cannot unconditionally say: it is impossible." Hahn wrote up the results for publication, concluding as follows:

> As chemists the experiments we have briefly described force us to substitute for the [heavy] elements formerly identified as radium, actinium, thorium the [much lighter] elements barium, lanthanum and cerium, but as "nuclear chemists" close to physics we cannot yet take this leap which is contrary to all experience of nuclear physics.[3]

On the other hand, Strassmann recalls that he had no such hesitations.[4] Hahn did not state explicitly that the presence of barium implied a break-up of the uranium nucleus into fragments, but this was clear to all who read the paper.

Shortly after receiving Hahn's letter, Meitner set off to spend Christmas with her nephew, the young physicist Otto Robert Frisch, with friends on the west coast of Sweden. Frisch was then working in Niels Bohr's Physics Institute in Copenhagen. In his memoirs, Frisch recalls that dramatic meeting:

> When I came out of my hotel room after my first night in Kungälv I found Lise Meitner studying a letter from Hahn and obviously worried by it. I wanted to tell her of a new experiment I was planning, but she wouldn't listen; I had to read that letter. Its content was indeed so start-

> ling that I was at first inclined to be sceptical. Hahn and Strassmann had found that those three substances [they had discovered] were not radium... [but] that they were isotopes of barium.
>
> Was it just a mistake? No, said Lise, Hahn was too good a chemist for that. But how could barium be formed from uranium? No larger fragment than protons or helium nuclei (alpha particles) had ever been chipped away from nuclei, and to chip off a large number not nearly enough energy was available. Nor was it possible that the uranium nucleus could have been cleaved right across. A nucleus was not like a brittle solid that can be cleaved or broken; George Gamov had suggested early on, and Bohr had given good arguments that a nucleus was much more like a liquid drop. Perhaps a drop could divide itself into two small drops in a more gradual manner, by first becoming elongated, then constricted, and finally being torn rather than broken in two? We knew that there were strong forces that would resist such a process, just as the surface tension of an ordinary liquid drop tends to resist its division into two smaller ones. But the nuclei differed from ordinary drops in one important way: they were electrically charged, and that [owing to mutual repulsion of the positive charges] was known to counteract the surface tension.
>
> At that point we both sat down on a tree trunk (all that discussion had taken place while we walked through the wood in the snow, I with my skis on, Lise Meitner making good her claim that she could walk just as fast without), and started to calculate on scraps of paper. The [positive] charge of the uranium nucleus, we found, was indeed large enough to overcome the effect of the surface tension almost completely; so the uranium nucleus might indeed resemble a very wobbly, unstable drop, ready to divide itself at the slightest provocation, such as the impact of a single neutron.
>
> But there was another problem. After separation, the two drops would be driven apart by their mutual electric repulsion and would acquire high speed and hence a very large energy... . Where could that energy come from? Fortunately Lise Meitner remembered the empirical formula for computing the masses of nuclei and worked out that the two nuclei formed by the division of a uranium nucleus together would be lighter than the original uranium nucleus by about one-fifth the mass of a proton. Now whenever mass disappears energy is created...and one-fifth of a proton mass was just equivalent to [the right energy]. So here was the source for that energy; it all fitted![5]

Frisch's account is too sober to make the reader grasp how staggering the result of their calculation turned out to be. He and Meitner used Einstein's famous equation $E = mc^2$ to calculate the energy equivalent to the loss of one-fifth of a proton from one atom of uranium.[6] This calculation showed that the splitting of one gram (1/28 of an ounce) of uranium would release as much energy as the burning of two and a half tons of coal.

Why had neither Hahn and Strassmann, nor Fermi, nor Curie and Savitch noticed this? They had merely inquired into the chemical nature of the new radioactive elements formed when uranium was irradiated with neutrons, and none of them knew as yet that they came from the isotope with 235 times the weight of a hydrogen atom, which makes up less than 1 percent of the bulk of uranium. Since only a tiny fraction of the atoms of the minuscule amounts of uranium used in their experiments had split, the violence of the splitting had gone unnoticed.

After his return to Copenhagen, Frisch set up an experiment designed to measure the force of the fragments shot out when he irradiated uranium with neutrons and confirmed that it was as great as predicted by his and Meitner's calculations. It was a situation, unusual in research, that fitted Karl Popper's ideas of the scientific method. The violence of the reaction had remained unnoticed without a hypothesis predicting it; and Frisch detected it by an experiment designed to falsify the hypothesis.

Meitner and Frisch sent two letters to the British journal *Nature*, one signed by both with their theoretical interpretation of Hahn and Strassmann's results,[7] and another written by Frisch alone describing his experiments.[8] They pointed out that the new radioactivities which Fermi and the Berlin group had attributed to elements heavier than uranium had, with one exception, been products of the splitting of uranium into lighter elements. The exception later turned out to be a precursor of plutonium. What Curie and Savitch had called lanthanum was a product of the radioactive decay of a barium isotope, but this possibility had been too far from their thoughts.

Meitner and Frisch coined the term fission for the new phenomenon, in analogy with the term biologists used to describe the spontaneous division of yeast cells. They did not suggest that the enormous energy released by the reaction might supply man with limitless energy for an almost unlimited time. Nor did they mention the possibility of making an atomic bomb, although this would have been clear to any physicist reading their papers. Besides, such a statement would have been frowned upon in those days as sensationalism unworthy of true scientists. The papers appeared in February 1939. Meanwhile, Niels Bohr described Frisch's experiment to a meeting of the American Physical Society in Washington, where it aroused so much excitement that even before Bohr had finished speaking, some physicists were hurrying back to their laboratories to repeat it. On February 7 Bohr, now at Princeton, sent a letter to the journal *Physical Review*,[9] attributing fission to the rare isotope uranium 235 rather than the abundant uranium 238.

The next, fateful step was a letter to *Nature* by Hans von Halban, Frederic Joliot, and Lev Kowarsky on 22 April 1939. They found that irradiation of uranium with neutrons causes emission, on average, of three and a half neutrons for each neutron absorbed. (In fact, later studies showed the correct number to be two and a half.) These neutrons would be absorbed by other uranium atoms and make them split, thus starting a chain reaction that could lead to an explosion. The question was how large a mass of uranium would be needed for such an explosion to occur, a gram or a ton?

In March 1940, Frisch and Rudolf Peierls, working at the University of Birmingham in England, calculated that no more than one kilogram (2.2 pounds) of uranium 235 would be needed to make an atomic bomb. They also indicated exactly how the rare isotope uranium 235 could be separated from the abundant uranium 238 and how it could be detonated. Their secret memorandum on their results set in motion the making of the bomb which destroyed Hiroshima.[10]

In the summer of 1941, Peierls engaged the German-born physicist Klaus Fuchs to help him with further theoretical work on the atomic bomb project. He did not realize until eight years later that Fuchs was a devoted Communist and had passed copies of all their work, including the Peierls-Frisch memorandum, to the Soviet Embassy in London, where it was collected by the NKVD case officer for technical intelligence, Vladimir Barkovsky. Barkovsky, now aged eighty-two, recounted his experience at a meeting held in Dubna, near Moscow, in May 1996, where he referred to Fuchs as "a hero who did the world great service," but in England people thought otherwise. Barkovsky stressed that Fuchs was not paid.[11]

Meitner was shattered when she realized that the hypothesis on the transuranes underlying her last four years of work in Berlin had now been disproved and that her departure from Germany had excluded her from the great discovery to which her work had led. The fact that she and Frisch had been the first to realize and publish the implications of Hahn and Strassmann's discovery was no consolation to her. Nor was she consoled when Americans tended to quote her and Frisch's papers in preference to Hahn and Strassmann's, perhaps because they were written in English. She wrote to Hahn: "Now Siegbahn will gradually believe...that I never did anything and that you also did all the physics in Dahlem," and to her brother: "Unfortunately I did everything wrong. And now I have no self-confidence, and when I once thought I did things well, now I don't trust myself." Perhaps she also reproached herself for having dismissed Curie and Savitch's experiment rather than following it up. Her fears were confirmed when the Nobel Prize for chemistry for 1944 went to Hahn alone. Having been locked up in the Nobel Committee's files these fifty years, the documents leading to this unjust award now reveal that the protracted deliberations by the Nobel jury were hampered by lack of appreciation both of the joint work that had preceded the discovery and of Meitner's written and verbal contributions after her flight from Berlin.[12,13]

Because of the war, the jury was also hampered by lack of communication with the outside world. Hahn and Meitner had already been nominated jointly for their earlier discoveries and were again nominated jointly for their discovery of nuclear fission. But the Nobel Committee for chemistry ignored these nominations and confined its attention to Hahn and Strassmann's two publications proving, by purely chemical methods, that irradiation of uranium with neutrons produced radioactive isotopes of barium. They never even considered including Strassmann, who had in fact done many of the experiments and introduced a crucial innovation while Hahn, a staunch anti-Nazi, was busy fending off Nazi attacks on himself and his institute. There was a proposal to award the physics prize to Meitner and Frisch at the same time as the chemistry prize to Hahn, but it went instead to the great theoretician Wolfgang Pauli, and later that proposal was crowded out by claims for other candidates.

Early in the war, Hahn told the young physicist Carl Friedrich von Weizsäcker: "If my discovery leads to Hitler obtaining an atomic bomb, I shall kill myself." On hearing the news of Hiroshima, he really did want to kill himself, but his internment in England robbed him of the means, and his friend Max von Laue managed to calm him down. When Meitner was invited to join the team at Los Alamos, she refused, wanting to have nothing to do with building an atomic bomb. In August 1945, she was enjoying a quiet vacation in the Swedish countryside when a reporter called to tell her of Hiroshima. Shocked beyond words, she walked alone for many hours. Her friends had never seen her so distraught. Worse, reporters pursued her since she had suddenly become a public figure who shared responsibility for the atomic bomb. Under a headline FLEEING JEWESS, one news story described how she escaped from Germany with the secret of the bomb and handed it to the Allies.

In September 1945 Meitner wrote to her sister:

> I feel like an impostor when American Jews...praise me especially because I am of Jewish descent. I am not Jewish by belief, know nothing of the history of Judaism and do not feel closer to Jews than to other people. And just now, when one wishes so strongly that all racial prejudices be eliminated from the world, isn't it unfortunate if Jews themselves document such racial prejudice?

All the same, Meitner enjoyed her fame and the recognition she received in the postwar years in both Germany and the United States, and she reestablished friendly relations with her former Berlin colleagues and other leading physicists. After Einstein's death in 1955 she wrote to Max von Laue:

> For all my great admiration and affection for Einstein during the Berlin years I often stumbled inwardly over his absolute lack of personal relationships... .Only later did I understand that this separation from individuals was necessary for his love and responsibility toward humanity.

My own experience, on the contrary, is that people develop a love for humanity in general not because they deliberately turn their backs on personal relationships, but because they are incapable of forming them. Peter Medawar told me about a colleague who loved all humanity, while the technician who worked for him could enter his room only at the risk of his life.

In 1960, aged eighty-two, Meitner moved to Cambridge to be with Otto Frisch and his family, and I was fortunate to get to know her. She showed no bitterness, and I admired her brilliance, her selfless passion for science, her warmth, and her sense of humor. In 1964, the U.S. Atomic Energy Commission invited me to nominate a candidate for the prestigious Enrico Fermi prize. I decided to nominate Meitner, but since I am not a physicist, I asked Sir Lawrence Bragg and Hans Bethe to support me. Hahn, who was also asked, nominated Strassmann. I was delighted when, in 1966, the Commission awarded the prize jointly to Meitner, Hahn, and Strassmann, which rectified to some extent the injustice of omitting Meitner and Strassmann from the Nobel Prize. Since Meitner was by then too frail to travel to Washington to receive the prize, Glenn Seaborg, the Commission's chairman and discoverer of the true transuranes, came to Cambridge and presented it to her in a short ceremony at my house.

Ruth Sime accuses Hahn of belittling or ignoring Meitner's contributions, but she does not give convincing evidence for her charges. Sime quotes Meitner's letters written in Stockholm after the award of the Nobel prize to Hahn in which she complains that he failed to mention their collaboration in his interviews with the press. This may have been true, but on the other hand, Hahn's printed Nobel lecture gives Meitner full credit for all their joint work; the curriculum vitae appended to the lecture stresses the years and the topics of their collaboration.[14] Hahn's autobiography gives details of their work together and quotes the full texts of the crucial exchange of letters between them in December 1938.[15]

Frisch's widow has assured me that in the years Meitner lived in Cambridge she never voiced anything but deep affection for Hahn. Manfred Eigen, a younger Nobel Laureate in chemistry, was struck by Hahn and Meitner's manifestly warm friendship when he spent time with them in Göttingen in the 1960s. Weizsäcker writes that he never met anyone more decent and benevolent than Hahn. When Sime criticizes Hahn for boasting that physics had nothing to do with his and Strassmann's discovery, she fails to take into account that chemists of their generation learned little physics, which left them feeling inferior to physicists, whom they needed to interpret their experiments. A desire to compensate for that feeling, rather than any wish to belittle Meitner's contribution, may have been the source of that boast. Sime also reports Hahn saying that he and Strassmann would not have made their discovery had Meitner stayed in Berlin. There is some truth in that because she apparently discounted Curie and Savitch's results as spurious and not worth bothering about.

Sime's accounts of Meitner's and her colleagues' scientific work are accurate, readable, and intelligible to anyone with a rudimentary knowledge of physics and chemistry. She gives a vivid picture of the life and personality of this remarkable woman, illustrating Peter Medawar's dictum:

> It is high time [laymen] recognized...the misleading and damaging belief that scientific inquiry is a cold dispassionate enterprise, bleached of imaginative qualities, and that a scientist is a man who turns the handle of a machine of discovery; for at every level of endeavor scientific research is a passionate undertaking, and the Promotion of Natural Knowledge depends above all else on a *sortie* into what can be imagined but is not yet known.[16]

Strassmann deserves to be mentioned as a quiet hero of the story and not only because he did many of the crucial experiments. When he was unemployed in the 1930s he preferred to starve rather than accept a job that would have required him to join the Nazi Party. During the war, he and his wife hid the Jewish pianist Andrea Wolffenstein in their apartment for several months at the risk of their lives. Wolffenstein survived and Strassmann was honored at the Israel Holocaust Museum over forty years later as one of the many Germans who refused to collaborate with the Nazis.

Meitner died in 1968, a few months after her old friend and colleague Otto Hahn.

The Man Who Patented the Bomb[*]

On August 13, 1940, Lt. Col. S.V. Constant of U.S. Military Intelligence reported: "ENRICO FERMI.... He is supposed to have left Italy because of the fact that his wife is Jewish.... He is undoubtedly a Fascist.... Employment of this person on secret work is not recommended." "MR SZELARD. [*sic*] He is a Jewish refugee from Hungary. It is understood that his family were wealthy merchants in Hungary and were able to come to the United States with most of their money....He is stated to be very pro-German.... Employment of this person on secret work is not recommended."

Fermi was not a Fascist, and Szilard lived in terror of the Germans; his family was neither wealthy nor had it come to the United States, but no matter. Had military intelligence been heeded, the atomic bomb would not have been built and the coming of the atomic age would have been at least delayed.

Leo Szilard was born in Budapest in 1898, the son of a successful Jewish civil engineer. He was mentally precocious but physically lazy, preferring to organize his playmates rather than take part in games himself; he made close friends and suddenly dropped them, habits that he was to maintain throughout his life. Szilard's mother, though originally Jewish and a nonbeliever, had worked out her own practical religion based on

*A review of *Genius in the Shadows: A Biography of Leo Szilard, the Man Behind the Bomb* by William Lanouette with Bela Szilard, foreword by Jonas Salk (Scribner's/A Robert Stewart).

Jesus' teachings, and she inculcated strong ethical values in her children. They were to guide one side of Leo's life; another was guided by the persistence and multiplication of his childhood terrors.

As engineering students at Budapest's Technical University in 1919, Leo and his brother Bela supported Bela Kun's short-lived Communist regime. After Kun's government was overthrown, they were hounded by the police and attacked by anti-Semitic students. Leo decided to continue his studies in Berlin. To mislead the border police, he took an excursion steamer going up the Danube to Vienna. As he sadly watched the Hungarian shores recede, an old Hungarian farmer returning to Canada after a home visit tried to console him: "Be glad, as long as you live, you'll remember this as the happiest day of your life!" Szilard would not have become famous if he had stayed at home, but he was to spend the rest of his lonely life as a vagabond in hotels and temporary lodgings.

In Berlin, Szilard enrolled as an engineering student at the Technical University as his father wished, but engineering soon bored him. On discovering the weekly physics seminars at the Friedrich-Wilhelm University attended by Albert Einstein and other eminent scientists, Szilard switched to physics. For his Ph.D. he wrote a brilliant mathematical thesis on a theory of fluctuations derived from thermodynamics which was later published, and he followed this with an ingenious paper on how entropy in a thermodynamic system can be reduced by the intervention of intelligent beings, showing that information is equivalent to negative entropy, i.e., to less disorder.[1] The second paper foreshadowed the information theory that was later developed by Claude E. Shannon and Warren Weaver.

These two papers were to remain Szilard's only major scientific publications. In his remaining years in Berlin, he published only two short notes on X rays, for which most of the credit probably belongs to his senior co-author, Hermann Mark. On the other hand, Szilard filed several ingenious patents, one with Einstein on an electromagnetic pump using fluid metals, which later proved useful in the Manhattan Project, and two on atom-smashing machines before other physicists began to build them: a linear particle accelerator in 1928 and a cyclotron in 1929. The patents are original and sound, but Szilard never developed them even to a preliminary experimental stage.

Max Volmer, one of Berlin's leading physicists, described Szilard as "one of the most capable and versatile people I have ever met. He unites in a rare fashion a complete understanding of the development of modern physics with a capacity for dealing with problems in all fields of classical physics and physical chemistry." Years later, the great zoologist Konrad Lorenz recalled Szilard as one of the most intelligent people he had ever encountered. Anna Kapitsa, widow of the famous Russian physicist Peter Kapitsa, said that ideas flowed from Szilard as water from a fountain, while Erwin Schroedinger, the founder of wave mechanisms, wrote that "what he had to say was always profound and original" and of a kind that "would not occur to anyone else."

What was it that inhibited this brilliant man from following up his ideas, except on one occasion when he was driven by sheer terror? Dostoevski used to write his novels when he had gambled away all his money, but lack of money did not induce Szilard to apply himself to the daily grind of research. He had an aversion to being tied to a piece of scientific work, a job, a home, or a woman, and remained a wanderer all his life. Viktor Weisskopf called him an intellectual bumblebee, which may seem derogatory but also implied that he fertilized many people's scientific work.

Szilard invariably foresaw the worst, and was often proven right. A few days after Hitler came to power in January 1933, he went to visit his family in Budapest and told them to emigrate because Hitler would soon overrun all of Europe. On March 30, 1933, he left Berlin by train for Vienna with his savings hidden away in his shoes. The very next day he read that Nazi guards had started to search passengers riding the same train. After that, Szilard always lived with two packed suitcases, in case he had to flee from wherever he happened to be.

In Vienna, Szilard made the first of several great contributions to help his fellow scientists. His deep concern about the German Jewish academics whom the Nazis had ousted from their jobs led him to initiate the Academic Assistance Council, a charitable organization financed largely by voluntary contributions from British academics, which helped displaced scholars to become re-established abroad. The council later changed its name to the Society for the Protection of Science and Learning, with its seat in London; it has helped thousands of academics persecuted for their race, religion, or political beliefs to find new homes and opportunities for work, and continues to do so to this day.

Thanks to warm recommendations from Berlin, Szilard could immediately have found jobs in Britain, but an account of a lecture by Lord Rutherford, the physicist and discoverer of the atomic nucleus, changed the course of his life. According to the account in the scientific journal *Nature*, which is the one Szilard is most likely to have read at the time, Rutherford concluded his lecture "with a timely word of warning to those who look for sources of power in atomic transmutations—such expectations are the merest moonshine." Szilard objected that even the great Rutherford could not know what someone else might invent. The previous year James Chadwick, one of Rutherford's young colleagues at Cambridge, had discovered the neutron when he bombarded a thin foil of beryllium with alpha-particles. Szilard recalled:

> As I was waiting for the light to change...to green and I crossed the street, it suddenly occurred to me that if we could find an element which is split by neutrons and which would emit *two* neutrons when it absorbed *one* neutron, such an element, if assembled in sufficiently large mass, could sustain a nuclear chain reaction. I didn't see at the moment just how one would go about finding such an element or what experiments would be needed, but the idea never left me.

At first Szilard thought only of the generation of power and filed a patent for his idea. It includes the following words:

> We shall now discuss the composition of the matter in which the chain reaction is to be maintained...(a) Pure neutron chains...are only possible in the presence of a metastable element...the mass of which is sufficiently high to allow the disintegration of its parts under liberation of energy. Elements like uranium or thorium are examples...

The patent was dated March 12, 1934, four and a half years before Otto Hahn and Fritz Strassmann discovered the fission of uranium by neutrons. They would have been unaware of Szilard's prescient patent, because he eventually realized its military implications and assigned it to the British admiralty with an injunction to keep it secret.

Happily for the world, Szilard never tried to bombard either uranium or thorium with neutrons, but first bombarded beryllium and later indium, neither of which undergo nuclear fission. After four years of intermittent and futile research, Szilard wrote to the Admiralty on December 21, 1938, from New York, asking them to withdraw his useless patent, only to hear a month later that Hahn and Strassmann had verified his ideas. On February 2, 1939, Szilard therefore wrote to the Admiralty again, asking them to keep his patent after all. At first he was delighted to have been proven right, but soon he became terrified that the Germans might build an atomic bomb and put it into Hitler's hands.

Szilard was on a scientific visit to the United States when he learned that on September 30, 1938, Chamberlain and Daladier had signed the Munich Pact with Hitler; fear of war made him decide not to return to the research that he was then doing in Oxford. He wrote to Professor Lindemann, later Lord Cherwell, that he would not be able to concentrate on it any longer:

> It seems to me that those who wish to dedicate their work to the advancement of science would be well advised to move to America where they may hope for another ten or 15 years of undisturbed work.

The idea that in the event of war British scientists would want to stay and do their duty by their country did not occur to him.

In New York, Szilard met Enrico Fermi, the Italian physicist who, when he had been in Rome, had started the irradiation of uranium with neutrons in the hope of generating elements heavier than uranium, then the heaviest one known, but had missed detecting its fission. Fermi had just collected the Nobel Prize for his work and used it as a pretext for leaving Italy and emigrating with his family to the United States directly from Stockholm.

Fermi doubted Hahn and Strassmann's results at first, but others quickly confirmed and extended them. In March 1939, Szilard and Walter Zinn found that the bombardment of uranium by neutrons was indeed accompanied by the emission of more neutrons, as he had feared. In Paris, Frédéric Joliot, Hans von Halban, and Lev Kowarski obtained the same result and published it as a letter to *Nature.* A month later they followed it with another letter, claiming that 3.5 neutrons were emitted for every uranium atom split (the correct number is between 2 and 3). After the fall of Paris in June 1940, Halban and Kowarski continued their experiments in the room immediately underneath my office in the old Anatomy School of Cambridge University. I have often wondered how heavily they irradiated me with their neutrons, but so far it seems to have done me no harm.

Their result implied that the construction of an atomic bomb was possible in principle, thus confirming Szilard's fears, but it was not known how large the piece of uranium would have to be before it exploded. Moreover, uranium was known to be a mixture of a major component of mass 238 and a minor one of mass 235, and it was not clear which was the fissionable one until Niels Bohr showed on theoretical grounds that it was the minor component, which makes up only 0.71 percent of natural uranium. All the same, the stage was set.

In the summer of 1939, Germany became the first country to establish a research unit for the study of the military applications of nuclear fission. Szilard suspected this, and he and his Hungarian colleague Eugene Wigner feared that Germany might attack Belgium in order to gain control of the uranium mines in the Belgian Congo. Knowing that Einstein was friendly with the Belgian King and Queen, they asked him to warn them. Together with Einstein, they drafted an appropriate letter, but Szilard had an uneasy feeling that it was being sent to the wrong address. He discussed the matter with Dr. Alexander Sachs, a Wall Street economist who knew President Roosevelt personally and offered to take a letter to him.

After weeks of agonizing over its text, Szilard and Wigner warned the President that "this new phenomenon would also lead to the construction of bombs, and it is conceivable—though much less certain—that extremely powerful bombs of a new type may thus be constructed." They recommended that the President should appoint someone to maintain contact with the physicists, keep government departments informed, and recommended action, especially to secure uranium supplies. That person should also use his contacts to secure private funds and equipment from industrial laboratories for experimental work on the chain reaction. In those days the idea of a government itself financing that research did not even seem to have occurred to Szilard, Wigner, or Einstein.

The letter to Roosevelt was dated August 2, 1939, but Sachs failed to get an appointment before October 11, 1939, six weeks after Germany's invasion of Poland. On hearing its message, Roosevelt asked Lyman J. Briggs, the head of the National Bureau of Standards, to set up an advisory committee to study the problem. The committee held its first meeting ten days later and was joined by Szilard, Wigner, and Edward Teller, who secured a promise of $6,000 for their uranium research.

Szilard then sent a ten-page report to Briggs, outlining the research needed to find out whether uranium could sustain a chain reaction, and he organized a drive to locate the necessary supplies of chemically pure uranium and graphite. He also recruited Fermi's help for the design of a pile made up of a lattice of alternate lumps of the two elements, but he had no job and received no pay for any of this work. At the end of 1939, he wrote bitterly to a rich friend who had lent him $2,000:

> Unfortunately, I have not earned anything during this year, as I was tied up with this work on uranium. It looks as though I shall not be able to earn anything next year either.

The promised $6,000 was not released until six months after the meeting of the Briggs committee, by which time Fermi's interest in the project had waned, while Szilard had become increasingly alarmed by reports of German progress. When the money finally arrived, it allowed them to buy the uranium and graphite needed for a crucial experiment. The carbon atoms in graphite were thought to make the neutrons emitted by uranium fission bounce backward and forward between them and slow them down without absorbing them. The slowed neutrons would induce fission in other uranium atoms, which would emit more neutrons. The experiment was to test whether this was true, but it required graphite free from the impurities of other atoms known to absorb neutrons.

Szilard succeeded in finding a manufacturer who made sufficiently pure graphite. He and Fermi then proved that such pure graphite absorbed few enough neutrons to sustain a chain reaction. Fermi wanted to publish their result, but Szilard restrained him, and this may have saved the world. As we now know, the Germans did the same experiment, but with impure graphite, and wrongly concluded that graphite absorbed too many neutrons to sustain a chain reaction—they therefore decided to use heavy water instead. This decision proved one of the main stumbling blocks that prevented them from ever establishing a chain reaction in their atomic piles. Had Fermi and Szilard published their result, the Germans would have realized their fatal mistake.

Censorship of papers on nuclear fission became an issue once more in May 1940 when the Princeton physicist Louis A. Turner sent Szilard a review suggesting that absorption of neutrons by uranium 238 might transform it into the fissile element 239, later called plutonium, which would be easier to separate from uranium than its fissionable isotope 235 and therefore better suited for a bomb. Szilard lacked the necessary authority to stop publication of Turner's paper, but his anxieties were relieved by the decision of Roosevelt's newly formed Defense Research Committee that all papers on uranium be censored.

That decision had an unforeseen and fateful effect. In 1942 George N. Flerov, a bright young Russian physicist who had begun his own research on nuclear fission before the war, got a few days of leave from the army in the town of Voronezh. He went to the library of the physics institute to see what had been published in his field since he had been called up. When he found that publication of American papers on this hot subject had suddenly been stopped, he wrote a letter to Stalin warning him that the Americans must be working on an atomic bomb. Stalin had apparently discounted previous reports on similar lines from his spies as too fanciful, but Flerov's letter convinced him, and he ordered work on a Russian bomb to be started forthwith. Hence the imposition of nuclear censorship in the U.S. may have deprived Germany of valuable information, but it also started the nuclear arms race with the Soviet Union.

In the U.S., the building of the nuclear pile began first at Columbia and later at Chicago under Arthur Compton. Fermi took charge of experimental work and did much of it with his own hands, while Szilard never dirtied his, but instead he poured out ideas and searched for supplies of pure graphite and uranium. After Pearl Harbor, Compton offered Fermi and Wigner, but not Szilard, high positions. Wigner recalls in his autobiography that when he accepted, he "was quite conscious of an immoral element in my action. But I was far more concerned with the moral failings of a man across the ocean: Adolf Hitler."[2] He describes how the entire team was driven by the constant nagging fear that one small error might yield the lead to the Germans.

Edward Teller and John von Neumann were two other Hungarian theoreticians drawn into the project. Wigner has written that Teller's imagination was more fertile than that of any of the others, or anyone he had ever known, and that he did not linger over elegant mathematical formulations as other theoreticians were apt to do, but studied the physical phenomena themselves with brilliant insight. One now wishes Teller's fertile imagination had not run amok in conceiving the futile and costly "Star Wars" program and foisting it on President Reagan. Wigner recalls that von Neumann contributed an expert knowledge of explosions. His precise calculations helped to convince Robert Oppenheimer and others that implosions (compression of the fissile material by conventional explosives all around it) would ignite the atomic bomb.

Von Neumann, Teller, Wigner, and Szilard, old friends from school days in Budapest, were part of an entire generation of brilliant Jewish Hungarian émigrés. Wigner attributes part of their success to the superb high schools in Budapest, which gave them a good start, and part to forced emigration and their need to establish their reputations in a new country.

In his autobiography, Wigner writes that Szilard, as the initiator of the Manhattan Project, expected a high position and preferably full control of the laboratory in Chicago, but believes that Compton was right not to give it to him because he was incapable of leading a group of scientists and was not a good enough physicist. Though he was a prolific source of original ideas, he expected others to work them out, no matter whether they were sound or crazy, and had no stomach for the daily grind of science, either in theory or experiment. Wigner denies that Szilard ever contributed any really good new idea to science, but this strikes me as unjustified in the light of the posthumously published collection of Szilard's scientific papers and patents and the many tributes from scientists who benefited from his casually thrown out ideas. All the same, Szilard was gradually pushed to the margins of the Manhattan Project and was never allowed to join the team at Los Alamos that finally designed and constructed the bomb.

Szilard's open disdain for authority and his erratic and undisciplined ways grated on General Leslie R. Groves, the commander-in-chief of the Manhattan Project. Apparently, Groves considered Szilard to be the villain in the Manhattan Project, a man of "doubtful discretion and uncertain loyalty," while Szilard considered Groves a fool and showed it. Eventually, Szilard infuriated Groves so much that Groves drafted the following letter to be sent to the Attorney General by the Secretary of State for War, Henry L. Stimson:

> The United States will be forced without delay to dispense with the services of Leo Szilard of Chicago, who is working on one of the most secret War Department projects.
>
> It is considered essential to the prosecution of the war that Mr. Szilard, who is an enemy alien, be interned for the duration of the war. It is requested that an order of internment be issued against Mr. Szilard and that he be apprehended and turned over to representatives of this department for internment.

When Stimson refused to sign, Groves put the FBI on Szilard's trail with orders to follow his every step. There is no mention of any of this vindictive persecution in Groves' autobiography. He must have become ashamed of it later.

Pushed to the sidelines, Szilard had time to think about the consequences of the bomb. Once more he enlisted Einstein's help to put his case before President Roosevelt. On March 25, 1945, he warned the President that "our 'demonstration' of atomic bombs will precipitate a race in the production of these devices between the United States and Russia and that if we continue to pursue the present course, our initial advantage may be lost very quickly in such a race." In an earlier draft, he wrote prophetically, "After this war it is conceivable that it will become possible to drop atomic bombs on the cities of the United States from very great distances by means of rockets." Szilard suggested delaying the bomb's use and called for international control, but Roosevelt died before Szilard's memorandum reached him. Szilard could not have known that the nuclear arms race had already started three years earlier and that Stalin was unlikely to stop it or accept international controls. Besides, his letter was preceded by one from General Groves promising Roosevelt that use of the bomb would end the war against Japan, as indeed it did.

From 1945 onward, Szilard led a frantic life of public appearances and private negotiations aimed at saving the world from his own invention. In 1961, broadcaster Edward R. Murrow said of Szilard that the bomb "has left him with one driving purpose, and that is to try to help dismantle the era of terror he helped to create." Szilard himself later declared on CBS television that he felt no guilt, but his denial is belied by his attempt in 1945 to stop publication of the Smyth Report, which described the scientific basis and the construction of the bomb together with the vast administrative organization and industrial effort behind it. Szilard feared that the report would brand him as a war criminal. After its publication he insisted, unsuccessfully, that the army should furnish him with a personal bodyguard.

In May 1946, Szilard called on Einstein to help him found the Emergency Committee of Atomic Scientists to raise money for public education on atomic energy. If the nuclear arms race was not controlled, Szilard recommended dispersing 30 to 60 million Americans from big cities all over the countryside.

Szilard campaigned against the anti-Communist witch hunt of the House Un-American Activities Committee of the late 1940s and early 1950s, and proposed that all academic staff take a one percent cut in their salaries to support their dismissed colleagues. Szilard was outraged that no one spoke up against the committee's activities, but I know at least one courageous member of the Harvard faculty who did. Biochemist John T. Edsall wrote letters of protest to *The New York Times* and to Lewis L. Strauss, the head of the Atomic Energy Commission. According to Edsall about two-thirds of the scientists called to testify at Oppenheimer's hearings spoke in his favor.

On October 13, 1951, Szilard finally married his lifelong friend, the Viennese physician Trude Weiss. After parting from her the next morning he was already overcome by terror at being confined by marriage ties and wrote to her: "I lost all hope of 'freedom' and I felt terrible; almost incapable to work at the lab., absentminded, sweats, and high pulse rate. This has been going on for three days." Unable to control his fears, he inquired about divorce proceedings in several states. His patient bride repeated earlier suggestions that he should see a psychoanalyst, psychiatrist, or marriage counselor, but to no avail. Eventually Szilard came to accept his new bondage.

Meanwhile, Szilard concocted plans for surviving the nuclear war that he feared to be imminent. He made plans to found a school in Mexico where his friends' children might go. In 1958, when Eisenhower sent American troops to Lebanon, he declared, "If a confrontation develops I shall leave the country." He really did pack his belongings into fifteen pieces of luggage and fled to Geneva together with his wife during the Cuban Missile Crisis, and he did not return until several months later. In Geneva Szilard walked into the office of Viktor Weisskopf, director of CERN, the European Laboratory for Nuclear Research, with the words, "I am the first refugee from the third world war." By then Szilard's interests had shifted to molecular biology. Realization of the great stimulus that CERN had brought to European physics led Szilard to suggest that an international laboratory run on similar lines would give a much needed boost to European molecular biology. His initiative led me and my colleagues to found the European Molecular Biology Organization, which financed traveling fellowships and workshops and later persuaded European governments to set up the laboratory that Szilard had envisaged.

Some years earlier, Szilard had concocted a scheme for limiting a nuclear conflict whereby the U.S. and Soviet Union would agree on "sister cities" designated for mutual destruction that each side would evacuate when the conflict started. He wrote two wordy, convoluted, repetitive articles for the *Bulletin of Atomic Scientists* on this subject and urged several other impractical schemes in "How to Live with the Bomb and Survive." Convinced that he was cleverer than the politicians, he decided to negotiate directly with Khrushchev, and actually managed to have a two-hour discussion with him when Khrushchev visited New York in September 1960.

In a memorandum translated into Russian, Szilard proposed that the U.S. and the Soviet Union should agree on a nuclear test ban whose observance should be monitored by independent citizens (Americans in the

U.S., Russians in the U.S.S.R., who should be paid $1 million for reporting any genuine violation). Khrushchev had declared a unilateral moratorium on nuclear tests two years earlier. Szilard blamed the Americans for wanting to continue testing for the sake of developing tactical nuclear weapons of which Khrushchev said he was not interested. Szilard's memorandum also recommended that America and Russia install direct telephone connections between their governments that would be readily available in an emergency. This appealed to Khrushchev, but it was done only after the Cuban Missile Crisis, when it would have been most needed. Szilard later told an interviewer that in Khrushchev he had found "a kindred spirit... a bold and intuitive strategist who was personally committed to controlling nuclear arms."

Szilard did not know that his kindred spirit had declared a unilateral nuclear test ban only because an explosion had destroyed the Soviet plutonium plant. When that had been rebuilt, Khrushchev's personal commitment to the control of the nuclear arms race ceased. The Soviet Union was to explode fifty weapons, including the new fifty-six-megaton hydrogen bomb, the largest every exploded, built under the supervision of Andrei Sakharov, for which Khrushchev and Brezhnev publicly embraced him in front of the entire Politburo.

In October 1960, Szilard went to Moscow to attend a meeting of the Pugwash Group, founded by Albert Einstein and Bertrand Russell in 1955 to bring American, West European, and Soviet atomic physicists together to discuss limitations to the nuclear arms race. Szilard stayed in Moscow for some time after the meeting, but despite persistent efforts he failed to get another appointment with Khrushchev.

Later that year, when Kennedy was elected President, Szilard installed himself and his wife in a Washington hotel. The same mixture of conceit and naiveté that had turned him into a self-appointed negotiator with Khrushchev now made him decide to advise the new administration on peace and disarmament, undeterred by being told that "people in government pay more attention to the advice they invite than to the advice they are offered." He also founded a movement of "Scientists for Peace" because "besides being cleverer than congressmen, scientists should lead this movement, because, unlike politicians, they have integrity and purity." Szilard may have been fortunate not to have lived long enough to have his faith in scientists' integrity and purity shaken.

In January 1962, he founded the Council for a Livable World, a pressure group financed by citizens interested in the control of nuclear weapons and general disarmament; it supports the election and activities of senators favorable to their cause and still continues its work today. Inspired by his postwar interest in molecular biology, he conceived the idea for the highly successful Salk Institute for fundamental research at La Jolla, California, where he and his wife set up his first home since he left Budapest forty-five years earlier. He died there three months later of a heart attack.

Was it a life well spent? It was lonely and haunted by fears, both rational and irrational, stoked by Szilard's vivid imagination. For example, when Szilard was at Oxford in his thirties, he made a date with a girl to go out on the river Isis, but when they arrived there he pulled the boat up onto the bank because he was terrified of even that sluggish river's shallow waters. He even refused to pull the chain in the bathroom, because he had never overcome his childish terror of the water rushing down. On one occasion this saved his life. When the Hungarian oncologist George Klein used the bathroom after him, he saw blood in Szilard's urine and diagnosed that Szilard suffered from cancer of the bladder. Klein recommended surgery. Szilard, after much vacillation, rejected this, studied the literature on the effects of radiation on tumors, prescribed his own radiation treatment, had it carried out at the Sloan-Kettering Memorial Hospital in New York, and was cured.

Szilard's great abilities as a scientist were largely wasted because he lacked the application needed for systematic research. All he needed, he thought, were ideas; their working out could be left to lesser mortals. Only when driven by panic fear that the Germans might do it first did he work hard on the verification of the neutron chain reaction, the behavior of graphite under neutron bombardment, and the design of the first atomic pile. Terror of the Germans drove him to start work on the atomic bomb and terror of its consequences ruled the rest of his life. Was Szilard's terror of the Germans justified? Release of the Farm Hall transcripts and other evidence shows that the German scientists neither succeeded in establishing a nuclear chain reaction in a uranium pile nor had any idea of the critical mass needed for a bomb.[3]

Was it justified to drop the bomb on Japan? It is sometimes said that there was no need because Japan was about to surrender, but the historian Gordon Craig informs me that there exists no convincing historical evidence that the Japanese military commanders were prepared to surrender. Had the bombs not been dropped, the U.S. Army would have had to invade Japan, with untold numbers of casualties that President Truman did not wish to risk. According to Laurens van der Post, who was a prisoner of the Japanese in Java, the Japanese army was under orders to kill all of the Allied prisoners in the event of an American invasion. These circumstances should be kept in mind before we condemn the decision to drop the bomb.

*Why Did the Germans Not Make the Bomb?**

Did the German physicists make no atomic bombs during the Second World War because they wouldn't or because they couldn't? This is a question finally answered by the recent publication of the secretly recorded conversations between Heisenberg and the other German atomic physicists interned at Farm Hall, near Huntingdon, in England in the summer of 1945.

Heisenberg's leading role among German physicists stems from the revolutionary mathematical theory which he formulated at the age of 24. Heisenberg was born in 1901 in Würzburg, the university town where Röntgen had discovered X rays a few years earlier and where Heisenberg's father was professor of Greek philology. Heisenberg shone at school, especially in mathematics and physics. True to the geneticist André Lwoff's dictum, "L'art du chercheur, c'est d'abord de se trouver un bon patron," he began his career in physics as the pupil of Germany's greatest teacher, Arnold Sommerfeld.

In the early 1920s, atomic physics was dominated by Niels Bohr's model of electrons circling like planets around the sun-like nucleus, their concentric orbits governed by Newtonian mechanics combined with Max

*A review of the books *Heisenberg's War: The Secret History of the German Bomb* by Thomas Powers (Cape) and *Operation Epsilon: The Farm Hall Transcripts* (introduced by Charles Frank; Institute of Physics).

Planck's quantum theory. Bohr's theory accounted for the spectra of the simplest atom with only one electron, hydrogen, but it ran into difficulties with the spectra of larger atoms and left many other observations unexplained. Heisenberg broke away from Newtonian mechanics and substituted a new kind of "quantum mechanics", which correctly predicted many hitherto uninterpretable phenomena and was immediately acclaimed by most physicists as a tremendous advance. Refinements of his theory led him to the formulation of the Uncertainty Principle, which says that it is impossible to measure simultaneously the position and momentum of an atomic particle. Ironically, it was the philosophical implications of this principle, based largely on its misconceived application to the macroscopic world, that brought Heisenberg fame comparable to Einstein's. For his discovery he was awarded the Nobel Prize for Physics in 1932. Quantum mechanics still forms the theoretical basis for much of present-day physics and chemistry—without it, for example, there would be no microchips, no computer technology, and no electronics industry.

I first saw Heisenberg soon after he received the Nobel Prize, when he lectured in Vienna, where I was a chemistry student. Knowing nothing about him, I expected to see a portly professor, but in came a slim young man who looked like one of us students and was quite without pomp. We were enormously impressed. I next encountered him in Cambridge in 1947, when he told my wife and me that he had never wanted to build an atomic bomb for Hitler. Fortunately, he did not say to us, as he did to a physicist in Oxford, a German refugee some of whose family had been murdered by the Nazis: "The Nazis should have been left in power longer, then they would have become quite decent." We believed Heisenberg until we read Samuel Goudsmit's book *Alsos*, which argued that his talk of not wanting to build an atomic bomb was merely an excuse invented after the war to explain Germany's atomic failure. (Shortly before his death, however, Goudsmit wrote to Heisenberg to apologise for having maligned him.) Powers quotes extensive documentary evidence to refute Goudsmit's accusation, confirming the conclusion already reached in David Irving's *The German Atomic Bomb*, that the German physicists wanted to build a reactor, but not a bomb. Their secretly recorded comments on Hiroshima have now provided further evidence of their reluctance. Heisenberg apparently regarded it as impracticable.

After Hans von Halban, Lev Kowarski, and Frédéric Joliot in Paris discovered in April 1939 that fission of one uranium atom, induced by the absorption of one neutron, led to the emission of more than two neutrons, physicists realised at once that this opened the way to a nuclear chain reaction with the release of enormous energy. Two German physicists at Göttingen told the Ministry of Education in Berlin that uranium fission might be employed in an energy-producing reactor, while Paul Harteck at Hamburg wrote to the German War Office that "it will probably make it possible to produce an explosive many orders of magnitude more powerful than the conventional ones... That country which makes first use of it has an immeasurable advantage over the others." Harteck's letter led to a conference at the Army Ordnance Department in September 1939, two weeks after Germany's invasion of Poland.

At this meeting it was decided to co-opt Heisenberg, who later reached an understanding with his colleague Karl Friedrich von Weizsäcker to build an atomic pile (they called it a uranium machine) as a post-war power source and as a way of protecting young German physicists from the call-up. However, in May 1940 Weizsäcker realised that the plutonium generated in such a pile could easily be separated from the uranium and would be suitable as an explosive, and sent a report to that effect to the Army Ordnance Department. Heisenberg, on the other hand, felt sure that even though this was possible in principle, a bomb could not be built in time for the war—a view he consistently expressed to the authorities. An elderly Jewish physicist, Fritz Reiche, who managed to reach the United States from Berlin in March 1941, brought a message from the nuclear physicist Fritz Houtermans that "Heisenberg will not be able to withstand any longer the pressure from the government to go earnestly and very seriously into the making of the bomb, despite the fact that Heisenberg tries to delay the work as much as possible." In fact he did withstand the pressure, simply by insisting that it would, as he rightly believed, take too long.

In June 1942, the physicists' lack of response to the critical state of the war appears to have induced Albert Speer, Hitler's newly appointed minister of armaments and munitions, to call a meeting between high army and air force officers and nuclear physicists in order to urge the building of atomic weapons. Heisenberg introduced the subject with a general talk about nuclear research. According to Speer's memoirs, Speer then asked him directly "how nuclear physics could be applied to the manufacture of atomic bombs." Heisenberg was aware of Hitler's order that all expensive projects which would take more than nine months to complete were to be cancelled, and replied that "the scientific solution had been found and that nothing stood in the way of building such a bomb, but the technical prerequisites for production would take years to develop, two years at the earliest, even if the programme was given maximum support." Heisenberg also reassured his audience that Germany need not fear an American atomic bomb before 1945 at the earliest. The physicists complained that their progress had been hampered by lack of money, but when Speer asked how much they needed immediately, Weizsäcker said 40,000 marks (about £4000), even then a very small sum. Twenty-two years later, Speer complained to *Der Spiegel* of the physicists' paltry demands, which suggested that their work had only just begun. Heisenberg himself recorded a government decision to continue work on a nuclear reactor on a modest scale.

The question was: what with? A nuclear reactor using natural uranium metal or uranium oxide as fuel requires a moderator, which must be made of a substance in which neutrons are slowed down by being bounced backwards and forwards without being absorbed. The only suitable substances seemed to be either pure carbon, as in graphite, or heavy water—i.e., water in which hydrogen is replaced by its heavier isotope, deuterium. Fermi used graphite for the first reactor he built in Chicago. On the other hand, the German physicist Walther Bothe found that the probability of neutrons being captured, as opposed to being bounced back by the nuclei of carbon atoms, was too high to make graphite a suitable moderator, however readily available. Bothe failed to realise that neutrons were being absorbed, not by the carbon atoms themselves, but by impurities in his graphite. On the basis of his erroneous result, the Germans decided to rely on heavy water. Their only source was a hydroelectric plant in Vermork, Norway, which used off-peak capacity for its production.

When the Norwegians informed British Intelligence of German demands for large quantities of heavy water, the British knew it could serve only one purpose. In 1942, an attempt by British paratroopers to attack the plant failed, with heavy loss of life. In February 1943, the Norwegian underground undertook a heroic and brilliantly executed raid which put the heavy-water plant out of action for several months without any loss of life. After it had been repaired, the U.S. Air Force destroyed the plant beyond repair in November 1943. The Germans ordered the remaining stocks of heavy water to be shipped to Germany, but a member of the Norwegian Resistance sank the ferry carrying it across a fjord, with the loss of 23 Norwegian lives. Thanks, in part, to that sacrifice, the Germans never had enough heavy water to establish a self-sustaining chain reaction in a nuclear pile and could not even begin to make a plutonium bomb. Their experimental pile was constructed with uranium cubes, but these also became unobtainable after an Allied air-raid on the plant that made them. The Germans realised that heavy water could be replaced by ordinary water if their piles were made of uranium enriched in the fissionable isotope 235, but they never obtained any because all their enrichment experiments failed.

After the meeting at Harnack House, Heisenberg asked Speer and Erhard Milch, the General in Command of the Luftwaffe, separately and discreetly, what they believed would be the outcome of the war. They both indicated that the Germans were likely to lose, even though this was several months before they lost the battles of Stalingrad and El Alamein. That same summer Heisenberg apparently agreed with Karl Friedrich Bonhoeffer (the brother of Dietrich, whom Hitler later had executed for his part in the July 1944 plot) and other Göttingen physicists that it would be a catastrophe if Germany were to win the war; he did not express views of this kind when paying visits abroad—he would have risked his life if any defeatist remark of his had been reported back to Germany. His reaction seems to have been to redouble his efforts to save German physicists. One of his senior colleagues, Walter Gerlach, actually obtained Goering's authority to cancel the call-up papers of any physicist needed for research on their "uranium machine." Gerlach deluded himself that after the war the victorious allies would need the experience gained by German physicists to show them how to generate nuclear power.

In fact, the Farm Hall Transcripts and the documents quoted by Irving and Powers show how far behind the Germans were in 1945—they hadn't even reached the point that Britain reached in 1941, before the Americans became involved. There was nothing in Germany to compare with the Peierls-Frisch Memorandum of March 1940 that contained the estimate of the critical mass of the uranium bomb, suggested how it could be designed, and predicted its likely effects, including its radioactive fallout. Nor is there anything to compare with the "Maud Reports" of July 1941 that contained detailed technical recommendations based on pilot experiments for the manufacture of uranium 235 at a rate sufficient to produce five atomic bombs a month. It is worth remembering that up to June 1941, when Hitler attacked the Soviet Union, Britain had been fighting alone, and first possession of atomic weapons would have given them a strong chance of staving off subjugation by Hitler.

The Farm Hall transcripts reveal the German physicists' bitter disappointment on realising that the American and British had left them far behind. "You are all second-raters," Hahn taunted his colleagues after hearing of Hiroshima. They did not realise that their enemies had succeeded in part thanks to the work of the many brilliant "non-Aryan" German, Austrian, and Italian physicists whom the Nazis had forced to emigrate. "I must honestly say that I would have sabotaged the war had I been in a position to do so." Hahn also remarked; and "I thank God on my bended knees that we did not make a bomb." Heisenberg said that he had been "absolutely convinced of the possibility of our making a uranium engine, but I never thought that we would make a bomb, and at the bottom of my heart I was really glad it was to be an engine and not a bomb. I must admit that." Weizsäcker said:

> I don't think we should make excuses because we did not succeed, but we must admit that we did not want to, but even if we had put the same energy into it as the Americans and had wanted it as they did, it is quite certain that we should not have succeeded, as they would have smashed up our factories.

Later he remarked, "If I ask myself for which side I would have preferred to work, I would say for neither." He may have forgotten his memorandum to the Army Ordnance Department of May 1940. Walter Gerlach, administrative head of one of the uranium projects, feared that "when we get back to Germany we shall have a dreadful time. We shall be looked on as the ones who sabotaged everything." "I went to my downfall with open eyes," he said at another point, "but I thought I would try to save German physics and in this I succeeded."

It is true that the Germans were reluctant to build an atomic bomb—Hahn, because he found the idea abhorrent, and Heisenberg because he was sure it could not be done in time—but they could not have done it even had they wanted to, partly thanks to effective Allied military action, partly because of their own scientific, technical, and organisational failures, and mainly because even the Americans with their much greater resources did not have their first atomic bomb ready before July 1945, more than two months after Germany's final defeat.

As a patriot, Heisenberg had at first hoped for a German victory. He deluded himself that the Nazis would restore order and selfless purpose to Germany and eventually abandon their persecution of the Jews. By 1942 he must have begun to change his mind, because he joined the Wednesday

Society, a debating club which included eight leading Germans later to be executed as conspirators in the July 1944 plot. Heisenberg held his breath, but escaped arrest.

British Intelligence was well aware of the lack of German progress. For example, in January 1944 the Directorate of Tube Alloys (the code name for the atomic bomb) reported: "All the evidence available to us leads us to the conclusion that the Germans are not in fact carrying out large-scale work on any aspect of TA. We believe that... the German work is now confined to academic and small-scale research." Leslie Groves, the American general in charge of the bomb, remained sceptical, however, and demanded that the Kaiser Wilhelm Institute in Berlin-Dahlem, together with Hahn and Heisenberg personally, be designated as prime targets for air raids. The Americans destroyed Hahn's Institute for Chemistry, but missed Heisenberg's Institute for Physics.

Some evidence has come to light of an American cloak-and-dagger plot to kidnap or assassinate Heisenberg. It began with a plan to kidnap him on a lecture visit to Zürich in November 1942 that was later abandoned. It reached its culmination with the dispatch of Moe Berg, former baseball champion, but at the time a member of the OSS, with orders to shoot Heisenberg when he lectured in Zürich on 15 December 1944. Berg did indeed attend Heisenberg's lecture with a loaded gun in his pocket and even got himself invited afterwards to the house of Paul Scherrer, the professor of experimental physics. At the end of the evening he walked back to Heisenberg's hotel with him, but he kept his gun in his pocket, apparently because he had heard nothing to suggest that Heisenberg was building a bomb.

During their long talks, Heisenberg had told Scherrer about his and his colleagues' work on a nuclear reactor. Scherrer, a prominent member of the European physics community, appears to have acted as a regular OSS informer and promptly passed this conversation on to them. At Farm Hall, Heisenberg said that he had his own informer at Scherrer's Institute, a man who told him what the Americans were doing. Judging by Heisenberg's surprise and disbelief on hearing of Hiroshima, that informer cannot have kept him very well informed. The German SS had their own informer at Scherrer's party who denounced Heisenberg for having admitted that the war was lost, even though Heisenberg had infuriated his hosts by adding that he wished Germany had won and denying all knowledge of the Holocaust.

Heisenberg was a patriot who saw Germany as a bulwark of European culture, oblivious of the Nazis being bent on destroying that very culture, even though they had attacked him viciously for continuing to teach the "Jewish" theory of relativity, surely one of that culture's supreme achievements. He had no sympathy for the Nazis. On the other hand, he lacked the sensitivity to imagine what it would be like to be one of their victims, which turned his former mentor Niels Bohr against him for the rest of his life. In 1941, just as the German armies were approaching Moscow, Heisenberg saw Bohr on an officially sponsored visit to German-occupied Denmark. He told Bohr that a German victory over Russia would be a good thing. He regretted the destruction of Poland, but argued that it was to Germany's credit that they hadn't destroyed France as well, a remark which infuriated Bohr. He also advised Bohr, whose mother was Jewish, to make contact with officials at the German Embassy, since they might protect him, which to Bohr seemed tantamount to treason. Finally, Bohr gained the impression that Heisenberg was asking him to act as a go-between in securing an agreement between German and American physicists not to build any atomic bombs, a proposal which Bohr rejected because he saw it as an attempt by Heisenberg to discover what the Americans were doing.

On a visit to German-occupied Holland in 1943, Heisenberg told a Dutch colleague, Hendrick Casimir, that Germany's historic mission had always been the defence of Western culture against the onslaught of the Eastern hordes, a matter in which neither France nor England had been determined enough; that a German-dominated Europe would therefore be a lesser evil; and that the Nazis might improve once the war was over. Casimir was outraged. Memory of such remarks hindered restoration of Heisenberg's old friendships after the war. Casimir later wrote:

> A genius is someone who can create things that are initially beyond his own understanding. In that sense, Heisenberg was certainly a genius, and this goes rarely with a special gift for understanding the feelings and the way of thinking of others. Heisenberg did not have that gift. Perhaps his greatest shortcoming was that he was unable to grasp the full measure of the depravity of what was then the ruling group in Germany.

There were occasions when physicists in neutral or occupied countries passed on to the Allies information about German atomic work that had been given to them by disaffected German scientists. It is a great tragedy that this was met with disbelief, and construction of the atomic bombs pressed ahead in the constant panic and, as we now know, groundless fear that the Germans might make them first. We must be thankful, at least, that Germany was defeated with conventional arms two months before the atomic bombs were ready, even though the chief architect of the American bombs, Robert Oppenheimer, is reported to have regretted this.

*Bomb Designer Turned Dissident**

There seem to have been two very different Sakharovs who hardly communicated with each other. The first was the cold-blooded inventor of the Russian hydrogen bombs; the second the fearless leader of the Russian intelligentsia's struggle for human rights. For twenty years, from 1948 until his dismissal in 1968, Sakharov masterminded the scientific groundwork for the development and perfection of ever more lethal atomic weapons, blindly and obsessively absorbed in work that he describes as a theoretician's paradise. His inventive genius was rewarded by election to full membership of the Academy of Sciences at the unprecedented age of 32, and by being decorated three times with the gold medal of Hero of Socialist Labour. In 1962, he attended a banquet in the Kremlin, seated between Khrushchev and Brezhnev who hugged him in front of the entire Politburo and Presidium of the Supreme Soviet, to thank him for his patriotic work, "which was helping to prevent a new war." This work was the design of a new "improved" hydrogen bomb of unprecedented power. There is no sign that the second Sakharov, once he had realised the true nature of Khrushchev's and Brezhnev's regimes, ever asked whether it was wise to put these terrible weapons of mass destruction into their hands. Or what need there was to develop these weapons at all?

Sakharov was born in Moscow in 1921. His father was a good physicist who became reasonably prosperous by writing popular science books

*A review of *Memoirs* by Andrei Sakharov (translated by Richard Lourie; Hutchinson).

for laymen. An intelligent, kindly and tolerant man whose favourite motto was "A sense of moderation"-would "proportion" be the correct translation?—"is the greatest gift of the Gods"—a motto which his son admired, but confessed he found hard to follow. The cruelty and terror in the midst of which he grew up in the 1930s left its mark on all the adults who lived through it. Roy Medvedev estimates that at least four to five hundred thousand people—above all, high officials—were shot and several million imprisoned. "The spiritual atmosphere of the USSR cannot be explained without harking back to this era," Sakharov writes, "to the crippling fear that first gripped the big cities and then spread to the population at large, and which has left its mark on us even today, two generations later. The repressions caused panic because of their pervasiveness and cruelty, and perhaps still more because of their irrationality; it was simply impossible to fathom how or why their victims were chosen." Yet he describes himself as a "painfully introverted" child, "totally immersed in his own affairs," and largely unaware of those terrible events. He made no friends at school nor during his first three years as a physics student at Moscow University. There, the only subject that gave him trouble was Marxism-Leninism, but this was because he found himself unable to memorise words without meaning. It never entered his head that Marxism-Leninism was not the philosophy best suited to liberate mankind.

In June 1941, during Sakharov's third year at university, the German invasion began. Panic broke out in October 1941 when Moscow itself was threatened, and "after a week of indescribable chaos" the university was evacuated to Ashkhabad in the Turkmen Republic. Sakharov spent the month-long journey silently learning quantum mechanics and relativity, rather than making friends. At Ashkhabad the worst thing he experienced was continuous hunger.

After four years of study and a final examination in theoretical physics, he graduated with honours and was sent to work in a distant munitions factory rather than being drafted into the Army. The journey there opened his eyes to the terrible suffering the war had brought; the trains were filled to overflowing with people who were worn out, burdened by worry and confusion, and talking endlessly, as if compelled to share the horrors that were haunting them. At the factory, men and women had to work 11-hour shifts on meagre rations; those on night shift often had to queue until midday to get their bread ration, and no one was allowed any leave. Here Sakharov's inventive genius found its first outlet.

Armour-piercing steel cores of anti-tank bullets were hardened by immersion in salt baths, a process that sometimes failed. Sakharov invented a fast, non-destructive method of testing the quality of the steel that was based on a brilliant piece of physical insight; for this he was awarded a patent and a cash payment of 3000 roubles. He married Klava, a girl employed at the factory as a laboratory technician, and moved in to live with her and her parents.

At the end of 1944, Sakharov's father secured for him a place at the Physics Institute of the Academy of Sciences in Moscow as a graduate student working with Igor Tamm, an outstanding theoretician who later won the Nobel Prize. For the next four years he worked in theoretical physics. He gained his doctorate and became a member of the lab's staff. In June 1948, Tamm told him that he was to join a special research group to investigate the possibility of building a hydrogen bomb, a bomb thousands of times more powerful than those that destroyed Hiroshima and Nagasaki. Sakharov writes that no one asked him whether he wanted to take part in such work, but the concentration, total absorption and energy that he brought to the task were his own. He realised the terrifying, inhuman nature of the weapon he was to design, but was convinced his work was essential and thought of himself as a soldier in a new scientific war. A war against whom? Sakharov does not say. He writes:

> The monstrous destructive force, the scale of our enterprise and the price paid for it by our poor, hungry, war-torn country, the casualties resulting from the neglect of safety standards and the use of forced labour in our mining and manufacturing activities, all these things inflamed our sense of drama and inspired us to make a maximum effort so that the sacrifices—which we accepted as inevitable—would not be in vain.

Sacrifices for what? To enable the Soviets to suppress the Hungarian uprising in 1956, safe in the knowledge that no one would dare to intervene? Sakharov never raises the question, even though he became increasingly aware of the callous brutality of the Soviet regime. He continued to justify his work on the grounds that strategic parity and mutual deterrence kept the peace, even though in 1973 he was to tell foreign correspondents that the Soviet Union was "a country behind a mask, a closed totalitarian society of unpredictable actions." The West, he advised them, must not let the Soviets achieve military superiority—the very superiority that until 1968 he himself had done his utmost to bring about.

In 1949, Beria, the dreaded chief of secret police whom Stalin had put in charge of the project, ordered Sakharov's team to move to the "Installation," the Soviet Los Alamos, a secret laboratory in the Urals that had been built by slave labour and was kept incommunicado from the world outside. By July 1953 everything was ready to test Sakharov's first hydrogen bomb at a site in the Kazakhstan steppe, when it occurred to someone that they had all forgotten about its radioactive fall-out. The physicists calculated that tens of thousands of people would have to be moved if they were not to receive doses of radiation of more than two hundred röntgens (600 röntgens would have killed half the population exposed). To the defence officials' fury, the test had to be postponed. Either the bomb would have to be dropped from a plane, which would have entailed many months' delay, or tens of thousands of people, including the elderly, the sick, and the very young would have to be moved long distances by army trucks over rough roads, with inevitable casualties. Evacuation was decided on all the same; people were told they could return in a month, but in fact were not allowed back in their homes until eight months after the test. Sakharov does not mention the high incidence of leukaemia that prevails among the Kazakhs as a result of that test, even though doubts about the morality of atmospheric testing gradually grew in his mind. Pavlov, the KGB general in charge of the Installation, tried to still them by telling him:

> The struggle between the forces of Imperialism and Communism is a struggle to the death. The future of mankind, the fate and happiness of tens of billions of people, alive now and yet to be born, depend on the outcome of that struggle. We must be strong in order to win. If our work and our testing are giving us strength for that battle—and they certainly are—then the victims of that testing, or any other victims, don't matter.

Marshall Vasilevsky told him, "There is no need to torture yourselves. Army manoeuvres always result in casualties—twenty or thirty deaths can be considered normal."

When the physicists said that the bomb's force had reached that predicted for a thermonuclear explosion, a celebration was held at which the defence offical in charge of the Installation congratulated Sakharov on his "exceptional contribution to the cause of peace." The physicists were lying, however. American analyses of the radioactive pollution caused by that first test have shown that no thermonuclear fusion took place; the explosion was caused purely by a fission reaction. Presumably Sakharov and his collegues did not dare to confess their failure because Malenkov, the Premier, had already publicly announced the Soviet Union's possession of the hydrogen bomb. Sakharov keeps this quiet. Only in 1955 did his "Third Idea" produce a true thermonuclear explosion.

One day a commission arrived to check the political reliability of senior personnel. They asked Sakharov whether he believed in the chromosome theory of heredity. This, as we know, was a political question. In the 1930s, Lysenko, an agriculturalist, persuaded Stalin that he could induce beneficial inheritable changes in crop plants simply by changing their environment. When geneticists disputed this, Mendelian genetics had been proscribed as an anti-Marxist heresy. When Sakharov replied that the theory seemed correct, the inquisitors exchanged significant glances, but said nothing. Another physicist gave the same answer, but his role in the project was less crucial and he was saved from dismissal only by Sakharov's and another leading physicist's intervention.

Sakharov's worries about the global effects of radioactive fall-out from nuclear explosions gradually increased. In 1959, Khrushchev announced a unilateral Soviet moratorium on nuclear testing. It seemed such a noble gesture that America and Britain soon followed under pressure of public opinion. In June 1961, Khrushchev told Sakharov and others that testing was to be resumed because the USSR had done fewer tests than its opponents. Sakharov replied that the scientists had little to gain from a resumption of testing which would favour the USA, while jeopardising the testban negotiations, the cause of disarmament, and world peace. Instead of replying directly, Khrushchev chose the subsequent banquet to proclaim in an angry and coarse tirade that Sakharov should keep his nose out of politics, which he did not understand. Everyone else sat frozen, not daring even to look in Sakharov's direction. Sakharov does not mention that Khrushchev's moratorium had been a ploy to stop American and British tests after a great fire at a nuclear-waste dump in the Urals had put the adjacent plutonium factory out of action, so that the Soviets could make no more bombs.

Sakharov's final defeat came over the duplicate tests in 1962. Following the example of the United States, which set up an atomic weapons laboratory at Livermore to compete with Los Alamos, Khrushchev set up a rival to the Installation. By 1962 each of the two laboratories had developed a more powerful and lethal thermonuclear weapon, but there were only slight differences between them. Sakharov calculated that worldwide radioactive fall-out from each of them was likely to cause hundreds of thousands of cancers, yet despite his warning both were to be tested in order to satisfy the two laboratories' competitive spirit and to show the Americans the USSR's nuclear strength. Sakharov used all his standing and influence to prevent this happening, ending with a telephone call to Khrushchev. Khrushchev gave an evasive answer and both tests went ahead. "A terrible crime was about to be committed and I could do nothing to prevent it," Sakharov writes:

> I was overcome by my impotence, unbearable bitterness, shame and humiliation. I put my face down on my desk and wept. That was probably the most terrible lesson of my life: you can't sit on two chairs at once. I decided that I would devote myself to ending biologically harmful tests. That was the main reason I didn't carry out my threat to quit the Installation. Later, after the Moscow Limited Test Ban Treaty was signed, I found other grounds for postponing my resignation.

Sakharov does not say what these were and, strangely, he did not consider the making and testing of even a single such bomb to be a crime in the first place.

He never did resign, but was dismissed in stages. He first incurred the Party's displeasure in 1964 when he publicly and courageously opposed the election of one of Lysenko's henchmen to membership of the Academy of Sciences. Lysenko, who was present, demanded that Sakharov be arrested and put on trial and Keldysh, the president, reprimanded Sakharov, but the younger corresponding members applauded his intervention loudly. Afterwards, Sakharov wrote to Khrushchev to explain the scientific case against Lysenko, but Khrushchev was furious and ordered the head of the KGB to collect material to compromise Sakharov. He was saved by Khrushchev's dismissal and Lysenko's final rout shortly afterwards.

This episode ends the first part of Sakharov's autobiography. Its chief interest lies in his sometimes vivid accounts of the personalities, attitudes, and events that guided the Soviet bid to win the nuclear arms race and, if necessary, also a nuclear war. His contacts with the highest circles of the Soviet hierarchy show them stripped of their public rhetoric, in their ruthless quest for power and barbaric indifference to the suffering and death of their own people.

In 1967 Roy Medvedev, the historian, set in train Sakharov's escape from what he calls his "hermetic world." His book *Let History Judge* opened Sakharov's eyes to Stalin's crimes, of which he had been largely unaware. In that same year he made the first of his many interventions on behalf of accused or gaoled dissidents; he wrote a private letter to Brezhnev in defence of Alexander Ginsburg and his associates. Brezhnev did not reply; instead, Efim Slavsky, the minister in charge of the Installation, on learning of Sakharov's letter, sacked him from his post as head of the theoretical department and reduced his salary from 1000 to 550 roubles a month. However, he remained Deputy Scientific Director of the Installation.

Sakharov calls this the turning-point in his life. After this he became increasingly concerned with human rights, the environment, ending the Cold War, and reforming the Soviet Union. He bought a short-wave radio and listened to the BBC and the Voice of America. He felt compelled to speak out on the fundamental issues of our age and put down his ideas in a passionate essay on *Progress, Peaceful Co-existence and Intellectual Freedom.* The essay, which begins by warning the world of the dangers it faces from thermonuclear war, calls for some of the reforms Gorbachev later put into effect, but is still firmly based on Marxist-Leninist ideology. Sakharov advocates wide social changes and an extension of public ownership in the capitalist countries, as well as preservation of public ownership of the means of production in the socialist countries. Like Martin Luther, who wanted to reform the Catholic Church rather than break away from it, Sakharov wanted to preserve socialism while freeing it from Party tyranny. In his autobiography Sakharov omits any mention of that profession of his socialist faith, which is a pity, because it would have shown the reader that his detachment from orthodoxy was slow and gradual. It certainly would not detract from the great moral courage and independence of mind which he showed in writing such an essay when he was still living among the hawks of the Installation, isolated from Moscow's intelligentsia.

When copies of his essay began to circulate in samizdat, Andropov, then head of the KGB, summoned the research director of the Installation and reprimanded him for the heresy that had emanated from his laboratory. After its full text was published in a Dutch newspaper, Efrim Slavsky hauled Sakharov over the coals and ordered him to make a declaration to the effect that the Dutch paper had published a preliminary draft of his essay without his permission, but Sakharov refused to recant and stuck to his guns. The Minister rejected Sakharov's warnings of nuclear war because he still believed in the possibility of a nuclear victory of the Soviet Union over the Imperialist powers, just as American hawks still believed that the United States could win such a war against Communism. That interview took place in July 1968 during the last weeks of the Prague Spring. The Minister assured Sakharov that the Central Committee had ruled out armed intervention, but on 21 August the entry of Warsaw Pact troops into Czechoslovakia destroyed whatever was left of Sakharov's faith in the Soviet system. Soon after that interview he was forbidden to set foot in the Installation, which was tantamount to dismissal.

The next blow came in March 1969 when his wife Klava died of cancer, leaving him for many months in a daze, and unable to do anything in science or in public life. He did not inform Klava's parents of her death, a strange omission which he later came to regret (his account of this episode has been deleted from the English translation).

In May 1969, the Minister transferred him back to the Academy's Institute of Physics in Moscow where his career had begun, with a modest salary to supplement his Academician's income. Sakharov describes his scientific work from then on as minor; most of his energy went into fighting injustice in all its forms. In 1970 an acquaintance invited him to help found a Human Rights Committee that would study and publicise human rights problems in the Soviet Union. At first Sakharov worried that this would arouse too many false hopes, that the Committee would be powerless to respond to the flood of calls for help it would elicit, but he joined all the same, and "not having been spoiled by an abundance of friends" he welcomed the human contacts of its weekly meetings, especially those with Elena Bonner.

Sakharov's inability to form close personal relations extended to his own family. He paints a cheerless picture of family life, devoid of warmth, affection, and humour. All he has to say about his relations with Klava and, after her death, with his children is that he "always avoided confrontations," that his affluence never brought him and Klava much happiness, that their lives had been empty and had brought little joy to their children. The less he concerned himself with these insoluble personal matters, he writes, the more productive was his life outside his home. Sakharov's cold, distant description of Klava reminded me of Einstein's heartless references to his wife in his correspondence with Max Born. After Klava's death he gave all the money he had won from various prizes—139,000 roubles—to charity, without a thought that his children might need it, like Leo Tolstoy who made over the royalties from his books to charity and left his family destitute. Perhaps all three men tried to make up for their failure to love those closest to them by trying to love all mankind.

Even tragic history has its ironic side, but Sakharov appears to have been blind to that of his own life. There is not a single joke in his 600-page memoir, which made me wonder if he ever laughed. Did he want to atone for his lethal inventions by becoming an angel of peace? His book contains no hint of regret, even when he advises the West that only when it deployed equivalent weapons as bargaining chips would the USSR be induced to eliminate its gigantic ICBMs and its medium-range missiles. Had he forgotten that these weapons were, in part, of his own making?

In 1972 Sakharov married Elena Bonner, of whom he writes with the greatest admiration. The second half of his autobiography recounts, on the one hand, the misdeeds of an all-pervasive, callous, underhand, corrupt, and cynical tyranny that poses as a theocracy devoted to the defence of the holy Marxist Grail, and, on the other, it relates the Sakharovs' heroic struggle for human rights, for détente and disarmament, for Jewish emigration, and against injustices of all kinds, in the face of continuous, vicious harassment and intimidation by the KGB. One of his first cases was that of Zhores (the Russian spelling of the French socialist's Jean Jaurès's name) Medvedev, the biologist and twin brother of Roy, who had been arrested and confined in a psychiatric hospital when his "scandalous" book *The Rise and Fall of Lysenko* was published in the West. Sakharov responded by walking into an international conference on biochemistry and genetics which was being held in Moscow, writing the news of Medvedev's imprisonment on the blackboard and appealing for signatures on a letter of protest. This led Academician Alexandrov, the physicist who succeeded Keldysh as president of the Academy, to tell Sakharov that he himself needed psychiatric treatment. The storm of international protest set in train by Sakharov and by Roy Medvedev led to Zhores' release soon afterwards.

Sakharov didn't fight injustice only when scientists or dissidents were implicated. In the early 1960s, two black marketeers were sentenced to 15 years' imprisonment. In the labour camp they gossipped about members of the élite who had been among their customers. The élite silenced them by passing an amendment to the criminal law that made such black marketeering a capital offence and had the men retried under the new law and sentenced to death. Sakharov protested—in vain—against punishing them under a law passed after the crime had been committed.

Sakharov's advice to the West not to let the Soviets achieve nuclear superiority, and his later appeal to the International Red Cross to demand the right to inspect Soviet prisons, labour camps, and psychiatric hospitals, unleashed a press campaign against him of the kind often used by the Politburo to prepare public opinion for a show trial. This time, he may have been saved by a telegram from Philip Handler, president of the Academy of Sciences of the USA, who wrote:

> Were Sakharov to be deprived of his opportunity to serve the Soviet people and humanity, it would be extremely difficult to imagine successful fulfilment of American pledges of binational scientific co-operation, the implementation of which is entirely dependent upon the voluntary effort and good will of our individual scientists and scientific institutions.

Even so, a Russian mathematician who issued an open letter in defence of Sakharov was dismissed from his job.

The Politburo finally decided to silence Sakharov by exiling him to Gorky, where a whole regiment of KGB personnel supervised and impeded his every move. I believed I had heard enough about the KGB before I read this book, but the degree of petty vindictiveness and mean cunning disclosed in Sakharov's memoirs exceeds anything I had imagined. The autobiography ends in December 1986 when two engineers and a KGB man came to install a telephone in the exiled Sakharovs' flat in Gorky. The next day, Gorbachev rang to invite them back to Moscow "so that Sakharov could continue his patriotic work." Twenty years earlier, this had been the official designation for his work on atomic weapons. Did Gorbachev have in mind instead Sakharov's campaign for an end to the Cold War?

Liberating France[*]

Successful people's autobiographies often follow the pattern of Charlie Chaplin's, who first delights us with his youthful acting talent, which raised him to fame and fortune from a childhood of penury and want, and then bores us with an enumeration of his movies and all the important people he has met. By contrast, François Jacob gives no hint that he became president of the Pasteur Institute in Paris, a Nobel Laureate, and one of the world's leading biologists, but presents us with a life symbolic of the tragedy and rebirth of France. Jacob was born in Paris in 1920, of well-to-do Jewish middle-class parents, and had a happy childhood. He attended school in Paris, intending at first to become a soldier like his maternal grandfather, a four-star Jewish general in the French artillery, a wise man full of vigor and courage, a patriot yet no chauvinist, a humane soldier who was the "statue within" of the title on whom young François tried to model himself. The obligatory stepping stone to a military career was the Ecole Polytechnique, but the Draconian teachers of the *lycée* that prepared boys for entry to that famous institution were so sadistic that Jacob quit and decided to become a surgeon instead.

Jacob does not present a sequential account of his life, but a selection of vivid images and episodes—like a jumbled show of lantern slides. He sees his life as "a series of different selves—I might almost say, strangers…

*A review of *The Statue Within: An Autobiography* by François Jacob (translated by Franklin Philip; Basic Books).

Would I recognize them if I passed them in the street?" Yet he made me feel as if their moods and fantasies had been my own, as if I had been an only child loved by his charming mother, had dissected corpses in his anatomy class, had been wounded in battle during World War II, or bungled my research, or as if I had encountered the hauteur of General de Gaulle ("the majesty of a Gothic cathedral"), or lived with the charm, brilliance, and arrogance of his colleague Jacques Monod.

When the young François admired Napoleon, his grandfather told him to idolize no one, neither great men, because they are no gods, nor gods, because they don't exist. When the general felt he was about to die, he told the boy not to believe in a life hereafter. He clasped his hand, looked into his eyes, and repeated: "There is nothing. Nothing. The void. So my only hope is you. You and the children you'll have." In place of religion, the general built up in the boy's mind faith in France's great institutions. Jacob writes:

> The Constitution, the authorities, the civil service, the army,... the Polytechnique were a little like the Pantheon, the Arch of Triumph, Notre Dame.... They formed the indestructible framework of our country, of our life.... I scarcely imagined that better ones could be devised.

Yet in the spring of 1940 the inconceivable happened—under the impact of Hitler's tanks the entire edifice collapsed. A few days before the Germans reached Paris, Jacob's mother died of cancer. Desolate and disillusioned, he fled to Bordeaux and boarded a ship for England. "Calm, assured, orderly, confident England," after chaotic, defeated, demoralized France. True to his grandfather's spirit, he was determined to fight the invaders. His medical studies, his family, his girlfriends, "all that must be put away in a box to be opened only on his return to France." But would he ever see it again?

On the next page we find him, in August 1944, an auxiliary medical officer in the Free French Army aboard a British landing craft crossing the English Channel; on the horizon he spotted the Promised Land, the coast of France. This was the moment he had lived for during four bitter years of lonely exile, but within a week of landing in Normandy his hopes for a triumphant return to Paris were shattered by a German bomb that smashed his right arm and leg and almost fatally injured his chest. He could have escaped injury by sheltering in a ditch when the bombers approached, but he responded to the desperate plea of a mortally wounded comrade to remain by his side. This heroic and compassionate deed cost him nearly a year in the hospital, his chosen career as a surgeon ("the finest profession in the world" it had seemed to him), and a life of chronic pain (which he conceals).

Jacob recalls his years as a soldier in the war in vignettes. He describes four nostalgic, lonely New Year's Eves, each spent in a different God-forsaken place in Africa, and a fifth spent alone in a bare room of a military hospital in Paris. In 1942 he took part in General Leclerc's grueling but victorious march from Chad in Central Africa to the Libyan coast of the Mediterranean across a thousand miles of desert: "When we reached the sea, it seemed that in the distance we saw the coast of France." He makes us experience the mixture of exaltation and fear before battle as the French army awaited a German attack in the Tunisian desert, and his frustration at having nothing to fight with but bandages. The ill-equipped, part-European and part-African French force stood firm, but the deadly, professional German attack would have annihilated them, had it not been for the bold intervention of the British Royal Air Force, whose planes the Free French followed "with eyes conscious of owing them more than they could ever repay." Jacob recalls his elation after the victory:

> Neighbors had come from France and Germany to kill each other on an uninhabited, lifeless land. A strange land, suddenly transformed for a few hours into hell and now recovering its peace, its impassivity. In the darkness that gradually blurred the surrounding shapes, the unity of the night seemed to testify to the unity of the world. I felt a new life being born. Like an escaped prisoner who, in the evening of a long march, reaches the summit of a mountain to find a land of welcome and liberty. The universe seemed to me full and mysterious, like a young animal. Beyond this sand, beyond the mountains, there was the sea. Beyond the sea, France: so green, so full of life. And, for the first time in three years, I knew physically, in every particle of my body, that the return to France was no longer simply a dream. That nothing henceforth, and no one, would prevent us from going home. Nothing save death.

Jacob's story is haunted by death. His book begins in Paris after the war with the visit of a one-legged comrade who intimates that Jacob should help to release him from life if his suffering becomes unbearable. Jacob pretends not to hear and feels a coward for ignoring his comrade's *cri de coeur*; it raises in his mind the specter of being made helpless by old age, of losing his mind like his once proud grandmother, of being at the mercy of others. We cannot help being born, he asserts, but we can choose the moment of death, provided we don't leave it until too late. But when is the right moment?

In the war in Africa, his narrowest escape came in 1943 on a moonless night when he was ordered to walk across the Tunisian desert, past the German lines, to take medical help to a French outpost. He walked alone. A German shell exploded behind him. He threw himself to the ground and lay there in a cold sweat, paralyzed by fear for what seemed like hours. A dog's bark stirred him, made him pull himself together, discover that he had lain there for only five minutes. As he proceeded in the dark, he found himself face to face with a German sentry. Should he run away? He carried on, expecting to be killed by a salvo in the back, but miraculously, the German spared him. Years later, at home next to his wife, he still awakes bathed in cold sweat, paralyzed, in terror, the bitter taste of death in his mouth, the faces of his dead friends in his eyes. With an effort, he arouses himself and tiptoes to his children's room, "watching them until their faces blot out those of the dead." "In the evening I hurried home to rejoin this beautiful woman with these magnificent children. . . . It was like the return of spring, the leaves once again on the trees, the sun, flowers. It was like a revenge on the war, on death."

Yet after his narrow escape from the grave in Normandy, such happiness eluded him for five lonely, bitter years. In 1944, some weeks after an ambulance train had brought him to Paris, his father found him at last in the Val-de-Grâce Hospital. The visit brought back the stabbing pain at the loss of his mother; a pain exacerbated when the acutely embarrassed father confessed in a tragicomic scene that he would presently remarry. Another visitor at this bedside was Odile, his first love, who told him she was soon to marry another man. On Armistice Day in May 1945, on the day of triumph, nine months after he was hit, Jacob was still in the hospital, facing another operation to extract a piece of shrapnel from his infected thigh. When at last he was discharged nobody wanted him. "Everyone went about his business as though I had not returned." Perfunctorily he completed his medical studies, but his injuries did not allow him to take up surgery, the only medical specialty that attracted him. Like a sulking child who refuses all toys if he cannot have his favorite one, he gave up medicine altogether.

He found France still dominated by the money-grubbing petit-bourgeois whose flirtations with fascism before and during the war he blamed for France's defeat. His disgust drove him to Communist meetings, but they grated on him: "the words used, the meanings they were given, the arguments from authority, the constant references to the sacred Marxist texts, the...speakers'...certainty of being right, of being in possession of the truth...both political and moral" repelled him.

> What could people in my generation believe in, if they were neither religious nor communist? Their youth had been stolen from them; their friends killed; their hopes, their enthusiasm dashed. What meaning, what substance could they now give to words like *honor, truth, justice*, and even *nation*?

Jacob paints a picture of himself having dreams of glory but no conception of his own ability, no sense of vocation, no woman friend, no proper home. He dabbled in making antibiotics, wrote movie scripts, worked in a bank for a few days, abandoning it in disgust, bought a book on law and put it aside after three pages. He was shiftless, interested in nothing, wholly disillusioned. This is the saddest and most poetic part of his book.

The turning point came when he met a young musician from a Jewish middle-class family like his own; her charming looks reminded him of his mother. "As always when meeting a girl I felt awkward and dim-witted." but she fell in love with him all the same. Soon after they married they had dinner with a cousin of hers, a young man who had served in the war like Jacob and was now doing research with the great biologist Boris Ephrussi. As Jacob listened to the rapturous account he gave of his work, a thought flashed through his mind: "If that man can do it why can't I?" He suppressed his fears of being good for nothing, and knocked at the doors of two professors of biology, asking to be taken on as an apprentice. They both turned him down, but Jacques Tréfouël, the director of the Pasteur Institute, saw beneath Jacob's mask of "arrogance and shyness," disregarded his professed ignorance of biology, and awarded him a research fellowship. From that moment Jacob's story is transformed.

Jacob still needed a teacher. He approached the great microbiologist André Lwoff, but Lwoff said he had no room for him. Jacob tried again. Lwoff refused. Jacob persisted. One day, he found Lwoff when he had just made a discovery and felt so cheerful that he said "Yes." After this, Lwoff treated him like a son, encouraged him and gave him confidence in himself. I was reminded of David Keilin, Lwoff's former teacher in Cambridge, who was to become my own teacher and to give me as much encouragement and affection as Jacob writes he received from Lwoff. Had Keilin set Lwoff an example?

Jacob quickly discovered that scientific research is "not the cold, studious, stiff,...slightly boring world one often imagines. But, on the contrary, a world full of gaiety, of the unexpected, of curiosity, of imagination. A life animated as much by passion as by logic." "That one could live, travel, eat and raise a family while spending the best part of one's time doing what one loves, that seemed like a miracle I still found hard to believe." I still do.

For Jacob, research in the laboratory is as dramatic as the battles in the desert, and it is spiced by bemused observations of his colleagues' varied characters and idiosyncracies. His catalog of *Homo sapiens scientificus* includes a number of well-known molecular biologists, beginning with James Watson:

> Tall, gawky, scraggly, he had an inimitable style. Inimitable in his dress: shirttails flying, knees in the air, socks down around his ankles. Inimitable in his bewildered manner, his mannerisms: his eyes always bulging, his mouth always open, he uttered short, choppy sentences punctuated by "Ah! Ah!" Inimitable also in his way of entering a room, cocking his head like a rooster looking for the finest hen, to locate the most important scientist present and charging over to his side. A surprising mixture of awkwardness and shrewdness. Of childishness in the things of life and of maturity in those of science.

Jacob began research on the genetics of bacteria and viruses, a field allowing him to formulate a hypothesis, design an experiment, test it, and find an answer next morning. Such work was ideally suited to his restless, Faustian spirit. ("Once I had obtained a result, it no longer interested me.") Within a few years Jacob, together with Elie Wollman, discovered a brilliant method of mapping the sequence of genes along the chromosome of the coli bacterium. These bacteria mate by fusing together in pairs, and during their fusion genes are transferred from the male to the female bacterium. Agitation in a kitchen blender tears them apart. By tearing them apart at successive intervals after mating, Jacob and Wollman discovered that the genes are transferred from the male to the female in linear sequence according to a strict timetable, revealing the sequence of their arrangement on the chromosome, which turned out to be a circular double helix of DNA:

> The three or four years spent studying bacterial conjugation, erotic induction, the *coitus interruptus*, was a period of jubilation. A time of

excitement and euphoria. But my memory of it is frozen. It has crystallized in articles and reviews, abstracts and lectures. It has lost its color, dried up in a story too often told, too often formulated. A story that has become so logical, so reasonable as to have lost all juice, no longer conveying the sound and the fury of the daily research. What gave it life has been swallowed up by time. Gone are the abortive trials, the failed experiments, the false starts, the misguided attempts. Forgotten are the fallacious arguments, the hesitations, the jabs of the sword in the water, the groundless joys, the spurts of rage against oneself or against others. Vanished are the hours spent counting the colonies, the anxieties, the uncertainties, the endless waiting. Everything has become smooth and polished. A fine story, very clear, with beginning, middle, and end. With well-oiled, well-articulated, well-arranged experiments, one following another, leading without fault, without hesitation, in seamless argumentation, to a well-established truth. The truth found in textbooks on genetics.

Occasionally other fragments of the past come back to light. Loom up, intact. Impressions. A warmth suddenly coming to my cheeks, for example, as I come upon an old photo of Jacques Monod, with his ironical little smile. Immediately I am back in his office, seated before him. A room in the middle of the hallway, on the ground floor. I have come down to tell him about a result that turned up that very morning. A still uncertain result. Obtained once only. But I have to talk about it, tell the story, share my excitement. To think, to forge ahead, I need to discuss. To try out ideas, to see them rebound. And no one plays this game better than Jacques. He listens to me. Looks at me. Holding his chin in one hand, digging into it with his finger. He asks a question. Rises. Goes to the blackboard. Sketches a diagram. Returns. Abruptly asks whether I have thought of doing a certain control experiment without which my result is worthless. I feel myself redden in confusion. I have forgotten this control. A faint ironic smile plays over Jacques's lips. The smile in the photo. I would like the earth to open up and swallow me.

Jacob's story ends on Christmas Eve 1960, when he and Monod completed their famous paper on "Genetic Regulatory Mechanisms in the Synthesis of Proteins," which opened a new era in our understanding of living cells. It was known that all chemical reactions in living cells are accelerated by enzymes and that all enzymes are proteins. It was also known that the structure of each protein is determined by the gene that codes for it. Most enzymes are made only in response to need, showing that there must be some mechanism controlling their synthesis, but it was not known what that mechanism is. Jacob and Monod discovered that there exist two kinds of genes: the ones that code for proteins and others that regulate the rate at which these proteins are made. These regulator genes switch the synthesis of proteins on or off in response to chemical stimuli. Jacob and Monod discovered these genes in coli bacteria and provided a largely correct picture of their operating mechanism. They suspected similar mechanisms to exist in all forms of life, whence Monod's famous dictum: "Anything found to be true in *Escherichia coli* must also be true for elephants." Almost! The discovery won Jacob and Monod, together with Jacob's much loved and admired teacher André Lwoff, the Nobel Prize for Physiology or Medicine for 1965.

Enemy Alien

It was a cloudless Sunday morning in May of 1940. The policeman who came to arrest me said that I would be gone for only a few days, but I packed for a long journey. I said goodbye to my parents.

From Cambridge, they took me and more than a hundred other people to Bury St. Edmunds, a small garrison town twenty-five miles to the east, and there they locked us up in a school. We were herded into a huge empty shed cast into gloom by blacked-out skylights thirty feet above us. A fellow-prisoner kept staring at a blank piece of white paper, and I wondered why until he showed me that a tiny pinhole in the blackout paint projected a sharp image of the sun's disc, on which one could observe the outlines of sunspots. He also taught me how to work out the distances of planets and stars from their parallaxes and the distances of nebulas from the red shifts of their spectra. He was a warmhearted and gentle German Roman Catholic who had found refuge from the Nazis at the Observatory of Cambridge University. Years later, he became Astronomer Royal for Scotland. In the spring of 1940, he was one of hundreds of German and Austrian refugee scholars, mostly Jewish and all anti-Nazi, who had been rounded up in the official panic created by the German attack on the Low Countries and the imminent threat of an invasion of Britain.

After a week or so at Bury, we were taken to Liverpool and then to an as yet unoccupied housing estate at nearby Huyton, where we camped for some weeks in bleak, empty, semi-detached two-story houses, several of us crowded into each bare room, with nothing to do except lament successive

Allied defeats and worry whether England could hold out. Our camp commander was a white-mustached veteran of the last war; then a German had been a German, but now the subtle new distinctions between friend and foe bewildered him. Watching a group of internees with skullcaps and curly side-whiskers arrive at his camp, he mused, "I had no idea there were so many Jews among the Nazis." He pronounced it "Nasis."

Lest we escape to help our mortal enemies, the Army next took us to Douglas, a seaside resort on the Isle of Man, where we were quartered in Victorian boarding houses. I shared my room with two bright German medical researchers, who opened my eyes to the hidden world of living cells—a welcome diversion, lifting my thoughts from my empty stomach. On some days, the soldiers took us out for country walks, and we ambled along hedge-flanked lanes two abreast, like girls from a boarding school. One day near the end of June, one of our guards said casually, "The bastards have signed." His terse message signified France's surrender, which left Britain to fight the Germans alone.

A few days later, tight-lipped Army doctors came to vaccinate all men under thirty—an ominous event, whose sinister purpose we soon learned. On July 3rd, we were taken back to Liverpool, and from there we embarked on the large troopship Ettrick for an unknown destination. About twelve hundred of us were herded together, tier upon tier, in one of its airless holds. Locked up in another hold were German prisoners of war, whom we envied for their Army rations. On our second day out, we learned that a German U-boat had sunk another troopship, the Arandora Star, which had been crammed with interned Austrian and German refugees and with Italians who were being deported overseas. More than six hundred of the fifteen hundred people aboard were drowned. After that, we were issued life belts.

Suspended like bats from the mess decks' ceilings, row upon row of men swayed to and fro in their hammocks. In heavy seas, their eruptions turned the floors into quagmires emitting a sickening stench. Cockroaches asserted their prior tenancy of the ship. To this revolting scene, Prince Frederick of Prussia, then living in England, restored hygiene and order by recruiting a gang of fellow-students with mops and buckets—a public-spirited action that earned him everyone's respect, so that he, grandson of the Kaiser and cousin of King George VI, became king of the Jews. Looking every inch a prince, he used his royal standing to persuade the officers in charge that we were not the Fifth Columnists their War Office instructions made us out to be. The commanding colonel called us scum of the earth all the same, and once, in a temper, ordered his soldiers to set their bayonets upon us. They judged differently and ignored him.

One day, I passed out with a fever. When I came to, in a clean sick bay that had been established by young German doctors, we were steaming up the broad estuary of the St. Lawrence River, and on July 13th we finally anchored off gleaming-white Quebec city. The Canadian Army took us to a camp of wooden huts on the citadel high above the town, close to the battlefield where the English General James Wolfe had beaten the French in 1759. The soldiers made us strip naked so they could search us for lice, and they also confiscated all our money and other useful possessions, but I forestalled them by dropping the contents of my wallet out of the window of the hut while we were waiting to be searched, and went around to pick them up the next day, when the soldiers had gone. Sometimes jewels are safest on a scrap heap.

In Canada, our status changed from that of internees to that of civilian prisoners of war, entitling us to clothing—navy jackets with a red patch on the back—and Army rations, which were welcome after our first two days, when we were left without food. Even so, the fleshpots of Canada were no consolation for our new status, which made us fear that we would remain interned for the duration of the war and, worse still, that in the event of England's defeat we would be sent back to Germany to be liquidated by Hitler. To have been arrested, interned, and deported as an enemy alien by the English, whom I had regarded as my friends, made me more bitter than to have lost freedom itself. Having first been rejected as a Jew by my native Austria, which I loved, I now found myself rejected as a German by my adopted country. Since we were kept incommunicado at first, I could not know that most of my English friends and scientific colleagues were campaigning to get the anti-Nazi refugees, and especially the many scholars among them, released.

I had come to Cambridge from Vienna as a graduate student in 1936 and had begun my life's research work on the structure of proteins. In March of 1940, a few weeks before my arrest, I had proudly won my Ph.D. with a thesis on the crystal structure of hemoglobin—the protein of the red blood cells. My parents had joined me in Cambridge shortly before the outbreak of war; I wondered when I would see them again. But, most of all, I and the more enterprising among my comrades felt frustrated at having to idle away our time instead of helping in the war against Hitler. I never imagined that before long I would be returning to Canada as a free man, engaged in one of the most imaginative and absurd projects of the Second World War.

Our camp offered a majestic panorama of the St. Lawrence and of the lush green country stretching away to the south of it. As one stifling-hot, languid day followed another, freedom beckoned from the mountains on the horizon, beyond the United States border. I remembered the Bishop's advice to King Richard II; "My lord, wise men ne'er sit and wail their woes, but presently prevent the ways to wail." How could I escape through the barbed wire fence? Suppose I surmounted that hurdle without being spotted by the guards, who stood on watchtowers with their machine guns trained on us? Who would hide me after my absence had been discovered at the daily roll call? How could I persuade the Americans to let me join my brother and sister there and not lock me up on Ellis Island? These questions turned over and over in my mind as I lay on my back in the grass at night, listening to the faint hooting of distant trains and watching the delicately colored flashes of the northern lights dance across the sky. Soon I began to dream of jumping on goods wagons in the dark or of fighting my way across the frontier through dense mountain forests—or just of girls.

As a Cambridge Ph.D. of four months' standing, I found myself the doyen of the camp's scholars, and organized a camp university. Several of my Quebec teaching staff have since risen to fame, though in different ways. The Viennese mathematics student Hermann Bondi, now Sir Hermann, taught a brilliant course of vector analysis. His towering forehead topped by battlements of curly black hair, he arrived at his lectures without any notes and yet solved all his complex examples on the blackboard. Bondi owes his knighthood to his office as chief scientist as Britain's Ministry of Defense, and his fame to the steady-state theory of the universe. This theory postulates that, as the universe expands, matter is continuously being created, so that its density in the universe remains constant with time. A universe like that need not have started with a big bang, because it would never have known a beginning and it would have no end. Bondi developed that ingenious theory with another Viennese interned with us—Thomas Gold, who, like him, was still an undergraduate at Cambridge, and who became professor of astronomy at Cornell University. The theory's third author was Fred Hoyle, the Cambridge cosmologist and science-fiction writer.

Theoretical physics was taught to us lucidly by Klaus Fuchs, the tall, austere, aloof son of a German Protestant pastor who had been persecuted by Hitler for being a Social Democrat. Klause Fuchs himself had joined the German Communist Party shortly before Hitler came to power, and fled to England soon afterward to study physics at Bristol University. After his release from internment in Canada, he was recruited to work for the atomic-bomb project, first in Birmingham and then at Los Alamos, and when the war was over he was appointed head of the theoretical-physics section of the newly established British Atomic Energy Research Establishment, at Harwell. Everywhere, Fuchs was highly regarded for his excellent scientific work, and at Harwell he was also noted for his deep concern with security. Then, in the summer of 1949, just before the explosion of the first Russian atomic bomb, the Federal Bureau of Investigation found reason to suspect that a British scientist had passed atomic information to the Russians, and the Bureau's description in some ways fitted Fuchs. After several interrogations, Fuchs broke down and confessed—in January of 1950—that from the very start of his work he had passed on to the Russians most of what he knew of the Anglo-American project, including the design of the first plutonium bomb. A few days after Fuchs' conviction for espionage, the Prime Minister, Clement Attlee, assured Parliament that the security services had repeatedly made "the proper inquiries" about Fuchs and had found nothing to make them suspect him of being a fanatical Communist. Neither had I gathered this during my contacts with him in Canada, but when I said so to an old colleague he told me that Fuchs and he had belonged to the same Communist cell while they were students at Bristol. "The proper inquiries" cannot have been all that searching.

Having no inkling of the tortuous mind that later made Fuchs betray the countries and the friends that had given him shelter, I simply benefitted from his excellent teaching. In my own lectures, I showed my students how to unravel the arrangement of atoms in crystals, and I spent the rest of my time trying to learn some of the advanced mathematics that I had missed at school and at the university.

The curfew was at nine-thirty. The windows of our hut were crossed with barbed wire. Its doors were locked, and buckets were put out. Stacked into double bunks, about a hundred of us tried to sleep in one room where the air could be cut with a knife. In the bunk above me was my closest friend from student days in Vienna. We had roughed it together in the mosquito-ridden swamps of northern Lapland and had almost suffered shipwreck on a small sealer in the stormy Arctic Ocean. These adventures had inured us to the physical hardships of internment, but the exhilarating sense of freedom that they had instilled in us made our captivity even harder to bear. Lacking other forms of exercise, we made a sport of reading our jailers' regulation-ridden minds. One day, the prisoners were told that each could send a postcard to his next of kin in England, but two weeks later all the postcards were returned—without explanation, at first. The camp seethed with frustration and angry rumors, but my friend and I guessed that after leaving the postcards lying around for a couple of weeks the Army censor returned them all because not every card carried its senders' full name. It took a month more before my card reached my parents in Cambridge, with the laconic message that Prisoner of War Max Perutz was safe and well.

In time, we learned through rumor that our scenic and efficient camp was to be dismantled and we were to be divided between two other camps. Would friend be separated from friend? By age or by the alphabet? It occurred to me that the pious Quebecois might divide us into believers and heretics—that is to say, into Roman Catholics and the rest—and my hunch was soon confirmed. Since my Viennese friend was a Protestant and I was a Roman Catholic, we were destined for different camps. Adversity tightens friendships. Our familiar Viennese idiom, my friend's keen sense of the ridiculous, and shared memories of carefree student days with girls, skiing, and mountain climbing, had helped us to escape from the crowd of strangers around us into our own private world. I decided to stay with the Protestants and the Jews, who also included many scientists, and soon found a Protestant who preferred to join the Catholics. Like Ferrando and Guglielmo, the handsome young swains in "Così Fan Tutte," we swapped identities. The false Max Perutz was sent with the faithful to the heaven of a well-appointed Army camp, while I, the real one, was dispatched with the heretics and Jews to the purgatory of a locomotive shed near Sherbrooke, Quebec. To start with, it had five cold-water taps and six latrines for seven hundred and twenty men.

Some weeks later, our comedy of errors was unmasked. The stern camp commander, though impressed by the purity of my motives, sentenced me to three days in the local police prison. Here was privacy at last—yet not quite. They locked me up in a cage resembling a monkey's in an old-fashioned zoo. It had no chair, no bed—only some wooden planks to rest on. Unlike the prisoner in Oscar Wilde's "Ballad of Reading Gaol," I did not look

> With such a wistful eye
> Upon that little tent of blue

> Which prisoners call the sky,
> And at every drifting cloud that went
> With sails of silver by,

because I never even saw the sky. But I had smuggled in several books inside my baggy plus fours, so I was not as bored as the poor soldier who had to march up and down on the other side of the iron grille to guard me. My reading was undisturbed and my sleep interrupted only by the occasional drunk; the little mites burrowed into my skin without waking me. Only when they had made themselves at home there during the weeks that followed did the scabies rash keep me awake at night.

Back in the Sherbrooke camp, where my spirits sagged at the prospect of wasted years, the camp commander summoned me again—this time to tell me that my release had been ordered by the British Home Office and that I had also been offered a professorship by the New School for Social Research, in New York City. He then asked me if I wanted to return to England or remain in the camp until my release to the United States could be arranged. I replied that I wanted to return to England, and this drew the admiring comment that I would make a fine soldier. I have never heard that said by anyone else, before or since, but what led me to my decision was that my parents, my girl friend, and my research were in England, and from the safe distance of Sherbrooke the U-boats and the blitz did not frighten me. My American professorship had been arranged by the Rockefeller Foundation as part of a rescue campaign for the scholars whom the foundation had supported before the war broke out, and in principle it would have qualified me for an American immigration visa, but I was sure that as a prisoner of war without a passport I would never get it. The camp commander raised my hopes that I would be sent home soon.

From our perch on the citadel of Quebec, we had been able to watch the ships go by on the St. Lawrence, but in the locomotive shed we could only watch the men line up for the latrines. In Quebec, we had had a room in a hut set aside for quiet study, but here among a milling, chatting crowd of men my assaults on differential equations petered out in confusion. Camp committees, locked in futile arguments over trivial issues, were chaired by budding lawyers fond of hearing themselves talk. In excruciating boredom, I waited impotently from day to day for permission to leave, but weeks passed and my captivity dragged on.

There was little news from home except for hints that my father, who was then sixty-three and had been an Anglophile from youth, had been interned on the Isle of Man. He shared that fate, I learned afterward, with a frail, meticulous old Viennese with sensitively cut features who was distraught at having his life's work interrupted for a second time. This was Otto Deutsch, the author of the then incomplete catalogue of Franz Schubert's collected works. He finished it in later years at Cambridge.

Early in December, I was among some prisoners destined for release from my and several other camps who were at last put on a train going east. From its windows, the snow-clad forest looked the same each day, so that we seemed to move merely to stay in the same place, like Alice running with the Red Queen. I had been sad at leaving my Viennese friend behind, but was overjoyed to find his father—whom he had feared drowned on the Arandora Star—among the prisoners on the train. Some weeks earlier, the father, on discovering that his son was interned in another Canadian camp, asked to be transferred there, and he was disconsolate that instead the Army had now put him on a train carrying him even farther away. The train finally dumped all of us in yet another camp—this one in a forest near Fredericton, New Brunswick. No one told us why or for how long.

In the Arctic weather, I contracted a bronchial cold that made the dark winter hours seem endless. My father had taught me to regard Jews as champions of tolerant liberalism, but here I was shocked to run into Jews with an outlook as warped and brutal as that of Nazi Storm Troopers. They were members of the Stern Gang, which later became notorious in Israel for many senseless murders, including that of the Swedish Count Folke Bernadotte, whom the United Nations had appointed as mediator in the Arab-Israeli conflict.

At Christmas, we were finally taken to Halifax, where we were met by one of Britain's prison commissioners—the shrewd and humane Alexander Paterson, sent out by the Home Office to interview any of the internees who wanted to return to Britain. His mission was stimulated by public criticism—"Why Not Lock Up General de Gaulle?" was one of the sarcastic headlines in a London paper that helped to make the War Cabinet change its policy. Paterson explained that it had been impossible to ship any of us home earlier, because the Canadians had insisted that prisoners of war must not be moved without a military escort, yet had refused either to release us in Canada or to escort us to England, on the ground that our internment was Britain's affair. The British War Office had now fulfilled the letter of the regulation by detailing a single Army captain to take us home.

Chaperoned by the one urbane captain, two hundred and eighty of us embarked on the small Belgian liner Thysville, which had been requisitioned by the British Army complete with its crew, including a superb Chinese cook. From this moment, we were treated as passengers, not prisoners, but I became fretful once again when days passed and the Thysville had not cast off her moorings; no one had told us that we had to wait for the assembling of a big convoy. As we finally steamed out to sea, I counted more than thirty ships, of all kinds and sizes, spread over a huge area. At first, Canadian destroyers escorted us, but we soon passed out of their range, and our remaining escort consisted of only one merchant cruiser—a passenger liner with a few guns on deck—and a single submarine, neither of them a match for the powerful German battleships Scharnhorst and Gneisenau, which, so our radio told us, prowled the Atlantic not far from our route. We steamed at only nine knots—the speed of the slowest cargo boat—and took a far-northerly course, trusting to the Arctic night to hide us. Both my Viennese friend and his father were on board.

Early in the voyage, I stood at the rail imagining a torpedo in every breaker. Like the Ancient Mariner,

> Alas! (thought I, and my heart beat loud)
> How fast she nears and nears!

But time soon blunted my fears and I began to enjoy the play of wind and waves, I slept in a warm cabin between clean sheets, took a hot bath, brimful, each morning, ate my meals from white table linen in my friends' company, walked in the bracing air on deck or retired to read in a quiet saloon. Toward the end of the third week, we were cheered by the sight of large black flying boats of the Coastal Command circling over us, like sheepdogs running round their flock, to keep the U-boats at bay. One gray winter morning, the entire convoy anchored safely in Liverpool Harbor.

On landing, I was formally released from internment and handed a railway ticket to Cambridge, and I was told to register with the police there as an enemy alien. When I presented myself at my friend's house near London that night, she found me looking so fit that she thought I must have returned from a holiday cruise, but then she admired the elaborate needlework by which I had kept my tweed jacket in one piece for all those months, so as not to have to wear the prisoners' blue jacket with the large red circle on the back. Next morning, at the Cambridge station, our faithful lab mechanic greeted me, not as an enemy alien, but as a long-lost friend; he brought me the good news that my father had been released from the Isle of Man a few weeks earlier and that both he and my mother were safe in Cambridge. That was in January of 1941.

Less than three years later, I returned to Canada as a representative of the British Admiralty and was accommodated in a suite in Ottawa's luxurious Hotel Château Laurier without being searched for lice. I owed that change of fortune to the remarkable Geoffrey Pyke, former journalist and amateur strategist, who enlisted me for a project that bore the mysterious code name Habakkuk. In 1938, I had taken part in an expedition to the Swiss Alps, where we found out how the tiny snowflakes that fall on a glacier grow into large grains of ice. It had never occurred to me that the expertise I gained there would be of any use to the war effort. When I returned from internment, my professor, W.L. Bragg, encouraged me to resume my peacetime research on the structure of proteins, with the continued support of the Rockefeller Foundation, and for a long time nobody wanted my help for anything related to the war except fire-watching on the roof of the laboratory at night.

At last, one day in the spring of 1942, an urgent telephone call summoned me to London. I was directed to an apartment in Albany, a building—owned by the eccentric Sir William Stone, who was also known as the Squire of Piccadilly—where wealthy Members of Parliament and writers like Graham Greene rented pieds-à-terre. There I was met by Pyke, gaunt figure with a long, sallow face, sunken cheeks, fiery eyes, and a graying goatee, who was camped out amid piles of books, journals, and papers, and cigarette butts lying scattered on oddments of furniture. He looked like a secret agent in a spy film and welcomed me with an air of mystery and importance, telling me in a gentle, persuasive voice that he was acting on behalf of Lord Louis Mountbatten, then Chief of Combined Operations, to ask my advice about tunnelling in glaciers.

Six months went by before Pyke called me again. This time, he sized me up with a volley of provocative remarks, and then told me, with the air of one great man confiding in another, that he needed my help for the most important project of the war—a project that only he, Mountbatten, and our common friend John Desmond Bernal knew about. When I asked him what it was, he assured me that he would willingly disclose it to me, a friend who had understood and appreciated his ideas from the first, but that he had promised to keep it to himself, lest the enemy or, worse, that collection of fools on whom Churchill had to rely for the conduct of the war, should get to hear about it.

I left excited and not much the wiser about what I was supposed to do, but Bernal, who had been my first research supervisor at Cambridge, told me a few days later that I should find ways of making ice stronger and freezing it faster, never mind what for. The project had the highest priority, and I could requisition any help and facilities I needed. Despite my glacier research, I was not sure exactly what the strength of ice was, and could find little about it in the literature. Tests soon showed that ice is at the same time brittle and soft, and I found no way of making it stronger.

Then, one day, Pyke handed me a report that he said he found hard to understand. It was by Hermann Mark, my former professor of physical chemistry in Vienna, who had lost his post there when the Nazis overran Austria, and had found a haven at the Polytechnic Institute of Brooklyn. As an expert on plastics, he knew that many of them were brittle when pure, but could be toughened by embedding fibres such as cellulose in them, just as concrete can be reinforced with steel wires. Mark and his assistant, Walter P. Hohenstein, stirred a little cotton wool or wood pulp—the raw material of newsprint—into water before they froze it, and found that these additions strengthened the ice dramatically.

When I had read their report, I advised my superiors to scrap our experiments with pure ice and set up a laboratory for the manufacture and testing of reinforced ice. Combined Operations requisitioned a large meat store five floors underground beneath Smithfield Market, which lies within sight of St. Paul's Cathedral, and ordered some electrically heated suits, of the type issued to airmen, to keep us warm at 0°F. They detailed some young commandos to work as my technicians, and I invited Kenneth Pascoe, who was then a physics student and later became a lecturer in engineering at Cambridge, to come and help me. We built a big wind tunnel to freeze the mush of wet wood pulp, and sawed the reinforced ice into blocks. Our tests soon confirmed Mark and Hohenstein's results. Blocks of ice containing as little as four per cent wood pulp were weight for weight as strong as concrete; in honor of the originator of the project, we called this reinforced ice "pykrete." When we fired a rifle bullet into an upright block of pure ice two feet square and one foot thick, the block shattered; in pykrete the bullet made a little crater and was embedded without doing any damage. My stock rose, but no one would tell me what pykrete was needed for, except that it was for Habakkuk. The Book of Habakkuk says, "Behold ye among the heathen, and regard, and wonder marvellously: for I will work a work in your days, which ye will not believe, though it be told you," but this failed to solve my riddle.

At one stage, Mountbatten sent Pyke to Canada in aid of Habakkuk with a personal introduction from Winston Churchill to Mackenzie King, the Canadian Prime Minister. King received Pyke with outstretched arms, saying, "Mr. Chamberlain has sent me such a nice letter about you." He was just one Prime Minister behind. While Pyke enlisted the Canadians' help, Mountbatten decided to demonstrate the wonders of pykrete to the British Joint Chiefs of Staff. For this show, Pascoe and I prepared small rods of ice and of pykrete which were exactly the same size. We could break the ice rods easily in our hands, but the rods of pykrete stayed in one piece, however we tried to break them. We also prepared large blocks of each material as targets to be shot at.

So secret was Habakkuk that no one was supposed to know even who I was, lest my nationality (Austria=mountains=glaciers=ice) or my research record betray it. Pascoe and I worked down in the meat store while on the upper floors burly Smithfield porters in greasy overalls carried huge carcasses of meat to and from the elevator. They never gave us any of it to supplement our meagre rations.

Who was to demonstrate pykrete to the Chiefs of Staff? Surely not a civilian and an Austrian—an enemy alien at that! It was decided to assign this task to Lieutenant Commander Douglas Grant, who had been an architect in peacetime, and who administered Habakkuk. He had not handled pykrete before, but he did have a uniform. I gave him our rods of ice and of pykrete, packed with dry ice into thermos flasks, and also large blocks of ice and of pykrete, and I wished him luck. I waited for news the next day, but none came.

Rationing had hit the small restaurants and tea shops in the City. Pascoe and I therefore used to take the bus down bomb-scarred Fleet Street to the palatial Combined Operations Headquarters, in Richmond Terrace, just off Whitehall, where we could get a square meal at an affordable price and could listen to the latest gossip. But that day the entertaining Pyke was still away in Canada, and everybody else seemed to avoid us. After lunch, I searched for the normally unruffled Grant and found him in a black mood. When he had handed our little rods of ice and of pykrete around, the old gentlemen had been unable to break either. Next, he had fired a revolver bullet into the block of ice, which duly shattered, but when he fired at the block of pykrete the bullet rebounded and hit the Chief of the Imperial General Staff in the shoulder. The Chief was unhurt, but Habakkuk was under a cloud. Worse was to come.

In Pyke's absence, an Admiralty committee headed by the Chief of Naval Construction had examined Habakkuk and sent an unenthusiastic memorandum about it to Mountbatten. When this news reached Pyke in Canada, it merely confirmed his disdain for the conservatism of the British establishment, which he epitomized in his derisive motto: "Nothing must ever be done for the first time." He sent back a cable headed "HUSH MOST SECRET. CIRCULATION RESTRICTED TO CHIEF OF COMBINED OPERATIONS ONLY." The message read, "CHIEF OF NAVAL CONSTRUCTION IS AN OLD WOMAN. SIGNED PYKE." The classification "Hush Most Secret" was normally reserved for operational matters, and was therefore treated with respect, but the contents of Pyke's cable quickly reached the ears of its victim—an admiral. Outraged at having his courage questioned by a mad civilian, he marched into Mountbatten's office and demanded Pyke's instant dismissal. Habakkuk seemed doomed. Then Pyke returned from Canada, elated by the success of his mission, and especially by the splendid performance of a prototype that the Canadians had succeeded in launching on Patricia Lake, in Alberta. Prototype of what?

Geoffrey Pyke was born in 1893, the son of a Jewish lawyer, who died when the boy was five years old, and left his family with no money. Geoffrey's mother seems to have quarrelled with all her relatives and made life hell for her children. She sent Geoffrey to Wellington, a snobbish private school attended mainly by sons of Army officers, and yet she insisted on his wearing the dress and observing the habits of an Orthodox Jew. This made him the victim of persecution, and bred in him a contemptuous hatred of the establishment. Though he never finished his schooling, it was possible for him in those days to start studying law at Cambridge.

When the First World War broke out, Geoffrey decided to stop his studies and become a war correspondent. Characteristically, he began his career by persuading the editor of the *Daily Chronicle* to send him to the enemy's capital, Berlin. He bought a passport from an American sailor and made his way to Berlin via Denmark, but he was soon caught, and was told that he would be shot as a spy. After some time in jail, he was put into an internment camp at Ruhleben instead. Less than a year later, the *Daily Chronicle* appeared with the banner headline "CORRESPONDENT ESCAPES FROM RUHLEBEN." By ingenious and meticulous planning, Pyke and another Englishman, Edward Falk, had made their way to Holland and back to England.

Confident now that he could solve any problem by hard thinking, Pyke devised an infallible system for making money on the commodities market. At first, he succeeded, and in 1924 he used the money to finance a startling new experiment in education. He founded the Malting House School at Cambridge, where children between the ages of two and five were to receive no formal teaching but instead were to be guided to discover knowledge for themselves in purposeful play—"discovery of the idea of discovery." For a time, the school flourished, and it became a laboratory where the great child psychologist Susan Isaacs studied the intellectual growth and social development of young children. Pyke's lawyer urged him to endow the school with the fortune he had made on the Metal Exchange, but he had more grandiose educational plans.

To finance these plans, he bought metals on credit through several different brokers, keeping each of them in the dark about the full extent of his operations. At one point, he cornered as much as a third of the world's supply of tin. Then the day came when Pyke's infallible graphs misled him; prices fell when they should have risen and Pyke went bankrupt. His school had to close, his marriage broke up, and his health collapsed. He tried journalism again, but no one would print his long articles, and he lived on the charity of friends.

In the mid-thirties, recovering, he organized a campaign for sending supplies to the Loyalists in the Spanish Civil War. Later, he raised a band of young English volunteers to conduct a clandestine public-opinion poll in Nazi Germany. Its results were to prove to Hitler that the Germans did not want to go to war, but Hitler forestalled an evaluation of Pyke's poll by the invasion of Poland.

Despite his failures, Pyke remained unshaken in his faith that he knew how to perform any job better than those whose profession that job happened to be, and as soon as the war broke out he became intent on telling the soldiers how to win it. Initially, no one would listen to him, but persistent campaigning and connections in high places brought him an introduction to Mountbatten.

In March of 1942, Pyke proposed to the Chief of Combined Operations that Allied commando troops be parachuted into the Norwegian mountains to establish a base on the Jostedalsbreen, the great glacier plateau, for guerrilla warfare against the German Army of Occupation—a base from which the commandos would be able to attack nearby towns, factories, hydroelectric stations, and railways. These troops should be equipped with a snow vehicle of Pyke's design that would allow them to move at lightning speed across glaciers, up and down mountainsides, and through forests. Pyke persuaded Mountbatten that such a force would be invulnerable in its glacier strongholds and would tie down a large German Army trying vainly to dislodge it. Despite Churchill's enthusiastic comment "Never in the history of human conflict will so few immobilize so many," the plan was dropped, perhaps because someone had found out that there are no towns, factories, hydroelectric stations, or railways near the Jostedalsbreen. The snow vehicle that Pyke had demanded for it was meanwhile built by Studebaker and named the Weasel. It proved its worth during the war in France and Russia, and afterward conveyed research expeditions safely to the South Pole.

While Pyke was in the United States organizing the manufacture of Weasels, he composed his great thesis on Habakkuk. From New York, he sent it in the diplomatic bag to Combined Operations Headquarters in London with a label forbidding anyone other than Mountbatten in person to open the parcel. Inserted opposite the first page was a green sheet of paper with a quotation from G.K. Chesterton: "Father Brown laid down his cigar and said carefully: 'It isn't that they can't see the solution. It is that they can't see the problem.'" In Pyke's accompanying letter, he wrote, "The cover name for this... project, because of its very nature, and partly because of you, is Habakkuk, '*parce qu'il était capable de tout.*'"

I cannot remember anyone ever revealing to me officially what Habakkuk stood for, but gradually the secret leaked out, like acid from a rusty can. Pyke foresaw that for several purposes air cover was needed beyond the range of land-based planes. Conventional carriers, he argued, were too small to launch the heavy bombers and fast fighters that would be needed for the invasion of any distant shores. Already, to extend air cover for Allied shipping over the entire Atlantic, floating air bases were needed; such bases would allow planes to be flown from the United States to Britain instead of being shipped. They would also facilitate the invasion of Japan. But what material could such islands be made of since every ton of steel was needed for ships and tanks and guns, and every ton of aluminum for planes? What material existed that was still abundant? To Pyke, the answer was obvious: ice. Any amount of it could be had in the Arctic; an island of ice melts very slowly, and could never be sunk. Ice could be manufactured for only one per cent of the energy needed to make an equivalent weight of steel. Pyke proposed that an iceberg, either natural or artificial, be levelled to provide a runway, and hollowed out to shelter aircraft.

Mountbatten told Churchill of Pyke's proposal. Churchill wrote to his Chief of Staff, General Hastings Ismay:

> I attach the greatest importance to the prompt examination of these ideas... The advantages of a floating island or islands, even if only used as refuelling depots for aircraft, are so dazzling that they do not at the moment need to be discussed. There would be no difficulty in finding a place to put such a "stepping-stone" in any of the plans of war now under consideration.
>
> The scheme is only possible if we make Nature do nearly all the work for us and use as our materials sea water and low temperature. The scheme will be destroyed if it involves the movement of very large numbers of men and a heavy tonnage of steel or concrete to the remote recesses of the Arctic night.
>
> Something like the following procedure suggests itself to me. Go to an ice field in the far north which is six or seven feet thick but capable of being approached by icebreakers; cut out the pattern of the ice-ship on the surface; bring the right number of pumping appliances to the different sides of the ice-deck; spray salt water on continuously so as to increase the thickness and smooth the surface. As this process goes on the berg will sink lower in the water. There is no reason why at the intermediate stages trellis-work of steel cable should not be laid to increase the rate of sinking and giving stability. The increasing weight and depth of the berg will help to detach the structure from the surrounding ice-deck. It would seem that at least 100 feet in depth should be secured. The necessary passages for oil fuel storage and motive power can be left at the proper stages. At the same time, somewhere on land the outfits of huts, workshops and so forth will be made. When the berg begins to move southward so that it is clear of the ice floes, vessels can come alongside and put all the equipment, including ample flak, on board.

Could an ice floe thick enough to stand up to the Atlantic waves be built up fast enough? It was to find the answer to this question that Pyke and Bernal first called me in, but without being allowed to tell me what the question was. As anyone knows who has tried to make a skating rink in his back yard, a long time is needed even in very cold weather to freeze a thick layer of water because the thin film of ice that forms at the top delays the transfer of heat from the underlying water to the cold air above. By Churchill's method, it would have taken about a year to build up an ice ship a hundred feet thick—and then only if the action of natural forces could somehow be prevented from causing it to disintegrate. Then, what about a natural floe? In the nineteen-thirties, a Russian expedition had discovered that even at the North Pole the pack ice was no more than ten feet thick. Atlantic waves can be as high as ninety feet, with a distance of more than fifteen hundred feet from crest to crest. Our tests showed that a slab of ice ten feet thick and suspended on two knife edges only eight hundred feet apart would snap in the middle. Besides, bombs and torpedoes would crack it, even if they could not sink it. And natural icebergs have too small a surface above water for an airfield, and are liable to turn over suddenly.

The project would have been abandoned in 1942 if it had not been for the discovery of pykrete: it is much stronger than ice and no heavier; it can be machined like wood and cast into shapes like copper; immersed in warm water, it forms an insulating shell of soggy wood pulp on its surface, which protects the inside from further melting. However, Pascoe and I found one grave snag: though ice is hard to the blow of an axe, it is soft to the continuous pull of gravity, which makes glaciers flow like rivers—faster in the center than at their sides, and faster at the top than near their beds. If a large ship of ordinary ice were kept at the freezing point of water, it would gradually sag under its own weight, like putty; our tests showed that a ship of pykrete would sag more slowly, but not slowly enough, unless it were to be cooled to a temperature as low as 4°F. To keep the hull that cold, the ship's surface would have to be protected by an insulating skin, and its hold would have to carry a refrigeration plant feeding cold air into an elaborate system of ducts.

All the same, plans went ahead. Experts drew up requirements, naval designers settled at their drawing boards, and committees held long meetings. The Admiralty wanted the ship to be strong enough to stand up to the biggest known waves—a hundred feet high and two thousand feet from crest to crest—even though such gigantic waves had been reported only once, in the North Pacific, after prolonged storms. It said that the ship must be self-propelled, with enough power to prevent its drifting in the strongest gales, and that its hull must be torpedo-proof, which meant that it had to be a least forty feet thick. The Fleet Air Arm demanded a deck fifty feet above water, two hundred feet wide, and two thousand feet long, to allow heavy bombers to take off. The strategists required a cruising range of seven thousand miles. The final design gave the bergship (as it came to be referred to) a displacement of two million two hundred thousand tons—twenty-six times that of the Queen Elizabeth, the biggest ship then afloat. Turboelectric steam generators were to supply thirty-three thousand horsepower to drive twenty-six electric motors—each fitted with a ship's screw and housed in its own separate nacelle—on the two sides of the hull. These motors were to propel the ship at seven knots—the minimum speed needed to prevent its drifting in the wind.

Steering presented the most difficult problem. At first, we thought that the ship could be steered simply by varying the relative speed of the motors on either side, like a plane taxiing along the ground, but the Navy decided that a rudder was essential to keep it on course. The problem of suspending and controlling a rudder the height of a fifteen-story building was never solved. Indeed, even today rudders cause problems in supertankers of only a tenth the bergship's tonnage: in 1978, failure of the rudder control caused the supertanker Amoco Cadiz to be blown onto the rocks off the coast of Brittany, spilling its oil onto the white beaches.

While plans for the bergship became more elaborate with each committee meeting, Pyke's mind raced ahead to work out how such ships should be used to win the war. He argued that the bergships would solve the difficult problems of invading hostile coasts because they would be able to force their way straight into the enemy's harbors. The defending troops would be petrified, literally, by being frozen solid. How? The bergships were to carry enormous tanks full of supercooled water—liquid water cooled below its normal freezing point—which could be sprayed at the enemy to solidify on contact. Afterward, more supercooled water would be pumped ashore to build bulwarks of ice, behind which Allied troops could safely assemble and make ready to capture the town. It was Pyke's best piece of science fiction. In reality, the cooling of liquid water below its freezing point is observed only in the tiny droplets that clouds are made of. Pyke could not have found reports in the scientific literature of anyone's making more than a thimbleful of supercooled water, but this did not diminish his enthusiasm for its use by the ton.

My own next problem was to find a site for the building of a bergship. How could we follow Churchill's sensible directive to let Nature do the job? Surveying the world's weather maps, I was unable to find a spot on Earth cold enough to freeze two million tons of pykrete in one winter. Nature would have to be aided by refrigeration. Eventually, we chose Corner Brook, in Newfoundland, where wood pulp provided by the local mills was to be mixed with water and frozen into blocks in a two-hundred-acre refrigeration plant. The problem of launching our Leviathan was to be circumvented by laying down the first pykrete blocks on wooden barges cramped together to form a large floating platform. This would gradually sink as the mass of pykrete was built up. The prototype was to be built in the winter of 1943–1944, to be followed by a fleet of bergships constructed on the North Pacific Coast the following winter, in time for the invasion of Japan.

One day, Mountbatten called me into his office to ask who should represent Habakkuk at a high-level meeting. I suggested Bernal as the only man who possessed the technical knowledge, the intellectual stature, and the persuasiveness to stand up to the war leaders. Bernal was the most brilliant talker I have ever encountered. The son of a wealthy Irish Catholic farmer, he soaked up knowledge from an early age like blotting paper, and became mesmerized by science. Once, he tried to generate X-rays by focussing the light from a paraffin lamp so as to see through his hand, and nearly set the farm on fire; he was beaten by his father. He was converted to Communism in 1922, when he was a student at Cambridge, and remained a faithful Party member all his life. (He died in 1971.) Bernal is mentioned in Andrew Boyle's recent book *The Climate of Treason* as one of the founders of the Cambridge Communist cell in the nineteen-thirties, but he made no secret of his allegiance and was never suspected of disloyalty to Britain. As a Cambridge undergraduate in the early twenties, he studied natural sciences and then took up X-ray crystallography, a physical method used for determining the arrangements of atoms in solids. When I joined him as a graduate student, in 1936, he was at the height of his powers, with a wild mane of fair hair (no beard), sparkling eyes, and lively expressive features. We called him Sage, because he knew everything from physics to the history of art. He was a bohemian, a flamboyant Don Juan, and a restless genius always searching for something more important to do than the work of the moment.

When war broke out, the authorities asked Bernal to assess the likely damage from aerial bombardment. He requested that his former research assistant be taken on to help him, but, to his astonishment, the request was refused on security grounds. Bernal ridiculed the decision and demanded to see the reason. When he was reluctantly shown the file, the papers stated that the man could not be trusted because he was associated with the notorious Communist Bernal.

Mountbatten, who liked to have unconventional people around him as counterweights to naval orthodoxy, appreciated Bernal's prodigious knowledge and his original approach to any kind of problem. Mountbatten himself impressed me greatly by his quick and decisive mind. The high-level meeting he was preparing for took place in Quebec in August of

1943, and was headed by Roosevelt and Churchill. Bernal staged a demonstration of pykrete which so impressed the war leaders that they decided to give Habakkuk the highest priority. Detailed plans for the immediate construction of a prototype were to be drawn up in Washington. The British team was ordered there forthwith—except for Pyke, whose mordant wit had enraged the American military to the point where he was forbidden to come.

When the people at the United States Consulate in London saw my invalid Austrian passport, they said that they were not allowed to issue visas to enemy aliens, however vital for the war effort. Mountbatten's Chief of Staff tackled this trivial obstacle by phoning the Home Office and telling the people there to make me a British subject within the hour. But, like a parson asked to perform a shotgun wedding without calling the banns, the Home Office insisted on at least the semblance of its customary naturalization ritual. That night, a detective called on me at my lodgings, in Holland Park. Would I give the names of four British-born householders who could vouch for my loyalty. Normally, the detective said, he would make careful inquiries from each of them, but in my case he wouldn't bother. What near relatives did I have in enemy territory? Normally, he would cross-check my answers, but in my case he wouldn't bother. Had I been convicted of any crime? Yes, of riding a bicycle in Cambridge without lights. Normally, he would check the police records, but in my case he wouldn't bother. After an hour of such banter, I signed his form. Supposing I had gone and betrayed the secrets of pykrete to the Eskimos. Would the Prime Minister have assured Parliament that "the proper inquiries" had been made at the time of my naturalization to ascertain that I had not been an Eskimo sympathizer from an early age? The next morning, I swore allegiance to the King before a justice of the peace; my wife merely had to sign a piece of paper at home in Cambridge. The following day, I was issued a shiny blue passport that described me as a British Subject by Certificate of Naturalization Issued 3 September 1943. You cannot become an Englishman, as you can become an American, but at least we were no longer enemy aliens liable to be interned, and my new passport solved the United States visa problem.

The other members of the Habakkuk team had already sailed to New York. To catch up with them, I was now sent there by air. First, a Sunderland flying boat took me from Bournemouth to Shannon, where the

British officers on board donned civilian clothes in deference to Eire's neutrality. From Shannon, Pan Am's Yankee Clipper flying boat ferried us to Newfoundland in fourteen hours, and thence to New York Harbor, where we landed thirty-four hours after leaving London—a record time. When the immigration officer read in my passport that my British nationality was of exactly four days standing, he decided to unmask this foreign agent whom the wily British were trying to foist on their unsuspecting ally, and subjected me to a sharp interrogration. When I had told him most of my life history, except for my involuntary sojourn in Canada, he began questioning me about relatives in the United States. A brother. What is his name? When was he born? What does he do? Where does he live? My heart thumped as I remembered that my brother's house had been searched by the F.B.I. when they found out that he had been in correspondence with a prisoner of war in Canada. Would this be in the immigration officer's file? If it was, he gave no sign, but continued. What other relatives? A sister. Where does she live? Prytania Street, New Orleans. Suddenly, his tense face relaxed into a broad grin. "But that's the street where I was born." And I was admitted. No one whose sister lived on Prytania Street, New Orleans, could be a spy.

On arriving in Washington, where I imagined the British team to be busy sixteen hours a day with the planning of the bergship's construction, I was surprised to find them all welcoming me at Union Station in the middle of a weekday afternoon. They wondered what the weather had been like in London when I left—a question that I diagnosed as an expression of home-sickness—and seemed in no hurry to get back to their desks. The next morning, when I reported for duty in a hut outside the Department of the Navy Building, on Constitution Avenue, I heard that Habakkuk was under scrutiny by the department's naval engineers, and that pending their report there was nothing we could do. Lord Zuckerman, another of Mountbatten's wartime scientific advisers, recently explained to me why no one paid much attention to us in Washington. Shortly after our arrival there, Mountbatten left Combined Operations to become Commander-in-Chief of the Allied Forces in Southeast Asia. Since he had been Habakkuk's principal advocate, its priority took a deep plunge.

So as not to idle away my time, I asked for permission to visit the Canadian physicists and engineers who had carried out tests on ice and pykrete parallel to ours and had built a model ice ship, complete with insulation and refrigeration, on Patricia Lake. It was on this trip that I reentered Canada as a free man, but I evaded my hosts' conventional question of whether this was my first visit.

Back in Washington, I took a room in the suburbs, where I listened to a Republican fellow-lodger's denunciations of Roosevelt as a greater menace than Hitler. I read in the Library of Congress or went rock climbing on the banks of the Potomac until the United States Navy finally decided that Habakkuk was a false prophet. One reason was the enormous amount of steel needed for the refrigeration plant that was to freeze the pykrete was greater than that needed to build the entire carrier of steel, but the crucial argument was that the rapidly increasing range of land-based aircraft was making floating islands unnecessary. This was the end of Pyke's ingenious project.

It was hard for a civilian to find a place on a ship back to England, but finally I was allocated a berth in a first-class single cabin on the Queen Elizabeth, England's newest and fastest liner. When I stepped into my cabin, I found that I shared it with five others. One, a tall, dignified old gentleman, introduced himself as Mr. Coffin, Moderator of the Presbyterian Church in the U.S.A., and proudly announced that he was going to London to have tea with the Queen. To belie his lugubrious name, he entertained the rest of us by day with a great fund of stories; he also kept us awake at night with his loud snoring.

The ship carried fourteen thousand American soldiers, sent to join the great armies that were to liberate France the following summer. Under big notices of "No Gambling," piles of dollar bills slid across mess tables in the lounge every few minutes as the great ship heeled over, steering its zigzag course to evade the U-boats. After six days, we steamed up the Firth of Clyde, where a large Allied battle fleet lay assembled in the gloomy winter morning, the sinister gray shapes anchored between the dark, cloud-covered mountainsides lending drama to a scene that looked like a Turner painting of a Scottish loch.

When I reported the demise of Habakkuk to my superior at the Admiralty the next morning, he was not surprised. Pyke was disappointed, but he was already busy on new schemes. One of them was the construction of a gigantic tube from Burma into China—much easier than building a road over the mountains, he argued. Through this tube, Allied men, tanks, and guns were to be propelled to China by compressed air, like the pneumatic post in department stores, to help Chiang Kai-shek defeat the Japanese Army there.

Another of Pyke's plans plotted the destruction of the Rumanian oil fields from which Germany derived most of its fuel. In the dark of night one squadron of planes was to attack the fields with high-explosive and incendiary bombs, while another squadron was to drop a force of commandos nearby, charged with destroying the fields on the ground. How could they penetrate the defenses? Disguised as Rumanian firemen, they should capture a fire station and drive into the oil field with its engines, pretending that they were on their way to extinguish the fires started by the air raid, but fanning them instead.

I had come to realize some months earlier that construction and navigation of the bergships might prove as difficult as a journey to the moon then seemed to me, yet Habakkuk was one of several apparently impossible projects conceived during the war; in each case the question was not so much one of absolute feasibility as of whether the strategic advantages to be gained by carrying out the project were in proportion to the manpower and materials required. In retrospect, it seems surprising that Mountbatten should have taken any of Pyke's projects seriously, but then Mountbatten was the youngest member of the Chiefs of Staff and headed an organization set up for unconventional warfare. Faced with that task, he liked to attract to his headquarters men who had not been to Staff College and whose ideas were therefore less likely to be anticipated by the enemy—never mind if they wore no socks.

In peacetime, most of Pyke's ideas would have been discarded as the science fiction they were, but Mountbatten relied for scientific advice on Bernal, without realizing that Bernal's one great failing was a lack of critical judgment. Pyke had the Cartesian's arrogant conviction that an intelligent human being could reason his way through any problem, rather than Francis Bacon's humble maxim that "argumentation cannot suffice for the discovery of new work, since the subtlety of Nature is greater many times than the subtlety of argument." I returned to Cambridge, sad at first that my eagerness to help in the war against Hitler had not found a more effective outlet, but later relieved to have worked on a project that at least never killed anyone—not even the Chief of the Imperial General Staff.

Until recently, I did not know how and why, several decades ago, the British government had decided to intern and deport many thousands of innocent German and Austrian refugees and Italians living in Britain, and to start releasing them again a few weeks or months later, long before the danger of a German invasion had receded. I have since read *Collar the Lot!*, Peter and Leni Gillman's history of the internment of aliens in Britain, which is based on a scholarly study of official documents that were released thirty years after the events and on interviews with many of the survivors.

The book reveals a disheartening story of official callousness, interdepartmental intrigue, newspaper hysteria, public lies, lies told to Parliament and to the governments of the Dominions, and, as John Maynard Keynes said of David Lloyd George, decisions taken on grounds other than the real merits of the case. The book tells also of human suffering, and of a few upright individuals whose compassion turned the tide.

The story begins in the autumn of 1939, when the Home Office and the War Office were anxious to avoid a repetition of the wholesale internment of nearly thirty thousand mostly harmless Germans in squalid prison camps which had taken place during the First World War. The Home Secretary, Sir John Anderson, therefore established tribunals that classified Germans and Austrians as refugees from Nazi oppression, and ordered the internment of only those thought to be loyal to the Nazi regime.

On April 9, 1940, German forces invaded Norway, supposedly helped by a Fifth Column of Norwegian Nazis and by German spies posing as refugees. A month later, the Germans invaded Holland and Belgium, and Winston Churchill replaced Neville Chamberlain as Prime Minister. Churchill held his first Cabinet meeting on May 11th. At the insistence of the Chiefs of Staff, the reluctant Anderson was asked to abandon his enlightened policy and to intern all male Germans and Austrians living near the coasts that were threatened by invasion.

A few days later, Sir Nevile Bland, the British Ambassador at The Hague, returned to London with alarming stories of treachery by German civilians in Holland. His photograph shows him supercilious and vacant, like a figure out of Evelyn Waugh's farcical novels about the British upper class. He realized that his important hour had come, and, at the end of May, he solemnly warned the nation in a radio broadcast, "It is not the German and Austrian who is found out who is the danger. It is the one, whether man or woman, who is too clever to be found out." Having pondered this profound truth, the Chiefs of Staff warned the Cabinet that "alien refugees [are] a most dangerous source of subversive activity," recommending that all should be interned. "The most ruthless action should be taken to eliminate any chances of Fifth-Column activities."

On May 24th, Churchill told the Cabinet that he was in favor of removing all internees from the United Kingdom. Newfoundland and St. Helena were two of the inhospitable places to which Churchill proposed we should be banished. General Jan Smuts managed to do one better by suggesting the Falkland Islands instead. On June 10th, when Italy declared war, Churchill ordered the Home Office to "collar the lot" of Italians living in Britain.

Among four thousand Italians interned during the succeeding two weeks, and among those supposedly most dangerous ones later selected for deportation overseas, were H. Savattoni, the banquet manager at the Savoy Hotel, who had worked there since 1906; D. Anzani, the secretary of the anti-fascist Italian League of the Rights of Man; Piero Salerni, an engineer urgently needed by the Ministry of Aircraft Production; Alberto Loria, a Jew who had come to Britain in 1911; and Uberto Limentani, a Dante scholar working in the Italian service of the British Broadcasting Corporation. All except Loria and Limentani were drowned on the Arandora Star. Limentani later gave the following (pp. 97–103) description of his escape from drowning:

E come quei che con lena affannata,
Uscito fuor del pelago a la riva.
Si volge a l'acqua perigliosa e guata.
Così l'animo mio, che ancor fuggiva,
Si volse a diretro a rimirar lo passo
che non lasciò giammai persona viva.

(Inferno, I.22–27)

(And as he, who with panting breath
Had escaped from the ocean to the shore,
Turns and stares back at the perilous waters
So my fugitive soul
Turned back to contemplate the pass
That no one has ever left alive.)

In the event most of the interned Italians were sent to the Isle of Man. As for me, on 30 June I was separated from the other internees and dispatched, together with a few dozen young men—bachelors aged about twenty-five or more—to Liverpool, where I was deposited in front of a great gray-painted transatlantic liner called the Arandora Star. I remembered having seen the same liner, then painted all in white, at anchor in I Giardini in Venice eight years before, when she had been on a cruise around the world. At that time, I said to myself, "How splendid it would be to go on a cruise in that ship!" Now, faced with just that opportunity, the prospect seemed a lot less appealing.

The ship was armed with two small cannons, veritable popguns, one in the bow and the other in the stern. There was barbed wire all over the place. I was not put in the hold but into a cabin two or three levels below deck. Outside this cabin, in which I had to sleep on the floor together with three other internees, there was an English sentry, armed with a rifle and fixed bayonet, who told me that we were being transported to Canada. During the night the ship weighed anchor. Late in the afternoon on the following day, we were allowed up on deck for half an hour for a breath of fresh air. I then saw that we were between Scotland and Ireland, at the point where both coastlines are visible. Looking round the ship, I noted that the lifeboats were in very poor condition: they had obvious holes in them. They had been neglected and did not inspire confidence. During that night, that is between 1 and 2 July, the ship must have rounded the northernmost headland of the Irish coast and headed out into the Atlantic. At six-thirty the following morning, I was dozing when there was a sudden and inexplicable crash. At once I felt that some disaster had occurred because there was a fearful clattering noise, as if everything that *could* overbalance had come hurtling down. Through the crack under the door, I could see that the electric light had suddenly gone out and therefore guessed that the generators were out of action. I asked myself what could have happened, and it occurred to me that the ship might have collided with an iceberg. In fact, we had been torpedoed by a German U-boat. I learned later that we were the victims of a famous U-boat commander, Captain Prien, on his way back from patrolling the Atlantic. Seeing our ship sailing without an escort, he had been unable to resist the temptation of firing a torpedo, which struck us full on, and he had then continued on his course.

There were about eighteen hundred souls on board—Italian, Austrian, and German internees, and, naturally, some hundreds of soldiers escorting us. My three cabin mates vanished at once. However, I remained for a few seconds, groping in the dark, because I recalled having seen some life belts hanging on the walls. I found one, put it on, and then somehow or other found my way up on deck. I could see that there was some panic, but I don't think that I lost my head because in the event there was simply no time to worry. I was always able to act with a certain coolness, that is to say, reflecting before taking each decision. The first thing I did was to climb up to the highest point that I could find, in order to determine whether the ship really was sinking. This showed me that it was tilting over more and more steeply to one side. I saw a sailor lowering a lifeboat into the sea and told myself that my best course would be to try to get aboard that lifeboat. But when I got down to the place, I realized that boarding the lifeboat would be like jumping from the fourth story, and I couldn't bring myself to do it. Only one person succeeded in jumping, and he fractured his skull (although he survived). So I gave up that scheme and made my way along the side deck to see if there was any other way of getting into the sea. Eventually, I found a piece of rope that I thought might suit my purpose, but not satisfied even with this, I went on searching and at last found a rope ladder.

At this juncture I decided to wait quietly for a while, thinking it opportune to put off getting into the sea until the last moment, because in the north Atlantic it can be extremely cold on a morning of cloud and rain even though it was July. After a bit I began to climb down the rope ladder, but on reaching the lower deck I thought that it might be better to stop and make sure that the ship really was sinking. Almost immediately I became aware that the end was imminent, so I went down into the sea. My immediate concern was to swim far enough away to avoid being sucked under with the ship.

The few boats that had been launched were for the most part filled with German sailors who had been interned after capture in South Africa and who knew how to lower lifeboats. In all, there were only five or six boats because, as I was told afterward, those positioned on the side opposite the direction in which the ship was tilting could not be lowered. In any event, no lifeboats were visible. There was already some wreckage in the sea, and I swam toward some object thinking that it might keep me afloat. There was another Italian hanging on to this bit of wood, and I said to him, "Help me to push this farther away from the ship so that we can save ourselves." I asked him his name: the poor chap was called Avignone, and I later found his name on the list of those who drowned. Many of those who managed to get down into the sea froze to death after a few hours in those icy waters.

In the meantime, the ship was sinking fast. Almost fascinated by the sight, I turned continually to look, yet I was anxious to get as far away as possible for fear of being dragged under. In fact, many of those who were too close—some good swimmers among them—were sucked under and not seen again. This great liner of about 12,000 to 15,000 tons listed more and more to one side, thereby throwing hundreds of people into the sea, mainly elderly people who had not attempted to save themselves. At that moment, the seawater clearly got into the boilers, because there was an explosion. Almost at once, as the stern sank, the bow lifted briefly above the waves, and with a frightening noise the liner slid obliquely into the sea, making the water boil over all round. There was wreckage everywhere, and corpses. More than once I became entangled in some floating debris that had either wire or metal spikes sticking out of it. There were also patches of diesel oil that had caught fire, and I therefore found myself in the midst of flames, although these quite naturally burned out very quickly. I reckon that I remained in the sea for about two hours.

At first I tried to clamber onto a sort of seat from the ship, which might have served me as a raft, but I was frustrated in this attempt because it overturned each time I got on it. After some time—possibly an hour and a half—I caught sight of a lifeboat a good way off, perhaps a mile or so. I was only able to catch glimpses of this lifeboat when I was lifted up by successive waves, but I determined to make toward it by clinging to some wreckage and propelling it with the help, once again, of another victim of the shipwreck—I think he was an Irishman, probably one of the soldiers who had been in charge of us. So for a while we helped each other, until finally he left me and swam directly toward the lifeboat without any support. I was never able to discover whether he made it. As for me, I told myself that that piece of wood was my only support and that it would be folly to leave it. Even now, I did not doubt for one moment that they would come and rescue us, and it was perhaps this conviction that kept my spirits high, in a manner of speaking. The curious regularity of the waves brought to mind some verses of Alessandro Manzoni's *Cinque Maggio,* and I repeated them to myself:

Come sul capo al naufrago
L'onda s'avvolve e pesa.

(As the shipwrecked's head
Is enveloped and weighed down by the waves.)

I thought how true it was that the waves broke over and submerged the castaway's head, and I reflected on the significance of the lines that followed:

L'onda su cui del misero,
Alta pur dianzi e tesa,
Scorrea la vista a scernere
Prode remote invan.

(The wave from whose crest
The doomed man gazed anxiously but vainly
For a glimpse of a distant shore.)

And then, I asked myself, how did it go on? Ah well, I should have to reread the text of *Cinque Maggio* when I got back home.

I struggled on with my piece of wreckage for a while and realized that I was growing feebler. Clearly I could not carry on in that laborious fashion. I should have to let go and swim for the lifeboat. I remember thinking that this was a brave decision, that piece of wood being my only secure hold on life. Now I made one last and prolonged effort, since I was still some way off, and managed to get nearer to the lifeboat. By now I was almost completely exhausted, and I made my one mistake of the entire adventure by shouting *Aiuto* ("Help" in Italian). I found out later that aboard that overloaded and waterlogged lifeboat, which was carrying some 110 to 120 survivors, there was a British army captain who had declared that there was no more room and that from then on only British soldiers should be rescued and taken aboard. However, this view was rejected, so I was told later, by the second in command of the torpedoed liner (the captain went down with his ship), a certain Mr. Tulip, at the helm of the lifeboat, who said, "No, we're at sea and we must rescue all survivors." He it was who ordered that they should take me aboard.

As a matter of fact, I managed, with help from those already in the lifeboat, to hoist myself up and realized then that my lungs were on the point of collapse and that my body had been tried to the limit. Squeezed in between the mass of survivors in the lifeboat and shaking with cold, I asked for something to cover myself with. In reply I was punched on the head and found myself sitting at the bottom of the boat with three or four people on my back. By a stroke of luck, reaching out my hands, I found a sailor's jacket and somehow managed to get it on. My position was extremely uncomfortable not only because of the crush of people above me but also because the level of seawater was slowly rising in the bottom of the boat. We should certainly have sunk in a few hours, as some of the German detainee survivors considerately observed. After two hours, making a great effort, I managed to move into a more comfortable position, and by raising myself I was able to breathe freely like the rest.

The lifeboat commander tried to stay in a spot not too far from the other four or five boats. These we could see, but there was no sign of rescue. I reckon that it was some six hours or so after the torpedoing that we saw a four-engined Sunderland seaplane, which was carrying out a search without yet having found any survivors. Then, after a minute or two, it saw us, fired a Verey light, and vanished. We knew now that help would arrive, but we still had to wait for two hours before, to our great relief, a motor torpedo boat bore over the horizon toward us—a Canadian boat called the St. Laurent. There were seven hundred survivors of the eighteen hundred passengers on the Arandora Star, and first we had to solve the problem of getting aboard the rescue ship, which was anything but easy. The warship took up a position in the center of the large area over which the survivors were scattered, and each lifeboat had to make its way toward it. When it came to transferring from the lifeboat to the rescue ship, the problem was that the swell caused the deck of the motor torpedo boat to be at one moment about 33 feet higher than the lifeboat and the next instant the opposite. And so, in order to get aboard the warship, we had to seize the exact moment when both vessels were on the same level. In my turn, I managed somehow to accomplish this, and I remember having to move along the deck as quickly as possible—I was of course barefooted—since at that point the deck was scorching hot, probably because it was just above the engine room.

Seven hundred is a great crowd on a small vessel like a motor torpedo boat, and although the sailors did what they could, we passed a very unpleasant night, packed as we were into the between decks, to which we had been forced to descend. I found myself sitting on a seat with dozens of other survivors in one of the sailors' sleeping quarters. I spent the night sitting like that, very hungry. I recall getting a cup of hot chocolate laced, if I am not mistaken, with rum. Otherwise, I was safe and sound. Before we were torpedoed, I had a cold, which must have disappeared during my involuntary swim because I don't remember having it afterward. It was certainly a disagreeable night, aggravated by the somewhat irrational fear which spread among us that we might be torpedoed again.

The next morning, 3 July, we arrived off the Scottish coast and disembarked at Greenock. Two or three of the shipwrecked survivors had died during the crossing, and others had to be taken to hospital. Just before disembarkation, we were, in fact, asked whether we needed medical treatment. At first I refused, thinking that I was in my normal healthy condition, but after running by chance into one or two of my cabin mates from the Arandora Star I was advised to go with them to hospital. I then noticed that my bare feet were rather swollen on account of the freezing conditions of the preceding twenty-four hours. So I took their advice and went along with them, which was very lucky for me, because those who did not go to hospital were embarked for Australia the very next day, and their ship was torpedoed somewhere on route. The ship did not sink, but it must have been a terrifying experience.

So there we were, on the deserted wharf of the port of Greenock, a wretched band of shipwrecked civilians with nobody to take care of us. I had on the sailor's jacket I'd found in the lifeboat, but I was still barefoot. After a while a sort of Red Cross hostel opened, but they could give us no more than a biscuit each. Little by little, the powers that be must have noticed our existence, because around midday some trucks arrived to take us along the Firth of Clyde to a hospital whose location was not known to us at the time but which we subsequently discovered to be the Mearnskirk Emergency Hospital in the vicinity of Glasgow. Covered as I was from head to foot in the naphtha that had spread on the sea after the Arandora Star had sunk, I needed above all a bath. Instead, I had to wash myself as best I could with a sponge. Then we were put to bed and could relax at last after the exhausting events of the previous day and night.

We stayed in that hospital for seven or eight days, well fed and cared for. We were the first patients of a hospital constructed for the very pur-

> pose of caring for victims of the war. The nurses were especially attentive and, apart from the facts that we had to stay in bed and that there was always a sentry at the door of the dormitory, I believe that we had no complaints about our treatment. We had lost all our personal belongings, and after a week we were given some clothes, which were frankly rather comical, being either too big or too small for us, and some equally useless shoes, as well as certain essentials like razors. On about 11 or 12 July, we climbed aboard a bus that took us to a new internment camp. We traveled right across Scotland, although we did not know then that the harsh building in which we were imprisoned, with its massive walls and surrounding barbed wire, was the Donaldson School Hospital on the outskirts of Edinburgh.
>
> After a few days I was permitted to write to the BBC, and only then did my colleagues in London learn that my name had been mistakenly included on the list of those who had drowned. The BBC had obtained at the outset an order for my release, and this was immediately put into effect, so that on 31 July I was freed—that is, I was escorted by a soldier in a streetcar as far as Princes Street Station and there put on the train for London. I reached London on the evening of 31 July, and on the following morning I resumed my work at the BBC. There I continued to broadcast for the next five years, until the end of September 1945, that is for the duration of the war.

Limentani later became professor of Italian at Cambridge University, where he gave me the above account of his experiences.

In obedience to the chiefs of staff's directive, the War Office ordered that those who had survived the torpedoing of the Arandora Star be reembarked a few days later on the Dunera, a ship bound for Australia. Among those guarding the internees as they boarded the Dunera at Liverpool harbor was a young soldier named Merlin Scott. That night he wrote a letter home.

> I thought the Italian survivors were treated abominably—and now they've all been sent to sea again," his letter said. "That was the one thing nearly all were dreading, having lost fathers, brothers, etc. the first time… . Masses of their stuff—clothes etc. was simply taken away from them and thrown into piles out in the rain and they were allowed only a handful of things. Needless to say, various people, including policemen! started helping themselves to what had been left behind. They were then hounded up the gangway and pushed along with bayonets, with people jeering at them… . Masses of telegrams came for them from relatives, nearly all just saying 'Thank God you are safe,' and they were not allowed to see them." the telegrams "had to go to a Censor's Office… . Some of them said they had no mail for six weeks.

Shortly after the Dunera left harbor, a German submarine fired two torpedoes at it, but the Dunera happened to change course, and the torpedoes missed the ship by about a hundred yards.

Merlin Scott's father was an Assistant Undersecretary at the Foreign Office. His son's letter made the rounds of the office and was shown to Lord Halifax, the Foreign Secretary. He forwarded it to Sir John Anderson, the Home Secretary, together with a memorandum expressing concern about the bad effect that such inhumanity would have on public opinion at home and in the United States. Halifax and Anderson won over Chamberlain, who until then had been the chief executor of Churchill's deportation policy, and on July 18th, only a week after Scott had written his letter, Chamberlain persuaded the Cabinet that "persons who were known to be actively hostile to the present regimes in Germany and Italy, or whom for other sufficient reasons it was undesirable to keep in internment, should be released." The Cabinet also agreed that the "internal management, though not the safeguarding," of the internment camps should be transferred from the War Office to the Home Office. The deportations were stopped.

The Canadian government at first stonewalled Paterson's proposal to release in Canada those refugees who did not want to return to England, and the American State Department refused to admit even those refugees who had already held immigration visas before they were interned. Early in 1941, Ruth Draper, the great diseuse, gave one of her heartwarming performances in Ottawa for the Canadian Red Cross. Afterward, the Prime Minister asked her what Canada could do in return. She told him, "There is a young innocent boy, whom I have known since he was a baby, being held in one of your internment camps behind barbed wire, without offense, without a trial." The Prime Minister ordered the boy's release, and his decision opened the door for others. When I paid my return visit to Canada, in October of 1943, the last internment camp had just been dismantled. The Gillmans' book shows that even in wartime one person's compassion can sometimes prevail against hardened politicians and the military.

As far as I know, historical research has found no substance in the ugly rumors of spying by Germans who posed as refugees, either in Norway or in Holland; nor was there ever a case of a German or Austrian refugee in Britain who aided the enemy. Merlin Scott, whose letter saved so many Italians in Britain from internment and deportation, was killed by the Italians in Libya during the first British advance, early in 1941.

He was the only child of Sir David Montague Douglas Scott, who was not told how his son met his death until forty-four years later, shortly after his ninety-eighth birthday, when he received this letter from a soldier who had served under Merlin.

> While remembering the 40th anniversary of V.E. [Victory in Europe] day on 8th May, I recalled the privilege it was to have served with Sir Douglas Montague Scott [Merlin], who was Platoon Commander of the bren-gun carriers, A Company, 2nd Battalion, Rifle Brigade.
>
> Sir Douglas and I were in a bren gun carrier and were called back from O.P. [outpost] duty to go into action near Hell Fire Pass, Egypt. He left my carrier to go into a signal carrier driven by Rifleman Savage. Sergeant Whiteman, who was Platoon Sergeant, travelled in my carrier. We went into attack in line and came under heavy fire and, on being given a signal from Sir Douglas, we had to withdraw. The carriers withdrew except for his carrier. Sgt. Whiteman and myself in our carrier went forward again to investigate, while still under heavy fire, and found that Sir Douglas's driver had been killed and Sir Douglas severely wounded in the chest. We coupled up his carrier with a tow chain to pull him back from the line of fire. In so doing his carrier went into a gun emplacement and we had to de-couple the tow chains. To do this we had to drive towards the enemy lines, turn round and recouple up to Sir Douglas's carrier to enable us to get back to our own lines.
>
> On enquiring the condition of Sir Douglas I was told that he had died on his way to hospital. The battalion Commander sent for Sgt. Whiteman and myself to thank us for what we had done and said that the action would be mentioned for a military medal, the result being that Sgt. Whiteman was awarded the D.C.M. It is sad to say that Sgt. Whiteman was himself killed a few weeks later.
>
> On the early morning of this action Sir Douglas was chatting to me and said that if it hadn't been for the war, he might never have met people like myself.
>
> The reason why I am writing this is that quite possibly I could be the only surviving person left from this action and for many times I have felt that I wanted to pass this first hand knowledge on to you.
>
> Sir Douglas Montague Scott was a very brave and courageous gentleman and it was a great honour and pleasure to have served under him.

Sir David told me that Merlin had been compassionate even as a boy. When Merlin's letter from Liverpool arrived, Sir David had been at the Foreign Office in charge of American affairs, which put him in a good position to warn the Foreign Secretary, Lord Halifax, of the bad effect that the maltreatment of the Italians would have on public opinion in the United States. When I visited him in September 1985, he was blind and chair-bound, but someone who had known him a few years earlier described him as the handsomest man she had ever met. Sir David died in August 1986, a few months before his one hundredth birthday. His wife told me that he never got over Merlin's death.

*The Threat of Biological Weapons**

Science without conscience is the death of the soul.

PETER ABELARD (1079–1144)

Men and women of good will talk of a common humanity, a common empathy with people everywhere who have hopes and sorrows similar to their own. Ken Alibek's book is about a common inhumanity, a gigantic effort that employed tens of thousands of people to find ways of inflicting the most excruciatingly painful diseases and deaths on millions of men, women, and children in the hated capitalist world on the pretense or in the belief that they threatened the beloved motherland, the Soviet Union.

Biohazard is a candid autobiography of an ambitious Kazakh physician who helped to create and direct the largest and most advanced biological warfare program in the world. In the "Oath of a Soviet Physician" he had sworn "to do no harm." He knew, he writes, "that the results of my studies could be used to kill people, but I couldn't figure out how to reconcile this knowledge with the pleasure I derived from research."

The prologue sets the scene:

> On a bleak island in the Aral Sea, one hundred monkeys are tethered to posts set in parallel rows stretching out toward the horizon. A muffled

*A review of the book *Biohazard: The Chilling True Story of the Largest Covert Biological Weapons Program in the World—Told from Inside by the Man Who Ran It* by Ken Alibek with Stephen Handelman (Delta).

thud breaks the stillness. Far in the distance, a small metal sphere lifts into the sky then hurtles downward, rotating, until it shatters in a second explosion.

Some seventy-five feet above the ground, a cloud the color of dark mustard begins to unfurl, gently dissolving as it glides down toward the monkeys. They pull at their chains and begin to cry. Some bury their heads between their legs. A few cover their mouths or noses, but it is too late: they have already begun to die.

At the other end of the island, a handful of men in biological protective suits observe the scene through binoculars, taking notes. In a few hours, they will retrieve the still-breathing monkeys and return them to cages where the animals will be under continuous examination for the next several days until, one by one, they die of anthrax or tularemia, Q fever, brucellosis, glanders, or plague.

These are the tests I supervised throughout the 1980s and early 1990s. They formed the foundation of the Soviet Union's spectacular breakthroughs in biological warfare.

During World War II and the early stages of the cold war that followed, the United States, the United Kingdom, and the Soviet Union developed and field-tested biological weapons; but the United Kingdom gave them up in 1950 and in November 1969 President Nixon unilaterally renounced them, and in February 1970 he also renounced the development of toxins for military purposes. The United States stated that it had destroyed its stockpiles, converted the biological weapons laboratories to purely defensive purposes, and converted the factories providing the weapons to peaceful uses. There is no evidence that it did not. Harold Wilson's British Labour government had by then proposed an international treaty banning the development, production, and possession of biological weapons. In 1972 this proposal led to the Biological Weapons Convention, which was signed and ratified by the United States and the Soviet Union. The convention entered into force in 1975, remains in force, and now has 143 "State Parties" adhering to it, including the United States, the Russian Federation, and Iraq. Unfortunately it has no efficient mechanism for verifying compliance.

The President's bold initiative had been stimulated in part by a forceful memorandum which the Harvard biologist Matthew Meselson sent to his friend and former colleague Henry Kissinger in September 1969. This contained cogent arguments that the security of the United States would be enhanced by a policy of renunciation; an American biological weapons program would pioneer a technology that would leak out, so that other states could duplicate it for sinister purposes of their own.

The Soviet authorities seized upon the Convention as an excellent opportunity to acquire a monopoly on biological weapons that could be added to the Soviet Union's already extensive stocks of conventional and nuclear weapons. They told their scientists that the Convention was no more than a screen behind which the United States would continue to develop its own biological weapons. Alibek writes:

> We were the victims of our own gullibility. I have come to believe that the most senior Soviet officials must have known all along that the Americans had no serious biological warfare program after 1969–after all, our intelligence agencies were among the best at their craft, and they had not come up with any real evidence. But the fiction had been necessary to instill in us a sense of urgency. The Soviet biological warfare program, born initially out of fear and insecurity, had long since become a hostage to Kremlin politics. This would explain why KGB Chairman Kryuchkov had been so willing to trade it away in 1990 and why bureaucrats like Kalinin and Bykov refused to give it up.

In December 1991, Alibek was a member of a Russian delegation sent to inspect suspected US biological warfare installations. Before he left Russia, his superior, Major General Kalinin, told him: "Whatever you see there, come back with evidence that the Americans are making [biological] weapons." He found none.

By the time the Soviet Union collapsed, the number of people employed there on research, development, and manufacture of biological weapons had grown to 60,000. Gorbachev's five-year plan, issued in 1985, included a $400 million virus production plant in Mordovia that contained a 170-gallon reactor for smallpox virus. In 1990 Soviet expenditure on biological weapons was close to $1 billion. At the same time, the Soviet authorities were unable to control their country's rapidly deteriorating economy. They were rescued from bankruptcy by foreign loans that raised external debt from $43 billion at the end of 1988 to $67 billion at the end of 1991.[1] Still, biological weapons spending continued.

The worldwide eradication of smallpox ranks as the greatest achievement of the World Health Organization, a heartening example of international cooperation to rid mankind of a terrible scourge. After this success, vaccination of babies against smallpox was discontinued throughout the world. The Soviets recognized that this made the new generation susceptible to smallpox and initiated a program for the manufacture of the virus. A Soviet medical team sent to India to help eradicate the virus was accompanied by a KGB man who returned home with a highly virulent and stable strain of it. Alibek writes that by adapting methods for growing human or animal cells in cultures on a large scale that had been developed in the West, a new production line perfected by 1990 was capable of manufacturing eighty to one hundred tons of virus per year in a vast atomic bombproof underground factory.

Some experts I have consulted regard that figure as improbably high because smaller amounts of virus would have been sufficient. An aerosol droplet containing as few as five smallpox virus particles suffices to infect a monkey; if this applied also to humans, then even one ton, if dispersed around the world, would be more than enough to kill every nonvaccinated man, woman, and child on earth. Alibek points out that the 12 million doses of smallpox vaccine now stored in the US would not stop an epidemic. He and his colleagues must have been aware that most adults in the West had been vaccinated so that their new weapon was aimed mainly at children.

In 1967 an animal keeper at the Behring pharmaceutical works in Marburg, Germany, contracted a virus from green monkeys sent there from Africa. It was named Marburg virus and is one of the most dangerous viruses known. The Ebola virus, described by Richard Preston in his best-selling book *The Hot Zone*, is closely related to it. In April 1988 Nikolai Ustinov, a scientist in the Vector Institute in the small Siberian town of Koltsovo, accidentally injected some Marburg virus into his thumb. Antiserum sent from Moscow proved useless. Alibek describes in heartrending detail Ustinov's terrible suffering during the two weeks before he died. Since normal burial of his body was too risky, it was covered with disinfectant, wrapped in plastic sheets, and placed in a sealed metal container, which was then fitted into a wooden coffin.

Alibek blames Ustinov's accident on long hours of work and enormous pressure to achieve quick results. What for? Alibek does not tell us. Before the burial, samples of the virus taken from Ustinov's organs were found to be more virulent and stable than those taken from the fermentors, and orders went out to replace that strain with Ustinov's. When bomblets filled with the virus were exploded over monkeys every one of them was dead three weeks later. By early 1990 the virus and the related Ebola virus were ready for approval by the Ministry of Defense.

On January 10, 2000 Judith Miller reported in *The New York Times* that American subsidies had helped to convert the Koltsovo laboratories to peaceful purposes. They are now manufacturing diagnostic kits, enriched milk for children, interferon (used for treating certain leukemias and lymphomas), antibodies, and cosmetics, and they are also trying to develop a drug against the Marburg and Ebola viruses. A photograph in *The New York Times* shows Ustinov's widow sadly leaning over a gravestone that bears his photograph at a ceremony marking the twenty-fifth anniversary of the laboratory's founding. She must wish the change of policy had come sooner.

Excessive prescription of antibiotics by doctors and their use by animal breeders has spread resistance to them among many disease-producing bacteria. Some bacteria have developed resistance against all major antibiotics. The genes responsible for the resistance are concentrated in circular strands of DNA called plasmids which are separate from the main bacterial chromosome and relatively easy to isolate. Alibek's colleagues managed to introduce such plasmids into *Yersinia pestis*, the bacteria responsible for bubonic plague; these could be suspended in aerosols and released over cities, and would resist all known forms of treatment. Alibek notes that death from plague is invariably painful. Another triumph of his organization was the creation of a strain of *Yersinia pseudotuberculosis*, a relative of the plague bacterium, into which they introduced the gene for a paralyzing toxin, so that infection would cause both fever and paralysis.

Valery Butuzov, a pharmacologist who became Alibek's close friend, specialized in developing toxins for assassinating individuals rather than for mass murder. One day in 1990 he came to ask Alibek's advice about a brilliant new invention, an ingeniously designed pocket assassinator. It consisted of a miniature battery, an amplifier, and a vibrating membrane, all in a box the size of a packet of cigarettes. A speck of dried powder on the membrane would form an aerosol when the device was activated. Butuzov told Alibek that he was thinking of something like Ebola virus. Alibek objected that this would also kill everyone around the chosen victim, but Butuzov thought that this wouldn't matter. One possible victim was Zviad Gamsakhurdia, the newly elected president of Georgia, who was seen as hostile to Russia.

Another of Alibek's colleagues, called Markin, served in the KGB "Division for Special Countermeasures Against Foreign Engineering Intelligence Services." He was a civilian engineer who was expert in disguising the voluminous flow of waste from the fermentation plants used for the production of bacteria and viruses. An unhappy marriage depressed him and made him ask for a leave of absence on the pretense that he needed to take care of his sick mother. A few weeks later he applied to retire to the collective farm near Gorky where his mother lived; he said that he was not a traitor, but wanted to live peacefully in a natural setting. But Markin knew too much. The local commander of the KGB complained that this had given him two headaches instead of one, because he had already had to keep watch on Andrei Sakharov, who had been banished to Gorky. After some time passed, Alibek's superior told him with glee that he now had only one headache: Markin had been drinking too much, went out for a swim, and never returned. The commander brushed off Alibek's question whether Markin had been murdered by saying that all that mattered was that he didn't have to worry about Markin anymore.

Tularemia is a bacterial disease transmitted by ticks or mosquitoes; if inhaled, it causes weeks of chills, nausea, headaches, and fever. The authoritative book *Health Aspects of Chemical and Biological Weapons*, published by the World Health Organization, states that

> A case fatality rate of approximately 25 percent could be expected if a virulent strain were used. If an antibiotic-sensitive strain of relatively low virulence were used, with the aim of incapacitating, as opposed to killing, deaths would be considerably reduced but still be appreciable in number.[2]

Alibek writes that after World War II the United States and Canada developed tularemia for use on the battlefield, realizing that it could immobilize an entire division if the soldiers had not been vaccinated. In 1942 Soviet troops apparently sprayed tularemia successfully among the German troops who were attacking Stalingrad. From an unspecified "leading international research institute in Europe," Alibek obtained a tularemia strain capable of overcoming immunity in vaccinated monkeys, and after several months' effort turned it into a weapon. Five hundred monkeys were brought from Africa and immunized before being tested on that island in the Aral Sea. Two generals took charge of the test team. Back in his laboratory, Alibek waited for the outcome, too excited to concentrate on his research. When the coded test results came back, they were better than expected. Alibek was thrilled that nearly all the immunized monkeys had died. He was showered with congratulations and decorated by his commanding officer, Major General Y.T. Kalinin, who was also deputy minister in the Ministry of Medical and Microbiological Industry. Alibek has no word of sympathy for the intelligent monkeys tortured to death for the sake of his advancement and the aggrandizement of his bosses.

In 1987 Alibek was promoted to the rank of colonel and deputy chief of the Biosafety Directorate for developing the large-scale production of a very virulent strain of the anthrax bacillus in a factory that could produce two tons a day. The process used, he writes, was as "reliable and efficient as producing tanks, trucks, cars or Coca-Cola." (A British weapons inspector who saw the plant told me that it could not have produced two tons of dried anthrax spheres a day, but only two tons of a suspension of spores in water.[3]) Alibek claims that this transformed the Soviet Union into the world's first and only biological superpower. By 1986, over nine hundred people worked at that plant, and more were added every month. The extreme pressure under which they were forced to work led to one or two accidents every week, but that was considered a small price to pay for their productivity.

In November 1979, a Russian magazine in West Germany reported that in April of that year an explosion in an army factory near Sverdlovsk in the Ural Mountains had released a cloud of anthrax bacteria that killed about a thousand people. The Soviet news agency TASS denied this violation of the 1972 Biological Weapons Convention and attributed the incident to an outbreak of anthrax among domestic animals in the Sverdlovsk region. In her excellent recently published book *Anthrax: The Investigation of a Deadly Outbreak*,[4] the American sociologist Jeanne Guillemin describes in fascinating detail her investigations in Sverdlovsk together with a team of scientists and doctors to discover the true cause of the outbreak. They succeeded in establishing that an explosion had taken place, but the details remained obscure.

Alibek's book supplies them. In the anthrax drying plant at Compound 19 of the biological arms production facility, anthrax was manufactured around the clock. In order to dry the anthrax spores the fermenting liquid had to be evaporated. The spores were then ready to be ground into a fine powder for suspension in an aerosol. To prevent the escape of anthrax spores, the exhaust pipes of the evaporators were fitted with filters, which, after each shift, were shut down for maintenance.

One evening in 1979, the outgoing technician of the afternoon left a note for his supervisor: "Filter clogged so I've removed it. Replacement necessary"; but the supervisor failed to enter his message in the log book. The evaporators were switched on again for the night shift and a cloud of anthrax spores escaped which the wind blew over the town. Alibek writes:

> No one wanted to set off a panic or to alert outsiders. Sverdlovsk residents were informed that the deaths were caused by a truckload of contaminated meat sold on the black market. Printed fliers advised people to stay away from "unofficial" food vendors. More than one hundred stray dogs were rounded up and killed, on the grounds that they represented a danger to public health after having been seen scavenging near markets where the meat was sold. Meanwhile, military sentries were posted in the immediate neighborhood of the plant to keep intruders away, and KGB officers pretending to be doctors visited the homes of victims' families with falsified death certificates.

Reports about the number of deaths differ between 66 and 105, far fewer than the thousand reported earlier. Alibek calls Sverdlovsk as serious a disaster as the one at Chernobyl. In fact, the number of deaths is surprisingly small for an urban area containing about a quarter of a million people, possibly because inhaled anthrax spores are not very infectious to humans.

To remove production from population centers, Brezhnev then ordered the transfer of biological weapons manufacture to Stepnogorsk, a town isolated in the desert of Kazakhstan. Alibek applied for the directorship of the new plant and got it. He and his family now enjoyed the privileges reserved for the higher echelons of the Soviet army. He was well on the way to becoming a general when disaster struck. In October 1989, Vladimir Pasechnik, the head of the Leningrad Institute of Ultra-Pure Biopreparations where cruise missiles were filled with bacteria and viruses, defected to Britain. Shortly afterward the Berlin Wall fell and the Soviet Empire began to collapse.

Margaret Thatcher faced Gorbachev with Pasechnik's evidence, but he denied it. Eventually agreement was reached on mutual inspection of Soviet and American biological weapons facilities, and the arrival of an inspection team was imminent. Alibek describes a meeting with Foreign Ministry officials who seemed to be unaware of the army's activities, and he claims that even Foreign Minister Shevardnadze, a member of the Politburo, was ignorant of Gorbachev's orders to step up the manufacture of biological weapons. Even before Pasechnik's defection some of the biological weapons were being made in mobile plants that would escape detection by foreign inspectors.

A British-American inspection team finally arrived at Sverdlovsk on January 15, 1991. Alibek was ordered to act as their host. He and his colleagues and their accompanying KGB men played a cat-and-mouse game with the inspectors about the work in the facilities they visited. He spoke no English, but he enjoyed his first contacts with scientists "in the enemy camp." Despite Alibek's determined attempt to fool the inspectors, they obtained clear evidence of a biological weapons program and informed the Soviets of their findings. Alibek then seems to have realized that the game was up. He gradually resigned from his various offices and also from the Communist Party, and escaped with his family to the United States, where he claims to have disclosed to US authorities all the secrets of the Soviet Union's biological warfare program.

In his introduction, Alibek writes that Russia's stockpile of germs and viruses has now been destroyed and that offensive research is no longer conducted, but near the end of his book he cites evidence that some facilities in Russia are still being used. In a recent letter to me, Matthew Meselson reports that three of the former Soviet facilities remain closed to foreigners. It seems clear that Biopreparat, the organization to which Alibek belonged, was the civilian façade of the biological weapons program which has indeed been dismantled. But this was only one arm of the program; the other was run by the army, and there is no evidence that it has been discontinued. Discussions among the Americans, the Russians, and the British had led to agreement on a program of reciprocal visits to one another's biological facilities; some visits took place but the agreement seems now to be in abeyance and the discussions have stopped.

Without such visits it is hard to evaluate the Russian program, but even if these visits were still taking place, experience in Iraq has shown that plants manufacturing biological weapons are very difficult to detect. Continuing suspicions, together with the deterioration of US–Russian relations, hamper joint efforts to avert a threat to both countries. On the other hand, the Republic of Kazakhstan, with American help and subsidies, has in fact dismantled the large biological warfare plant at Stepnogorsk.

Alibek tells us that the initiative for the Soviet program had come not from the government or from the army, but from a scientist, Yuri Ovchinnikov, a biochemist of international repute, vice president to the Soviet Academy of Sciences, and the only scientific member of the Central Committee of the Communist Party. A wrestling champion and actor in his youth, he was strongly built, an outstandingly able scientist and manager, and a shrewd opportunist who exploited his connections in the army and the Party to gain strong support for his institute. He enjoyed his prominent position in the Soviet hierarchy. His chauffeur drove him in a large

Zil limousine along the middle lanes reserved for Party bosses, and he was an admired leader of the mostly women scientists in his laboratory.

His research concentrated on naturally occurring peptides and toxins with obvious military applications, such as the toxin incorporated in *Yersinia pseudotuberculosis*. Before he died of leukemia at the age of fifty-three, Ovchinnikov had obtained the equivalent of about $100 million for building a gigantic new institute of bioorganic chemistry on the outskirts of Moscow. In the 1960s and early 1970s, the Western literature on molecular genetics had made him aware of its possible military applications, and he persuaded the skeptical military leaders of its potential for creating deadly biological weapons; he found an ally in Leonid Brezhnev and persuaded him to use the international Biological Weapons Convention as a screen for turning the Soviet Union into a biological superpower. An Interagency Scientific and Technical Council was set up for coordinating the various organizations involved in this ambitious program.

What was the purpose of producing tons of deadly viruses and bacteria engineered to enhance still further the suffering and agonizing deaths they would cause? This was not just an enterprise designed to provide Soviet forces with additional weapons. It was a program pursued with the same feverish intensity as the construction of the first atomic bombs at Los Alamos in the 1940s. The physicists at Los Alamos, however, were driven by fear that the Germans under Hitler would make atomic bombs first. The Soviets ought to have known through their secret services that neither the United States nor Britain was making biological weapons. They were already equipped with thousands of nuclear missiles, enough to extinguish all life on earth, so why did they need biological weapons sufficient to kill all humans and higher animals on earth several times over as well?

Did the Politburo and the General Staff plan a preemptive strike against the West, even though nuclear retaliation would have made this suicidal? Was the program driven by paranoid fear that the USSR would be attacked? Or by mindless obsession with the accumulation of weapons? Or simply by members of an elite haunted by fear that power might slip from their hands? And what was in the minds of the scientists who provided them with the means to commit crimes against humanity?

One of the most disturbing features of Alibek's book is his obliviousness to the potential consequences of his work. He is a technocrat pure and simple, concerned only with the solution of the problem in hand and with his own career and a mind closed to all else. He writes fondly of his family, but it never occurs to him that the weapons he manufactures would subject thousand of other families to appalling suffering and death. And what for? The question does not occur to him.

For many years, Andrei Sakharov did nothing but design bigger and better hydrogen bombs, but eventually he gave it up in favor of working for human rights and an end to the cold war. Alibek merely changed to the winning side, saying he now felt he had a mission to warn the United States about biological terrorism. Thanks to his pioneering work, he writes, it has become cheap and easy to make the biological weapons, and former colleagues of his have offered their expert services to Iraq, Iran, and other hostile regimes to equip their missiles with deadly pathogens, but he provides no evidence to support this statement. In the US, the fear that these weapons might be deployed has been used to justify an antimissile defense, but according to a recent article in *Scientific American* such defenses are easy to penetrate by cruise missiles, decoys, and other means.[5] The authors of the article do not believe that the problems can be solved even by expenditure of many billions of dollars. Argument continues whether a limited missile defense is technically feasible or politically desirable. (It is also possible that enemies of the US could secretly assemble the components of nuclear and biological weapons within the US and take them to their targets in a truck, as did, for example, the suicide bombers who attacked US embassies in Kenya and Uganda.)

In an editorial last year in *Science*, the physicist Philip Abelson mentions a new instrument to protect against biological attacks. It is a miniature version of the mass spectrometer, a sophisticated, normally very large machine used to identify chemical compounds, which would identify the attacking organisms within minutes.[6] In 1984 Kary Mullis invented another ingenious chemical procedure, the polymerase chain reaction, which makes it possible to identify single genes and amplify them many millionfold. These methods would give the medical services time to treat victims with the appropriate vaccines or antibiotics, always assuming that sufficient stores of vaccines against virtually all possible bacteria and viruses are kept in all major cities.

This approach would also cost billions of dollars and would require elaborate training of police, ambulance, and hospital personnel, but it is unlikely that more than a small fraction of the thousands of victims of such an attack could be saved. It may also prove difficult to keep thousands of people on the alert for years in order to deal with a danger which, one hopes, will never materialize. While preparing to spend billions of dollars on defense against biological weapons, the US government also opposes a protocol to the Biological Weapons Convention which would cost little and would give the convention added effectiveness by providing for routine inspections for biological weapons activities. These would be analogous to the inspections for chemical weapons activities that have been routinely practiced under the 1993 Chemical Weapons Convention; but the US administration would prefer a protocol providing for inspections on challenge.

According to press reports, the opposition of the US government is a response to pressure by pharmaceutical firms which fear that inspections might reveal their commercial secrets to foreign competitors. It is also argued that the technologies for peaceful applications of biotechnology are the same as those for weapons and that inspections or surveillance will be unable to establish the intent of the producer. The distinguished biologist Joshua Lederberg has commented that there may be no technical solution to the problem of biological weapons, only an ethical solution, by which he apparently means that nations should join in accepting a moral obligation not to use them.

According to a recent analysis by Paul Schulte, the UN special commission charged with ridding Iraq of weapons of mass destruction found itself frustrated "not by sensationally detected cheating, but by endless, persistent, shameless—and boring—lies and obstruction. They got away with that, because they judged, and still judge that they could successfully defy a divided Security Council." It appears, he writes, that

> any determined state which is sufficiently rich and/or powerful to pose a serious proliferation threat can, in the long term, even after losing an aggressive war against a widely based UN sanctioned coalition, expect to escape the consequences of non-compliance (even if the disarmament requirement was a strict cease-fire condition) by exploiting its diplomatic and commercial leverage to divide the world community into dropping sanctions or other enforcement measures.[7]

All the same, Matthew Meselson, a biologist knowledgeable about the threat posed by chemical and biological weapons, believes that international criminal law offers a promising new means of preventing biological warfare and terrorism. He wrote:

> Recently, interest has developed in the possibility of a convention to create international law that would hold individuals criminally responsible

> for acts that are prohibited to states by the biological and chemical weapons conventions. Such a convention, which would be patterned on existing conventions that criminalize aircraft hijacking, nuclear theft, and other crimes that pose a threat to all, would make it an offense for any person, regardless of official position, to order, direct or knowingly render substantial assistance to the development, production, acquisition, or use of biological or chemical weapons. A person who commits any of the prohibited acts anywhere would face the risk of prosecution or of extradition, should that person be found in a state that supports the proposed convention.
>
> International law that would hold individuals criminally responsible would create a new dimension of constraint against biological and chemical weapons. Such individuals would be regarded as *hostes humani generis*—enemies of all humanity. The norm against chemical and biological weapons would be strengthened; deterrence of potential offenders, both official and unofficial, would be enhanced; and international cooperation in suppressing the prohibited activities would be facilitated.[8]

It has been argued that it may be unjust to brand all those involved in biological weapons programs as criminals, because totalitarian governments may have given them no choice, forcing them to work under threats to them and their families. Even so, Meselson's proposal deserves to be explored.

Before 1969, both the United States and the Soviet Union had large biological weapons programs. But then, to his credit, President Nixon decided that "mankind already holds in his hands too many seeds of its own destruction" and abandoned it, while the Soviet Union continued it clandestinely and illegally even after acceding to the Biological Weapons Convention. Reading Alibek's book, I wondered whether he tried to make his revelations more sensational by painting a grossly exaggerated picture of the Soviet biological weapons program. Colleagues who worked on the British-American team that inspected the Soviet facilities in 1991 have told me the book contains many inaccuracies but that Alibek has not exaggerated the scale of the Soviet effort. On the contrary, he had detailed knowledge of only part of it. Even if only a fraction of his allegations are correct, his book uncovers a moral abyss and a perversion of science that threatens to nullify medicine's hard-won conquest of infectious disease.

How likely are biological attacks? An attack by a foreign power, say by Iraq or North Korea, would provoke retribution so destructive as to make it suicidal. While the British press thrives on scaring the public with the imagined threats posed by genetically manipulated foods, American publishers are now making much of the dangers of bioterrorism. Is the alarm justified?

Terrorists might grow bacteria and viruses in a plant that supposedly makes vaccines, but it would require large technical and scientific resources to turn them into effective weapons. Henry Sokolski, Director of the Nonproliferation Policy Education Center in Washington, D.C., has written a balanced assessment of the risks themselves and of their exaggerated perception that seems worth quoting at length. He writes:

> Last year President Clinton announced the US would spend $10 billion on countering terrorism, including biological and chemical threats, for fiscal year 2000. Would there be better things to spend such large sums of money on? As for biological attacks worldwide, seventy have occurred in the last century causing nine deaths, but only eighteen of these seventy attacks were made by terrorists. There are risks not only in underestimating the chemical and biological domestic terrorist threat, but in overestimating it as well. These include:
>
> - Raising public consciousness about the possible threat in a manner that emboldens criminals and terrorists to attempt precisely what the government and public want to avoid.
> - Reassuring the public about the preparedness of government such that any government shortcoming is likely to be magnified to politically fatal levels.
> - Preemptively undermining U.S. civil liberties in the name of enhanced homeland defense....
> - Encouraging an "America first" siege mentality and a retreat from foreign commitments critical to our nation's security.
>
> The downside risks listed are at least as likely as the domestic biological and chemical terrorism threats that might generate them. The technical challenges of terrorists using traditional biological agents to produce massive fatalities are no less daunting. Biological agents are lethal only if inhaled, and particles larger than ten microns are likely to be blocked before they reach the lungs. On the other hand, agent particles approaching one micron are likely to be exhaled and so will not remain in the lungs.[9] Operationally, particles sized between five and ten microns are optimal. Spreading biological agent in particles of that precise size, however, is difficult. The only organizations that have done so are states. Sunlight, moreover, kills or denatures most biological agents (making night-time dispersal imperative), and wind patterns and humidity can reduce the lethality of an anthrax attack 1,000-fold.

> As for dealing with domestic biological terrorism, the US is blessed with a massive health care system. The country spends nearly four times as much on its public health and medical system as it does on its entire military. Include the fire-fighting services and police, and it is clear that these civilian institutions (and the Centers for Disease Control) are the ones best positioned to respond to domestic terrorism.
>
> The point here is not to dismiss the possibility of any particular chemical or biological threat, but rather to weigh how much attention each one deserves. Assuming we are not foolish enough to demand 100 percent protection against all attacks, our medical system, federal and local governments, and military should be able to ensure against a lasting, strategic calamity. The key to success, however, will be the same as it was a decade ago in Desert Shield, which is to avoid focusing on the most horrific scenarios at the expense of preparing for the most likely ones.
>
> In general agreement with these views, delegates at a conference on "The New Terrorism" held at the Chemical and Biological Arms Control Institute in Washington, D.C., in April 1999 agreed that terrorism causing casualties by the use of chemical or biological agents was unlikely in the near future.

A detailed historical study of known terrorist attacks in the collection *Toxic Terror* leads the authors to similar conclusions:

> Crude delivery methods are likely to remain the most common forms of chemical and biological warfare terrorism. They are potentially capable of inflicting at most tens to hundreds of fatalities—within the destructive range of high-explosive bombs, but not the mass deaths predicted by the most alarmist scenarios. Although the devastating potential of a "catastrophic" event of chemical and biological warfare use warrants examination, history suggests that the most probable terrorist use of chemical and biological warfare agents will be tactical and relatively small-scale.
>
> To date, there are no known cases of state-sponsored chemical and biological warfare terrorism (at least in the public domain), probably because of the likelihood of severe retaliation against the sponsoring government if its involvement were to become known.[10]

In view of the kinds of irrational behavior described in *Biohazard* and the ease of acquiring and propagating bacteria and viruses for biological weapons, there is no reason for complacency about their dangers. But the dangers should be seen in the perspective of other threats to human life. In 1995, the last year for which official statistics are available, the number of people killed by tobacco in the United States was 502,000, of whom 214,000 were aged between thirty-five and sixty-nine. On average, each of those could have expected to live twenty-three years longer.[11] In view of these alarming numbers, it seems to me that the still-prospering tobacco industry poses a proven threat to health and life that is many thousand times greater than the potential threat of bioterrorism.

How to Make Discoveries

High on Science[*]

Peter Medawar was a great biologist whose research helped to make possible the transplantation of human organs. He also thought profoundly about the methods, the meaning, and the values of scientific research, and he published his thoughts in books and essays that are models of clarity, style, and wit.

Born in 1915 in Brazil of a Lebanese father and an English mother, he received his education in England and made his career there. He became a full professor at thirty-two, a Fellow of the Royal Society at thirty-four, a Nobel Laureate at forty-five, and head of Britain's largest medical research laboratory at forty-seven.

At fifty-four, when his intellectual powers and capacity for work seemed inexhaustible, a cerebral hemorrhage destroyed the right half of his brain, but it did not impair his determination, his vitality and optimism. Three years later he was back at his research and literary work, and he lectured around the globe. In 1980 a cerebral thrombosis set him back severely. Again he recovered and wrote more papers and essays as well as a hilarious autobiography which makes even his tragedy an occasion for

*A review of the books: *A Very Decided Preference: Life with Peter Medawar* by Jean Medawar (Norton); *The Threat and the Glory: Reflections on Science and Scientists* by P.M. Medawar (Harper Collins/A Cornelia and Michael Bessie Book); and *Peter Brian Medawar: 28 February 1915–2 October 1987* by N.A. Mitchison (reprinted from *Biographical Memoirs of the Royal Society*, 1990, volume 35).

laughter.[1] In 1985, another series of strokes robbed him of his ability to speak clearly and of most of his eyesight, and in 1987 they finally killed him.

Medawar's first great discovery was that the rejection of skin grafts from a donor is an immunological reaction, a reaction mediated not by antibodies, but by white blood cells, which James Gowans later identified as lymphocytes. Medawar wrote that these cells behaved

> like the chorus in a provincial production of Gounod's *Faust;* lymphocytes in the bloodstream at any one moment disappear behind the scenes and re-enter by another route.

Medawar asked how lymphocytes distinguish between self and non-self. In the mid-Fifties this led him and his colleagues R.E. Billingham and L. Brent to their second great discovery. They found that animals can be made tolerant to grafts of foreign tissue if they have been immunized by an injection of such tissue while still in their mother's womb. The mere fact that animals can be made tolerant to foreign grafts raised the hope of inducing such tolerance in humans and encouraged experiments on the transplantation of organs.

At first this was possible only between identical twins. Later, Medawar and his colleagues felt encouraged that steroids delayed the rejection of skin grafts. They then tried to immunize rabbits with mouse lymphocytes, and injected the immune serum of the rabbits back into the mice. This suppressed the lymphocytes' attack on foreign grafts, at least for some weeks or months.

Medawar's first attempts to develop this method for human use were overtaken by the discovery that cyclosporin, a compound isolated from a fungus at the Sandoz Laboratories in Switzerland, had a powerful suppressive effect on lymphocytes in animals and a low toxicity. Roy Calne in Cambridge, England, pioneered its use and that of other immunosuppressive drugs in humans. He was able to tell Medawar before his final series of strokes that one of his life's great aims, to make human beings tolerant of transplanted organs, had at last been achieved.

Medawar never wanted to die, severely crippled though he was. In the last of his essays, now collected and published under the title *The Threat and the Glory,* he pours scorn on the crumbling baroque edifice of Freudian psychoanalytic theory that postulates a death instinct, "the most deeply unbiological explanatory concept in Freud."

> The tenacity of our hold on life and the sheer strength of our preference for being alive whenever it is an option is far better evidence of a life instinct than any element of the human behavioral repertoire is evidence of a death instinct. It is odd, then, that nothing in modern medicine has aroused more criticism and resentment than the lengths to which the medical profession will go to prolong the life of patients who need not die if any artifice can keep them going... . Charity, common sense, and humanity unite to describe intensive care as a method of preserving life and not, as its critics have declared, of prolonging death.

Medawar was tall, with "the pride of bearing that comes from good looks known to be possessed"[2] and powerfully built, good at tennis and cricket. He was outgoing, vivacious, sociable, debonair, brilliant in conversation, approachable, restless, and intensely ambitious. From his student days he was determined that none of human knowledge should be beyond his grasp; for example, he declared Bertrand Russell's formidable monograph on mathematical logic, the *Principia Mathematica*, to have been the book that influenced him most when he was a student at Oxford.

Later he fell under the spell of the philosopher Karl Popper, whose book *Conjectures and Refutations* taught him the scientific method that he adopted and popularized in his writing. According to Popper, scientists do not derive general laws from observations, but they formulate hypotheses which they test experimentally. This method leads them gradually closer to the truth. Popper's view that imagination comes first had a strong appeal for Medawar, since it implied that a scientist is not a robot who turns the handle of discovery, but a creative spirit on a par with artists and writers. Medawar called Popper's "the hypothetico-deductive method."

Popper gave explicit form to a method already used by the greatest scientists in the past. For example, in 1856 Michael Faraday wrote about the propagation of light waves through the supposed ether: "I have struggled to perceive how far... experimental trials might be devised, which... might contradict, confirm, enlarge, or modify the idea we form of it, always with the hope that the corrected or instructed idea would approach more and more to the truth of nature."[3]

Medawar allowed nothing to deflect him from the pursuit of knowledge, except possibly laughter. He rarely relaxed, did not believe in holidays, and continued his research even while heading an institute of several hundred people. He maintained that a lot of his work was "as good as a rest," and he prided himself on wasting no time, but neither Russell's logic nor Popper's hypothetico-deductive method saved him from wasting several of his best years on experiments that proved to be futile.

He had formulated bold hypotheses about the spread of pigments in animals' skin, which is a fundamental problem related to growth and differentiation. He then devised ingenious experiments to test his hypotheses, but he blinded himself to the possibility that he might be looking at the wrong kind of cells. As I have learned to my own cost, one can become so enamored of one's ideas that doubts aroused by inconsistent results are stilled with far-fetched explanations rather than being allowed to overturn basic premises. Medawar preached that research is a passionate enterprise, but he did not warn scientists that that very passion can lead them into a trap.

What did it feel like to be married to such a man? Like Enrico Fermi's widow Laura, who began her light-hearted biography of the great physicist while he was still alive,[4] Jean Medawar's biography, called *A Very Decided Preference,* draws an affectionate and bemused cartoon of Peter instead of chiseling him in marble. He told his young bride, pretentiously, that she had first claim on his love, but not on his time, made her buy her own wedding ring and often also her own Christmas presents. So preoccupied was he with his work that Jean had to be both father and mother to their four children. He had no patience with real people's emotional problems, but was spellbound when he heard them transformed into music in Wagner's operas. Wotan bidding Brünhilde goodbye in *Die Walküre* stirred him more than his own daughter leaving home for months. When Jean reproached him for having hardly noticed that she had gone, he explained that "his emotions were stirred by art," the sort of priggish remark that Mr. Casaubon in *Middlemarch* might have made to Dorothea.

Medawar's emotions were stirred by Wagner and Verdi (hardly by Mozart), but apparently not by great painters until he reached middle age and paid his first visit to the Frick Museum in New York. In his review of the fifteenth edition of the *Encyclopedia Britannica,* he scoffs at the entries for Sisyphus, Tantalus, or Leda as superfluous, oblivious that much of the visual arts and literature was inspired by classical mythology.

Young Medawar proudly told a friend, "My mind, you know, never lets me rest," but old Medawar wrote modestly that "It is a natural tendency of the mind to come to and remain at a complete standstill. This is a principle of Newtonian stature." In hospital, books "are crucially important for keeping the mind in working order. Some serious works should therefore be among them. Remember, however, that if you didn't understand Chomsky when you were well, there is nothing about illness that can give you an insight into the working of his mind." Medawar excelled in such sallies, and Jean tells us that he would laugh out loud as he wrote them.

Her biography begins with the tragic episode in Exeter Cathedral, when he fell "from a pinnacle of achievements into a state very near to death." He had been invited to give the presidential address at the annual meeting of the British Association for the Advancement of Science, and he chose that honorific occasion for a passionate declaration of his faith in science. He named his lecture after Francis Bacon's New Atlantis, an island kingdom where the Merchants of Light establish "the noblest foundation that ever was on earth" whose object is "the knowledge,... and the secret motions; and the enlargement of the bounds of human empire: THE EFFECTING OF ALL THINGS POSSIBLE." The purpose of his address, Medawar said, was ""to draw certain parallels between the spiritual and philosophic condition of thoughtful people in the seventeenth century and in the contemporary world." The Thirty Years War on the continent of Europe and the Civil War in England brought "a period of questioning and irresolution and despondence,... a failure of nerve." Besides, people believed that the Biblical apocalypse was at hand, much as we fear, with more reason, man's destruction by nuclear war. "Then as now the remedy for discomforting thoughts was less often to seek comfort than to abstain from thinking" (a typical Medawar aside).

In the second half of the seventeenth century the belief that the rational pursuit of science could improve the human condition engendered a new spirit of optimism.

> The repudiation of the concept of decay, the beginning of a sense of the future, an affirmation of the dignity and worthiness of secular learning, the idea that human capabilities might have no upper limit, an excellent recognition of the capabilities of man—these were the seventeenth century's antidote to despondency.

Medawar calls for a similar antidote in our time. "We wring our hands over the miscarriages of technology and take its benefactions for granted." "The real trouble is our acute sense of human failure and mismanagement, a new and...oppressive sense of the inadequacy of man." Nevertheless "to deride the hope of progress is the ultimate fatuity, the last word in poverty of spirit and meanness of mind."[5]

Jean relates that he had worked on this lecture for months, and had moved her to tears when he tried it out on her at home. It still is one of the most eloquent and erudite pleas for science.

Two days after delivering that address Medawar read the lesson to a large congregation of the British Association assembled in Exeter Cathedral. He chose a passage from chapter seven of the Wisdom of Solomon:

> For Wisdom, the fashioner of all things, taught me, for in her there is a spirit that is intelligent, holy, unique, manifold, subtle, mobile, clean, unpolluted, distinct, invulnerable, loving the good, keen, irresistible, beneficent, humane, steadfast, sure, free from anxiety, all powerful, overseeing all, and penetrating through all spirits that are intelligent and pure and most subtle.

He started to read more slowly. Gradually he spoke as if the words were costing him a colossal effort.

> For wisdom is more mobile than any motion; because of her pureness she pervades and penetrates all things.

At that moment his speech began to slur, and he collapsed with a cerebral hemorrhage.

It was a fall from the heights of a brilliant career to helpless dependence on life-support machinery, doctors and nurses, and above all on Jean, whose sheer determination to restore Peter to full activity fortified his own "very decided preference for staying alive."

The title of Peter Medawar's collected essays, *The Threat and the Glory*, is taken from his review of several books on genetic engineering that appeared in the *New York Review* in October 1977 under the title "Fear of DNA." These fears have not diminished, even though there has not been a single mishap to substantiate them, and the benefits have multiplied. Medawar disposes of the fear that gene technology will make it possible to change people's characters and personalities in order to create a nightmarish Brave New World of brainy masters and stupid slaves. He argues that this could have been embarked upon at any time in the past thousand years "merely by applying the most powerful of all forms of biological engineering—Darwinian selection—to a population—mankind—known by its open breeding system, lack of speciation, and rich resources in inborn diversity to be perfectly well able to respond to the empirical acts of the stockbreeder."

> If these enormities have not been perpetuated or even seriously attempted hitherto by the comparatively straightforward and empirically well-understood methods available for their execution, why should we now

> begin to fear that enormities as great or even greater will be executed by the much more costly and technically more difficult procedures of genetic engineering?

Besides, Medawar points out that the technology needed to fill the mind with untruth, with a resistance to new learning and to anything that might conduce to improvement has been known for 5,000 years or more and is known as "education."

In another essay, "Biology and Man's Estimation of Himself," Medawar disputes that the propounding of Darwin's theory of evolution was a great affront to human dignity, but I doubt that he was right. Noel Annan's *Life of Leslie Stephen* relates the profound shock that Victorians experienced when they felt their religious beliefs shaken by the *Origin of Species;* I also remember the outrage that Jacques Monod's book, *Chance and Necessity,* recently caused among the French, even though Monod merely restated Darwin's theories in molecular form.

Medawar's review of the sociologist Harriet Zuckerman's *The Scientific Elite* contains observations on the great scientists who did and those who did not win the Nobel Prize. I had believed American laureates to be, like most other scientists, of working-class or lower middle-class origins, but in fact 90 percent of their fathers were professional men, proprietors, or managers. Half of them had worked either as students, post-doctoral fellows, or junior collaborators under older laureates, but they owed the prize to the older men's good teaching rather than their teachers' sinister influence on the Nobel committees in Stockholm. Medawar writes that "the laureates who would have pleased Nobel are those to whom the award gave a terrific fillip to proceed with their research" (like Medawar himself) but Nobel would have been dismayed if "the benefaction intended to put its recipients beyond the reach of want had, in reality, deprived them of any further incentive to continue with scientific research." These people spend their time traveling from conference to conference, with platitudinous titles like "Man and the Universe," "Science and Culture," or "Environment and the Future." Fortunately they are in a small minority, but it has not escaped my notice that the sly organizers of such meetings flatter Nobel Laureates by persuading them that their presence is essential for the survival of mankind, and then exploit their promised attendance to extract funds from gullible benefactors.

Medawar's bêtes noires are psychoanalysts and traders in IQ. He never misses an opportunity of deriding Freud and his faithfuls. In his essay on Darwin's illness he ridicules the psychoanalysts who uncovered "a wealth of evidence that unmistakeably points" to the idea that Darwin's chronic indigestion, cardiac symptoms, and lack of physical energy were "a distorted expression of the aggression, hate and resentment felt, at an unconscious level, by Darwin towards his tyrannical father… As in the case of Oedipus, Darwin's punishment for the unconscious parricide was a heavy one—almost forty years of severe and crippling neurotic suffering… ." Medawar comments sarcastically:

> These deep and terrible feelings found outward expression in Darwin's touching references towards his father's memory, and in his describing his father as the kindest and wisest person he ever knew: clear evidence…of how deeply his true inner sentiments had been repressed.

The real cause of Darwin's chronic malaise was probably a parasitic infection that he contracted in his travels through South America.[6]

In Popper's view, psychoanalytic and Marxist theories are equally unscientific because both are self-fulfilling and neither can be falsified experimentally. What would Medawar have made of R.C. Lewontin's dogmatic and unprovable assertion that "Darwin's theory of evolution by natural selection is *obviously* nineteenth-century capitalism writ large, and his immersion in the social relations of a rising bourgeoisie had an overwhelming effect on the contents of his theory."[7] Marxism may be discredited in Eastern Europe, but it still seems to flourish at Harvard.

In his review of W. Broad and N. Wade's book *Betrayers of Truth*, Medawar relates the findings of the English psychologist Sir Cyril Burt that working-class children have lower IQs than children of the professional and managerial classes. For many years Burt's publications influenced educational policies until Leon Kamin and Oliver J. Gillie discovered after his death that he had faked his data. Medawar believes that "there was no effective check of Burt's findings, because he told the IQ boys exactly what they wanted to hear." Medawar asks how the young scientists with obvious intelligence and ability mentioned in *Betrayers of Truth* had supposed they would get away with sensational results that were based on fraudulent experiments. Efraim Racker, a professor of biochemistry at Cornell University, was himself a victim of fraud committed by one of his graduate students. He describes him as outstandingly able and believes that he and others like him were mentally unbalanced, suffering from a kind of schizophrenia that closed their minds to the likely consequences of their actions.[8]

Once I became the victim of scientific fraud, committed by a student who was told by his professor what result we expected him to find, and promptly found exactly that. I doubt that the student was schizophrenic, but probably just naive and anxious to please. Medawar remarks sarcastically that if among the thousands of scientists there are just a few odd crooks, then he would draw "a clear distinction between the scientific profession and the pursuit of mercantile business, politics, or the law, professions of which the practitioners are inflexibly upright all the time." Medawar's sharp pen punctured obscurantism, pomp, mysticism, cant, pseudoscience, and empty high-flown verbiage posing as profundity, as in Pierre Teilhard de Chardin's book *The Phenomenon of Man.*[9] In an innuendo about semantics he writes: "The innocent belief that words have an essential or inward meaning can lead to appalling confusion and waste of time. Let us take it that our business is to attach words to ideas and definitions, not to attach definitions to words."

He lashed out at "geneticism" as "an application to human affairs of the genetic understanding which is assumed to be much greater than it really is." He dismissed eugenics as lacking any scientific foundation and would have indignantly rejected the recent claim that criminal behavior is inherited as just another example of geneticism.

*Deconstructing Pasteur**

There is a real world independent of our senses; the laws of nature were not invented by man, but forced upon him by the natural world. They are the expression of a rational world order.

MAX PLANCK, *The Philosophy of Physics*

Louis Pasteur was the father of modern hygiene, public health, and much of modern medicine. He was born in 1822 at Dole, halfway between Dijon and Besançon in eastern France, where his father owned and ran a small tannery. He attended school in nearby Arbois, obtained his first science degrees in Besançon, and in 1847 graduated with a doctorate in science from the Ecole Normale Supérieure in Paris.

Scientists at that time believed that the fermentation of grapes, or the souring of milk, or the putrefaction of meat, were processes unrelated to microorganisms. The causes of infectious diseases were unknown. Malaria was believed to arise from "miasmas" emanating from swampy ground, and outbreaks of plague were attributed to unfavorable constellations, to comets, to the wrath of God, or to the poisoning of wells by Jews (who often paid for it with their lives). The "animalcules" first observed in the seventeenth century by the Dutchman Anton van Leeuwenhoek were believed to arise spontaneously in decaying meat or vegetable matter; they had not as yet been connected with disease. In the eighteenth century,

*A review of *The Private Science of Louis Pasteur* by Gerald L. Geison (Princeton University Press).

Edward Jenner had introduced vaccination against smallpox with liquid drawn from the pustules of pox-infected cows, but the infectious agents involved were unknown, and vaccination against other diseases did not exist.

Pasteur revolutionized science by proving that fermentation and putrefaction are organic processes invariably linked to the growth of microorganisms; that these never arise spontaneously from inanimate matter but only by reproduction of their own kind; that they are ubiquitous in the environment, but can be killed by subjecting them to heat, the process now known as pasteurization. He showed that infectious diseases of silkworms, animals, and human beings are caused by microorganisms and he devised ways of preventing them by vaccination. His discoveries inspired Joseph Lister in London to introduce antiseptics into surgery, which reduced mortality to a fraction of what it had been. Shortly before his death in 1895 two of Pasteur's pupils discovered that bubonic plague is caused by bacteria which are transmitted by fleas from dead rats to man, a discovery that helped to eliminate plague from much of the world.

Pasteur led a simple family life and devoted all his time to research. To generations of Frenchmen and to many others, Pasteur's has been the image of the selfless seeker after the truth who was intent on applying his science for the benefit of mankind. In *The Private Science of Louis Pasteur,* Gerald L. Geison, a historian of science, claims to have deconstructed Pasteur, and to have produced "a fuller, deeper and quite different version of the currently dominant image of the great scientist." I propose to deconstruct his deconstruction and restore the rightly dominant image.

Geison analyzes Pasteur's major discoveries: the asymmetry of biological compounds; fermentation; the vaccines against anthrax and rabies; and his demonstration that life is not generated spontaneously from nonliving matter. By a painstaking comparison of Pasteur's notebooks with his publications, Geison claims to have found him guilty of deception, of stealing other peoples' ideas, and of unsavory and unethical conduct. Some of these claims are scientifically flawed, while others defy common sense.

Geison's argument follows the line laid down by certain social theorists who assert that scientific results are relative and subjective, because scientists interpret empirical facts in the light of their political and religious beliefs, and under the influence of wider social and cultural pressures. They allege that instead of admitting their preconceptions, scientists misrepresent their findings as absolute truths in order to establish their power. What truth is there in these assertions?

Pasteur transformed medicine, but he started out as a chemist and devoted the first ten years of his career to the study of a seemingly recondite problem, the relationship between the crystalline forms of certain salts of tartaric acid, a compound found in wine gone sour, and the effects of solutions containing these compounds on polarized light transmitted through the solutions. Acute observation and brilliant reasoning led him to discover that tartaric acid can exist in two alternative forms which are chemically indistinguishable, but which have their component atoms arranged asymmetrically in space so that they are mirror images of each other, like left and right hands. Since such asymmetries had never been observed in compounds synthesized in the laboratory, Pasteur reasoned that the capacity to produce them must be an intrinsic property of the living cell. This was one of the great discoveries in chemistry and immediately established Pasteur's reputation. Since tartaric acid is a product of fermentation, the discovery led him to the study of fermentation as such and to research on disease.

Geison insinuates that Pasteur cheated because the effects on polarized light of his right- and left-handed salts of tartaric acid, which according to his interpretation should have been exactly equal and opposite, were in fact slightly different. Geison writes:

> Pasteur minimized the difference—in effect, he explained it away—by pointing to the difficulty of completely separating the two… forms [of crystals]. The deviation would "probably be the same for very well-chosen crystals" he now claimed.

I challenged Geison to go to the chemistry department at Princeton University, where he is a professor of history, and to repeat Pasteur's experiment. He would have a hard time getting as close an agreement between the two measurements as the very skillful Pasteur did. Later experiments by others proved Pasteur's explanation of the small discrepancy exactly right; but it seems that, because references to right and wrong would imply the existence of objective truth, they have been eliminated from the vocabulary of Geison's school of sociologists of science.

Geison also accuses Pasteur of concealing the guidance he had received from his teacher, Auguste Laurent, because such acknowledgment might have implied sympathy with Laurent's radical political views and could have been damaging to Pasteur's career. But Geison's own book shows that Laurent's ideas on the relationship between the crystalline forms of tartaric acid and the transmission of polarized light were misleading, confirming Pasteur's statement that under Laurent's guidance he "was enveloped by hypotheses without basis." Politics had nothing to do with the judgment, and Geison's accusation of opportunism is unjustified.

Pasteur's turn from molecular asymmetry to fermentation has often been attributed to his having connections with the brewing industry while he worked at Lille, but Geison writes that Pasteur's notebooks confirm his statement that "the 'inflexible' internal logic of his work" led him to it. Fermentation of grapes produces an alcohol that does not affect polarized light; but Pasteur discovered that products of fermentation included another alcohol that did. Since he associated that property with living organisms, he concluded that fermentation must be an organic process performed by microorganisms rather than a purely chemical one, as the great chemists Jöns Jakob Berzelius and Justus von Liebig maintained, and he demonstrated this in a series of brilliant experiments which convinced everyone except Liebig.

Pasteur next asked if these microorganisms arose spontaneously from inanimate matter, as was then widely believed. He answered this question in a brilliant lecture in the grand amphitheatre of the Sorbonne before a distinguished audience. After linking spontaneous generations to the kind of materialism in which there is no need for a divine creator, a doctrine abhorrent to himself, and, as he knew, also to the Church and the royal family, he stressed that "neither religion, nor philosophy, nor atheism, nor materialism, nor spiritualism has any place here.... It is a question of fact. I have approached it without any preconceived idea." Geison disputes this and alleges that his "approach to the question... was strongly conditioned by an intertwined set of philosophical, religious, and political interests." But he does not give any clear evidence for the claim.

The experiments Pasteur described in that lecture were stimulated in part by Félix-Archimède Pouchet, a biologist in Rouen who claimed that living eggs are generated spontaneously by a "plastic force" in dead plant and animal debris, and that microorganisms arise spontaneously in liquid extracts (or "infusions") made from boiling hay, even when they are exposed to chemically produced, hence sterile, oxygen. By contrast, Pasteur demonstrated that sugared yeast water, boiled briefly, would not ferment when exposed to sterile air. As a final demonstration, he took sterile sugared yeast water to the Mer de Glace above Chamonix and opened the bottles there. As he expected, the air was germ-free and no fermentation occurred. To disprove him, Pouchet took his boiled-hay infusions up a glacier in the Pyrenees and found that they fermented. Pasteur dismissed this finding as the result of sloppy work in preparing the boiled hay—unjustly, as it later turned out.

Pasteur's predecessor, the great naturalist Georges Cuvier, had already disputed the idea of spontaneous generation as unproven and had associated the idea with the philosophers responsible for the French Revolution on the ground that it denied the divine creation of life. Later, spontaneous generation became associated with materialism and also with Darwinism. Against this background, Geison castigates Pasteur's conduct in his controversy with Pouchet over fermentation, because some of Pasteur's own attempts at preventing fermentation of sugared yeast water had not succeeded, and many of his other experiments had also failed. According to Geison, he disregarded these failures because his religious and political views prejudiced him against taking them seriously. Geison implies that Pasteur acted dishonestly by not repeating Pouchet's experiment with hay infusions, since, according to the orthodox scientific method, a single disproof of a hypothesis invalidates all previous supporting evidence. For his part, Pasteur once remarked wisely: "In the observational sciences, unlike mathematics, the absolutely rigorous demonstration of a negation is impossible." Geison dismisses this as unscientific.

In fact, scientists rarely follow any of the scientific methods that philosophers have prescribed for them. They use their common sense. Having convinced himself by the most rigorous possible experiments that fermentation will not take place under sterile conditions, Pasteur could be confident that any contrary evidence was the result of error and he wasted no time searching for it. He felt sure that the source of the error was bound to emerge eventually. It did indeed fourteen years later when Pouchet's boiled-hay infusions were found to have contained heat-resistant bacterial spores which boiling would have failed to kill. Geison attributes Pasteur's immediate victory over Pouchet to his gift for persuasive advocacy rather than to the intelligent judgment of his audience, which Geison dismisses as a scientific elite in league with Pasteur's quest for power. Now that we have seen that the complexity of life on the atomic scale is vastly greater than that of non-living matter, the idea of spontaneous generation seems even more absurd than it did in Pasteur's time. But to Geison; the line of thought with which he is associated may define such generation as just an alternative paradigm.

Pasteur's discovery that all fermentation was caused by the actions of microorganisms convinced him that the same must be true of contagious diseases. He observed that animals which had recovered from a disease became immune to reinfection by the same disease. From there it was only a short step to the idea that virulent microorganisms somehow reduced in potency might serve as vaccines to make animals immune to infection by potent forms of the same organisms.

Pasteur and his young collaborator Emile Roux first put this idea into practice in a vaccine against a cholera that affects chickens and other domestic birds. Pasteur published this work in 1880, and his cholera vaccine for poultry became available soon afterward. Pasteur's and Roux's next target was anthrax, which was then decimating French sheep and cattle. Their vaccine contained live anthrax bacilli, whose effect was "attenuated" so as to render them non-infectious. This was done by either exposing cultures of virulent anthrax bacilli to air at 42°–43°C (i.e., "oxidizing" them) or subjecting them to "passage," a procedure by which one animal not susceptible to the disease—say, a mouse—is injected with a small quantity of bacteria; after they have multiplied, they are injected into a second mouse, and so on.[1] Two of Pasteur's collaborators, Charles Chamberland and Emile Roux, used instead potassium bichromate, an oxidizing agent, to produce similar attenuating effects, but faster and perhaps more drastic. All these treatments probably induced genetic mutations that weakened the bacteria without killing them. Pasteur's competitor Jean-Joseph Henri Toussaint tested anthrax-infected sheep blood that was heated or treated with carbolic acid, the disinfectant which Joseph Lister in London had introduced to kill bacteria. Toussaint's vaccines produced variable results, as indeed Pasteur's did initially.

In 1881, Pasteur's first publications about his air-oxidized anthrax vaccines drew a challenge for a public trial from veterinarians who were angry that a chemist was poaching on their preserve. Twenty-four sheep, one goat, and four cows were given two successive protective vaccinations before the trial; another twenty-four sheep, one goat, and four cows were left unvaccinated. On May 31 all animals received injections of virulent anthrax bacilli. By the day of the public trial, on June 2, all the unvaccinated sheep and the goat were dead and the cows were very sick, while the vaccinated animals were alive and healthy, except for one ewe which died the following day. A post-mortem showed that it carried a fetus that had died about two weeks earlier.

Geison tells us that according to Pasteur's notebooks this triumph was achieved not with Pasteur's own air-oxidized vaccine, but with Roux's and Chamberland's bichromate-oxidized one, which Chamberland had attenuated further by three passages through mice. However, after the trial, Pasteur continued to develop the air-oxidized vaccine; it was soon used successfully by farmers throughout the world. By 1894, 3,400,000 sheep had been vaccinated and mortality from anthrax had fallen to 1 percent, compared to 9 percent for unvaccinated sheep.[2]

Pasteur did not make a public statement that the trial vaccine had been oxidized by bichromate; but neither did he claim that it had been oxidized by air. All the same, Geison accuses him of having "actively misrepresented the nature of the vaccine actually used" and of "a significant and undeniable element of deception." (I wonder what difference there is between "misrepresenting" and "actively misrepresenting.") Geison also accuses Pasteur of taking credit that belonged to his competitor Toussaint, because Toussaint was the first to use carbolic acid, an antiseptic, to treat the sheep blood used for vaccines, and Geison regards the potassium bichromate used by Pasteur as just an alternative antiseptic.

Both accusations are based on misconceptions. Carbolic acid (phenol, as it is now called) kills bacteria, while Chamberland's bichromatic treatment kept them alive, as shown by their subsequent passage through mice. One of the empirical findings of immunology, probably discovered by Pasteur himself, is that live attenuated vaccines are more effective than dead ones. (This is why Sabin's polio vaccine has proved more effective than Salk's.)

Pasteur also pointed out that Toussaint's vaccine would be extremely difficult to adapt for practical use because, unlike his own, it could not be propagated in cultures. Therefore Pasteur owed no intellectual debt to Toussaint. There is, moreover, no qualitative difference between attenuating mutations induced by exposure to bichromate or air, because both are oxidants. Under the pressure of the public trial, Chamberland and Roux apparently decided that the bichromate vaccine was safer, but Pasteur later preferred his own, air-oxidized vaccine. The charge of "active misrepresentation" is ridiculous, especially since Pasteur's air-oxidized vaccine was used successfully until long afterward.

Pasteur's next vaccine, against the rabies virus, was also air-oxidized. In July 1885, a couple from Alsace brought to Pasteur's laboratory their nine-year-old son Joseph Meister, whose hands and legs had been severely bitten fourteen times by a rabid dog. Rabies has a long incubation period, so that a vaccine given soon after infection still stood a good chance of success. Pierre-Victor Galtier's earlier transmission of the infectious agent from rabid dogs to rabbits led Pasteur to the idea of attenuating it by repeated passages through rabbits. Pasteur's young collaborator Emile Roux then thought of attenuating the virulence by exposing strips of fresh spinal marrow taken from a rabbit that had died of rabies to dry, sterile air for various lengths of time. A small piece of marrow ground up and suspended in sterilized broth was then used as a vaccine.

Pasteur first gave Joseph Meister an injection of strips of spinal cord that had been dried longest, and he followed this by less and less attenuated strips that had been dried for successively shorter periods. They saved Joseph Meister's life and also that of a fifteen-year-old shepherd boy, Jean-Baptiste Jupille, who had been severely bitten while trying to save other boys from being attacked.

Pasteur realized later that Roux's method actually killed a progressively greater proportion of the rabies virus instead of merely attenuating it; all the same, the vaccine proved effective. In 1985, Dr. Hilary Koprowski, formerly president of the Wistar Institute in Philadelphia, wrote:

> Young Meister's treatment took place on 6 July 1885... By 12 April 1886, 726 people had been treated, 688 after having been bitten by dogs and 38 by wolves. There were four deaths. By 31 October, 2490 had been vaccinated, and since then the Pasteurians have had the last word... Despite many modifications, confidence in the original product was so strong that it was used until 1953 when the last person was vaccinated at the Pasteur Institute in Paris with Pasteur's original preparation.[3]

Nevertheless, Koprowski considers it hard to say even today whether Pasteur's vaccine was completely safe and whether some of the rabies infections after Pasteur's original vaccination were from the animal bite or from the vaccine. A vaccine made from rabies virus grown in cultures of human fibroblast (skin cells) has recently proved both efficacious and safe. It is known as the "human diploid vaccine" and is manufactured at the Mérieux Institute in Lyon for use throughout the world.

Dr. Michel Peter and other physicians accused Pasteur, a mere chemist, of having used an insufficiently tested vaccine. Geison eagerly takes up their case and accuses Pasteur of unethical conduct because "boldly, even recklessly, Pasteur was willing to apply vaccines in the face of ambiguous experimental evidence about their safety or efficacy." Pasteur had indeed obtained variable results when he tried to immunize dogs, beginning with the most briefly dried, and therefore least attenuated and most virulent, strips of spinal cord, and then followed that treatment with injections from spinal cords dried for progressively longer periods. But in a subsequent trial on dogs begun forty days before his vaccinations of Meister, Pasteur had reversed that order, and none of the dogs he treated contracted rabies, despite the virulence of the final injection. Twenty-seven days elapsed between that final injection and the first injection he gave Meister—a period that would have been long enough for the dogs to develop rabies symptoms if Pasteur's second procedure had been faulty. In the face of this evidence, Pasteur would have been timid and heartless to refuse the desperate appeal by Joseph Meister's mother. Geison's accusation that Pasteur's successful attempt to save Meister's life was unethical is without foundation.

In Pasteur's day there was no way of being sure either that a suspected dog really had rabies or that it had infected its victim. Pasteur's enemies benefited from these uncertainties, but in 1888 they seem to have been silenced by the report of an English commission which repeated and confirmed the successful vaccinations of dogs and concluded:

> From the evidence of all these facts, we think it certain that the inoculations practiced by M. Pasteur on persons bitten by rabid animals have prevented the occurrence of hydrophobia in a large proportion of those who, if they had not been so inoculated, would have died of that disease. And we believe that the value of his discovery will be found much greater than can be estimated by its present utility, for it shows that it may become possible to avert, by inoculation, even after infection, other diseases besides hydrophobia.[4]

Pasteur had no medical degree and therefore could not carry out the injections of Meister and Jupille himself. They were done not by Pasteur's own medical collaborator, Emile Roux, but by two other doctors. From this fact Geison speculates that Roux refused to give the injections because he considered them unsafe, and that he fell out with Pasteur over this issue. But Geison cites no documentary evidence for this claim.

According to Geison, Pasteur's colleagues supported him against the accusation of Peter and other physicians because Pasteur's treatment of

Meister "was a symbolic rallying point in a wider struggle for cultural authority and power... Critics of Pasteur's treatment for rabies... were... pushed aside in pursuit of a larger project: to secure the cultural domination of modern 'professional science.'" Might they not have been pushed aside because Pasteur's vaccine worked?

It is remarkable that Pasteur achieved his phenomenal practical successes while his theoretical concepts were still far from accurate. At first he thought his live, attenuated vaccines caused immunity by consuming the nutrients in the host, leaving none for virulent bacteria. Later he changed his mind and believed that the attenuated bacteria released a toxin that stopped further bacterial growth, but in 1890 Emil Behring and Shibasaburo Kitasato in Berlin discovered that the toxin was released not by the bacteria but by the defenses of the host animal in the form of antibodies.

Yet Geison has little to say about the correct explanations for the efficacy of vaccines, or for other phenomena, perhaps because his ideological approach denies their very existence. According to him, "Pasteur shared with many of his peers a rather simpleminded and absolutist notion of scientific truth, rarely conceding the possibility of its being multifaceted and relative." According to Geison, Pasteur's "scientific beliefs and modus operandi were sometimes profoundly shaped by his personal concerns, including his political, philosophical and religious instincts" while "...the real individual scientist... tries to navigate a safe passage between the constraints of empirical evidence on the one hand and personal or social interests on the other."

Had Michael Faraday's discovery of electromagnetic induction been "multifaceted and relative," there would be no electric power; had Albert Einstein's concept of the relation between mass and energy and James Chadwick's discovery of the neutron been relative, there would be no nuclear power and no atomic bombs; had Erwin Schrödinger's wave equation been relative, there would be no computers. Nor is there a shred of evidence that any of these scientists, or Ernest Rutherford, or Alexander Fleming, or James Watson, or Francis Crick, had to "navigate a safe passage between the constraints of empirical evidence on one hand and personal and social concerns on the other." Such concerns may have made Galileo and Darwin hesitate to publish their revolutionary ideas, and they may affect some of today's scientists who try to disentangle the respective influences of nature and nurture on human behavior, but these are exceptions.

I cannot think of any Nobel Prize–winning discovery in physics, chemistry, or medicine that was based on anything other than empirical evidence or mathematical insight.

According to Geison, it is now a commonplace among historians and sociologists of science that science, no less than any other form of culture, depends on rhetorical skills. I have known scientists who possessed great rhetorical skills which failed to conceal the shallowness of their research from their peers. On the other hand, Alexander Fleming's lectures put everyone to sleep, while his discovery of penicillin made him one of this century's most famous scientists. Good research needs no rhetoric, only clarity. The entire approach emphasizing "relative" truth seems to me a piece of humbug masquerading as an academic discipline; it pretends that its practitioners can set themselves up as judges over scientists whose science they fail to understand.

Toppling great men from their pedestals, sometimes on the slenderest of evidence, has become a fashionable and lucrative industry, and a safe one, since they cannot sue because they are dead. Geison is in good company, but he, rather than Pasteur, seems to me guilty of unethical and unsavory conduct when he burrows through Pasteur's notebooks for scraps of supposed wrongdoing and then inflates these out of all proportion, in order to drag Pasteur down. In fact, his evidence is contrived, and does not survive scientific examination.

Pasteur may have been domineering, intolerant, pugnacious, and, in his later years, a hypochondriac who searched every slice of bread for bacteria before eating it; but he was courageous, compassionate, and honest, and his scientific achievements, which have much reduced human suffering, make him one of the greatest benefactors of mankind. Joseph Meister became the proud janitor of the Pasteur Institute in Paris. In 1922, the French Ambassador to the United States, Jules Jusserand, said in a speech: "In the course of its history, France has produced many great men. There is no one of whom we are prouder than Pasteur.... Some years ago, before the war, a newspaper organized a kind of plebiscite and asked its readers who in their view were France's greatest sons. 2,300,000 replies came, and in this militaristic nation of ours... the emperor Napoleon came seventh and Pasteur came first."[5]

The Battle Over Vitamin C[*]

Albert Szent-Györgyi was a flamboyant Hungarian biochemist, famous for having isolated vitamin C and for other important discoveries. He was born in Budapest in 1893, lived in Europe through two world wars, and then spent the remainder of his long life at Woods Hole on Cape Cod, where he died in October 1986. His name is the Hungarian for Saint George, whom he tried to emulate when he attempted single-handedly to save his country from the Nazi and Soviet dragons.

Albert Szent-Györgyi was the son of a landowner who spent his time "thinking about the sheep, the hogs and manure" and of a sensitive, musical mother who was descended from a family of distinguished academics. Albert was a mediocre pupil at first, but at sixteen he began to read widely and decided to follow his uncle, the physiologist Mihaly Lenhossek, into medical research. His uncle's mistrust of his ability proved one of the spurs to his career. Szent-Györgyi told his biographer that from his earliest days he recognized in himself an intuitive, almost mystical ability to hear the voice of nature, something akin to a poet's inspiration. This ability was to guide him to success at first; it became a recipe for self-deception in later life.

In 1914 Szent-Györgyi was drafted into the Austro-Hungarian army and sent to fight the army of the czar. After three gruesome years he

*A review of *Free Radical: Albert Szent-Györgyi and the Battle over Vitamin C* by Ralph W. Moss (Paragon House).

"became increasingly disgusted with the turpitude of military service." "I could see that we had lost the war… . The best service I could do for my country was to stay alive."[1] He shot himself in the arm so that he could be discharged and complete his medical studies. Szent-Györgyi's horrifying experiences in the First World War made him fight for peace for the rest of his life. Soon after the end of the war, Szent-Györgyi left Hungary with his young wife and infant daughter to do research abroad, He had no grant, only six hundred pounds sterling from the sale of his father's estate. This proved insufficient to supplement his meager earnings at a succession of Czech, German, and Dutch universities. He and his family lived under hardships so great that he developed hunger edema, the swelling that comes from malnutrition. Yet he was determined to pursue his own ideas rather than work at his professors' bidding: "The real scientist is ready to bear privation, if need be starvation, rather than let anyone dictate to him which direction his research must take" (A. Szent-Györgyi in "Science Needs Freedom," 1943).

During the Twenties and Thirties the chemical mechanism of the oxidation of nutrients, the process from which animals get their energy, posed one of the great unsolved problems of biology. Most of our nutrients, like starch, proteins, or fats, are large molecules. They are first broken down to smaller ones in the digestive tract, and these small molecules then enter the bloodstream. They are made up of carbon, nitrogen, oxygen, hydrogen, and sometimes sulfur. To provide energy, these compounds are broken down further until the carbon is oxidized to carbon dioxide and the hydrogen to water. This breakdown proceeds in our tissues in a series of chemical reactions, each catalyzed (speeded up) by a different enzyme. In 1920, the steps involved in this process were largely unknown. The first of Szent-Györgyi's papers made a promising start on unraveling them. They were published in 1925 and caught the attention of the founder of English biochemistry, Frederick Gowland Hopkins, who invited him to his laboratory in Cambridge and helped him obtain a Rockefeller Fellowship. Szent-Györgyi was overjoyed to have an adequate salary at last and to find himself among brilliant young people in one of the world's best biochemistry schools.

Szent-Györgyi told his biographer that in 1926 he moved into an "ancient cottage" at 35 Oldstone Road. I moved into the same house ten years later. The road is actually called Owlstone and, like all the other houses in that road, no. 35 is small, plain, suburban, and semidetached, built in 1913. Szent-Györgyi romanticized it as part of what his biographer describes as "his appealing, self-dramatizing myths... which he had created and perpetuated for sixty years." Szent-Györgyi also told his biographer that he never talked about science to Hopkins, who shunned people and with whom it was hard to communicate. This memory conflicts with Szent-Györgyi's own expression of his "deepest gratitude" for Hopkins' "extreme kindness and helpfulness" at the end of his paper on the isolation of what later proved to be vitamin C. In fact, Hopkins was the most approachable of great men; he regularly wandered around his laboratory for friendly chats with the young scientists about their work.

Before Szent-Györgyi came to Cambridge he had found that the adrenal cortex contains a chemical factor that bleaches a brown solution of iodine, reducing the iodine to iodide. He wondered what the function of that factor might be, but failed to isolate it. In Cambridge he crystallized it in chemically pure form and showed that it was an acid, related to sugars, that also occurs in oranges and in cabbage. The chemical names of sugars all end in "-ose." Not knowing what kind of sugar it was, he first called it "ignose," when the editor of the *Biochemical Journal* objected to that flippant name, he changed it to "godnose," whereupon the incensed editor gave it the prosaic name "hexuronic acid," because it contained six atoms of carbon.

After describing the process of isolating it and analyzing its properties, Szent-Györgyi wrote: "The reducing substances of plant juice [i.e., the hexuronic acid] have repeatedly attracted attention, specially from students of vitamin C,"[2] but he did not test whether hexuronic acid actually was vitamin C, even though he could easily have done so at the Medical Research Council's nutrition laboratory, which had opened in July 1927 and was only two miles from the Biochemistry Department. Had he carried out such a test, his claim to the discovery of vitamin C would never have been disputed. In retrospect, Szent-Györgyi attributed this failure to his disdain for applied research, but that does not tally with his triumphal lecture tours once the identity of hexuronic acid and vitamin C had been established.[3]

This happened in 1932 after Szent-Györgyi had been appointed professor of biochemistry at the Hungarian University of Szeged. In 1931 Joseph Svirbely, a young American Ph.D. of Hungarian descent, arrived there. Having done his thesis on the isolation of vitamin C from lemons with Charles King at the University of Pittsburgh, he asked Szent-Györgyi to let him find out whether hexuronic acid was capable of curing guinea pigs of scurvy. It did, and the dose of hexuronic acid needed was the same as that of vitamin C extracted from lemons. With Szent-Györgyi's agreement Svirbely wrote this news to King in March 1932 (the exact date is not clear). On April 1 a letter by C.G. King and W.A. Waugh appeared in the American magazine *Science* to announce that vitamin C from lemons had chemical properties similar to those that Szent-Györgyi had described for hexuronic acid; hence the two compounds must be identical.[4] Svirbely's and Szent-Györgyi's announcement of the same discovery appeared sixteen days later in the British magazine *Nature*.[5]

Szent-Györgyi was terribly upset at having been scooped, needlessly because the scientific world saw that he had done the pioneering chemistry of purifying and characterizing the new compound, while King and Waugh had merely repeated a few of his tests to demonstrate the identity of their crystals with his. Besides, Szent-Györgyi had communicated Svirbely's and his results to the Hungarian Academy of Sciences twelve days before King and Waugh's letter appeared. However, Szent-Györgyi's anger was heightened when *The New York Times* and other American papers hailed King's discovery without mentioning his name.[6]

In 1937, when Szent-Györgyi received the Nobel Prize, the American press accused him of having stolen the discovery from King and abused the Swedes for not having awarded the prize to King and Szent-Györgyi jointly. I wondered why King was excluded, and asked the Nobel Committee for Physiology and Medicine if I could look at their files, which are open to inspection fifty or more years after the event. The Nobel committees do not themselves nominate candidates for the prize, but each year they solicit nominations from universities, academies, and individuals worldwide and appoint referees to report on their merits. I found that Szent-Györgyi had been nominated by scientists from Hungary, Czechoslovakia, Germany, Switzerland, Belgium, and Estonia (not by Hopkins, to my surprise), but that no one had nominated King.

In 1934, the committee had asked the Swedish chemist Einar Hammarsten to act as a referee; he wrote a seven-thousand-word report concluding that the discovery of vitamin C and its identity with hexuronic acid deserved a Nobel Prize, that Szent-Györgyi's role had been outstanding, but that the sum of the contributions made by several others had been equal to or greater than his. Given that no more than three people can share the prize, he could not recommend an award. Hammarsten cites King and Waugh's papers, but not as prominently as others do.

Meanwhile, Szent-Györgyi continued to work on the problem closest to his heart, the oxidation of nutrients. He discovered that fumaric acid and three other acids whose role in living tissues had been enigmatic represented successive steps in a chain of chemical reactions taking place during oxidation. After the publication of that work, Hammarsten and the biochemist Hugo Theorell wrote reports to the Nobel committee recommending that Szent-Györgyi be given the Prize for Physiology or Medicine primarily for that great advance, and he received it in 1937 "for his discoveries in connection with biological combustion processes, with special reference to vitamin C and the catalysis of fumaric acid." His Nobel lecture is a model of lucidity, liveliness, and scientific rigor.[7]

Szent-Györgyi's Nobel Prize made him a national hero in Hungary, but at that very moment his research on oxidation was overtaken by Hans Krebs, a young German refugee at the University of Sheffield. Krebs showed that Szent-Györgyi's four acids took part in a cycle of chemical reactions, since known as the Krebs cycle, in which carbon dioxide, water, hydrogen, heat, and chemical energy are abstracted from breakdown products of nutrients in successive small steps. At first, Szent-Györgyi felt disheartened by Krebs' success, but then he turned to another great problem, the contraction of muscle. The chief component of muscle was known to be a fibrous protein called myosin, but its role in contraction was unclear.

The Russian biochemists Vladimir Engelhardt and his wife M.N. Ljubimowa had just shown that myosin is also an enzyme because it splits the compound that carries energy within the living cell (adenosinetriphosphate, or ATP). During the early 1940s, Szent-Györgyi went one step further. He added ATP to fibers extracted from muscle and showed that it makes them contract. His young collaborator Bruno Straub then discovered that these fibers contained an additional protein, which he named actin because it activated contraction in the presence of ATP. In 1953, these important findings laid the foundation for the discovery of the mechanism of muscular contraction by H.E. Huxley, Jean Hanson, A.F. Huxley, and R. Niedergerke, who showed that neither myosin nor actin filaments change in length; instead they become interlocked like the fingers of one hand between the fingers of the other hand and slide relative to each other when muscle shortens. Each actin filament is surrounded by three myosin filaments for part of its length. When contraction is turned on, the actin filaments are drawn deeper into the spaces between the myosin filaments.

Szent-Györgyi recounts how unpopular he had made himself among his autocratic Hungarian colleagues by cultivating the informal relationships between professor and students that he had come to know at Cambridge. Contemporaries confirm this, yet I remember an episode suggesting that subtle differences remained between Cambridge and Szeged. An old friend of mine ran into Szent-Györgyi in a mountain hut high up in the Alps. After dinner, he found Szent-Györgyi dictating a scientific paper to a student with a typewriter carried all the way up the mountain by a porter for that purpose. In Cambridge, taking one's student to the mountains as a secretary would have been inconceivable; and the urge to prove one's ceaseless creativity would have been tempered by the sobering thought that after a hard day's climb fatigue might cloud the clarity of one's mind. The student was Straub, who later became the president of Hungary.

Szent-Györgyi and Straub did most of their work on muscle in Szeged during the Second World War. The slaughter of Hungarian troops in Russia and the persecution of Jews at home induced Szent-Györgyi to join a party that opposed Admiral Horthy's authoritarian regime and to sign a courageous public manifesto calling for democracy at home and Hungary's withdrawal from the war. This manifesto became known in England and earned him great credit. Early in 1943 he went to see the ostensibly profascist prime minister Kállay and proposed to travel to Istanbul, on its face to give scientific lectures, but in fact to ask Britain and the United States for a separate peace. Kállay gave his blessing, and Szent-Györgyi managed to put his bold plan to the head of the British Secret Service in Istanbul. He received some mild encouragement and returned to Szeged with instructions to set up a clandestine radio station for contact with Britain. With characteristic flourish he was to write later: "I had the whole fate of the war in my hands. I was to be the connecting link between the Prime Minister and the English Government, waiting for a chance to bring Hungary over to the right side."

Unfortunately all Szent-Györgyi's contacts and plans were betrayed. In a stormy interview with Admiral Horthy, Hitler himself demanded Szent-Györgyi's head, and Szent-Györgyi spent the rest of the war in a dangerous game of hide-and-seek with the Gestapo. In view of the strategic situation that prevailed in 1943, I wonder why he addressed his overtures to Britain rather than to the Soviet Union. He did take that initiative later, in 1945, when the Russians were already occupying part of Hungary. He planned to seize a Hungarian plane, fly it across the battle lines, and negotiate a surrender with Russia, but this plan was also betrayed. On Molotov's orders, he and his family were finally rescued by the Russians, probably at Engelhardt's initiative, and accommodated in luxury as General Malinovsky's guests.

Szent-Györgyi emerged from the war as a leading political figure, an idealist bent on encouraging the cooperation of East and West through the influence of a Hungary where science and the arts would flourish; but he was soon disillusioned by the Russians' heavy hand. He sought an interview with Stalin, but found himself rudely shouted at by one of Stalin's underlings instead. When an industrialist friend who had financed his research was arrested and viciously tortured, Szent-Györgyi together with his wife and daughter fled to the United States, and in September 1947 they settled at Woods Hole.

Two years after his arrival in America, Szent-Györgyi made his last important scientific observation. For experiments on muscular contraction, scientists used to free muscle from most substances other than myosin and actin by washing it with water, but such muscle quickly lost its ability to contract. Szent-Györgyi extracted muscle fibers with a mixture of glycerol and water instead and then stored them at –20° Celsius. Such fibers maintain their ability to contract and have since become widely used in muscle research.

Szent-Györgyi had a strangely divided personality. The Saint was an original thinker, an inspiring evangelist of science, a rigorous investigator, and a fearless, radical advocate of democracy and peace. He fought fascism, anti-Semitism, McCarthyism, nuclear tests, and the war in Vietnam. He was an internationalist who once said: "An Indian or Chinese scientist is closer to me than my own milkman." George, on the other hand, could not distinguish fact from fantasy, sometimes came near to megalomania ("I am always several steps ahead of everyone else"), made false claims to solicit money for his research, and surrounded himself with people who would not contradict him.

Before the Second World War the Saint prevailed; afterward George seems to have become increasingly dominant. In 1950, he wrote to the Rockefeller Foundation: "I am approaching the solution of rheumatic fever, hypertension and myasthenia [gravis]," and asked peremptorily for support; later he often claimed to have discovered the cause of cancer and to be on the verge of discovering how to cure it.

I attended one of Szent-Györgyi's lectures in Cambridge after the war, eager to hear the great man, but disappointed when he proclaimed that proteins conduct electricity because I knew that they are insulators. Later he asserted that "proteins are built to a great extent of *free radicals*,"[8] molecules made reactive by the loss or gain of single electrons or hydrogen atoms. This seemed equally wrong. Another of his claims was that tissues contain "charge transfer complexes," small molecules packed together so tightly that electrons can jump easily from one to the other. Only one such complex has been found in living cells, in a compound that Szent-Györgyi specifically excluded from his theory. He distinguished between two states of living matter, which he called α and β. The α-state prevailed before the appearance of oxygen in the Earth's atmosphere, when cell division is supposed to have proceeded uncontrolled. After the appearance of oxygen the β-state evolved, in which cell division is supposed to be controlled by the chain of enzymes that transfer electrons from nutrients to oxygen. According to Szent-Györgyi, cancer is a reversion from the β- to the α-state. This is all science fiction.

An outsider might think that scientists take no notice of far-fetched theories that run counter to firmly established knowledge, however eminent their source, but this is not what happened. Szent-Györgyi's theory of electrical conduction in proteins was claimed to have been confirmed experimentally by English chemists, and his ideas about free radicals were apparently confirmed by a Russian group who reported their presence not in proteins but in DNA. An experiment is an experiment and calls for an explanation, but it can be hard to discover the explanation for someone else's spurious results, and without it they cannot be convincingly disproved. A scientist at Bell Laboratories in New Jersey repeated the Russians' experiment but, however he prepared his DNA, he found no sign of free radicals. One day he played a fast game of squash; afterward he wrung out his shirt, and added a drop of the liquid to the DNA. Immediately there was a strong signal of free radicals, which made him realize that the Russians' result had come from their sweaty fingers. The electrical conductivity of proteins detected by the English chemists was eventually tracked down to contamination with traces of salt.

Living tissues do produce free radicals, not usually as part of protein molecules as Szent-Györgyi believed, but as toxic side-products of chemical reactions or under the influence of ionizing radiations. Most of these toxic radicals are either molecules of oxygen that have gained an electron or molecules of water that have lost an atom of hydrogen. White blood cells actually produce them as weapons against invaders, but when they react with DNA they can give rise to cancer. Animals have evolved mechanisms to scavenge for and inactivate toxic radicals, and one of the most important scavengers is vitamin C. The chemical reaction of vitamin C with free radicals is the same as that involved in the bleaching of iodine, which first brought its existence to Szent-Györgyi's attention; but by the time this became known, Szent-Györgyi was set on proving his own theories and never seems to have noticed the true connection between free radicals, cancer, and vitamin C that has given his original discovery of the vitamin added importance.

Peter Medawar wrote that "a senior scientist... should always hear behind him a voice such as that which reminded a Roman emperor of his mortality, a voice that should now remind a scientist how probably he may be, and how often he probably is, mistaken." What deafened Szent-Györgyi to that voice? Being scientific top dog in Hungary? A life of high adventure, hobnobbing with the powerful during the war? Presiding over his own isolated research institute at Woods Hole where no one contradicted him? Perhaps it was just old age.

*A Mystery of the Tropics**

Some years ago, I listened to an emotional appeal by a director of the World Health Organization to fight against the parasitic infections that kill millions of children in the Poor World every year. His speech fell flat, partly because the audience regarded it as a well-rehearsed and often repeated performance, and partly because mere numbers fail to arouse people's emotions. By contrast, the opening of Desowitz's book on tropical diseases strikes to the heart with the story of the illness and unnecessary death of a single child in India. Unnecessary, because the mother could have saved her child had she or the Indian government been able to afford $15 worth of medicine.

The disease was a parasitic infection known as kala azar. Desowitz's book tells the story of the search for the organism that causes it and the way that organism is transmitted from patient to patient. Its heroes are the physicians of the British Indian Medical Service, many of them amateurs in science who, beginning in the late nineteenth century, pursued their research as a hobby with primitive means and often with more devotion than competence. They had many patients with the symptoms of the disease, among them enlarged spleen and liver, irregular fever, and anemia, but could not find out what caused it. The first clue came at the turn of the century when Dr. William Leishman in London found microscopic egg-shaped bodies in the spleen of a soldier who had died from kala azar; later

*A review of *The Malaria Capers: More Tales of Parasites and People, Research and Reality* by Robert S. Desowitz (Norton).

Charles Donovan in Madras found the same bodies in the spleen of a live patient and recognized them as protozoa, single-celled organisms a little larger and more complex than bacteria. They were aptly named *Leishmania donovani.*

How were they transmitted from patient to patient? First suspicions fell on the ubiquitous bedbugs. Dr. W.S. Patton in Madras spent five years patiently encouraging bedbugs to feed on his kala azar patients, hoping that the bedbugs would suck up the parasites, and was encouraged when he found some in their intestine. Later, Mrs. Helen Adie, a researcher in Calcutta, claimed enthusiastically that she had actually seen them in the bedbugs' salivary glands. Her findings were hailed as a breakthrough until others proved that she had looked at the wrong protozoa. People continued to probe in the dark until one man had a bright idea. This was Major John Stinton, the only man known to hold both the Victoria Cross for his bravery in battle and the fellowship of the Royal Society for his scientific achievements. Stinton took a map of India and compared the incidence of parasitic diseases with the distribution of various biting insects. His map showed a coincidence between the prevalence of kala azar and that of a tiny silvery sandfly, *Phlebotomus argentipes.*

Stinton published his findings in 1925, but it took another seventeen years until he was proved right. In 1923 a kala azar commission had been set up in Assam under the English parasitologist Henry Edward Shortt. He found that even the first problem, the breeding of the sandflies in the laboratory, needed years of research before it could be solved. When that had been accomplished, his sandflies died after their first blood meal, before they could transmit the disease to a healthy host. Shortt became so depressed by years of failure that he recommended that the commission be closed down, but the next day a hamster that had been bitten 1,434 times by infected sandflies over a period of seventeen months was found to be infected.

Even after this event, infected sandflies failed to transmit the disease to human volunteers, until in 1939 the physician-entomologist R.O. Smith observed that the sandflies needed refreshment to make the parasites grow. When he fed them on raisins, the *Leishmania* inside them multiplied so prodigiously that they plugged the flies' throats. Smith thought that they might infect their next host in their desperate efforts to cough out these accumulations of protozoa. In 1942 Shortt, together with an Indian scientist and physician, C.S. Swaminath, and L.A.P. Anderson, tested Smith's idea by feeding infected and "raisined" sandflies on six volunteers and found final proof when three of them contracted kala azar.[1] By then it could be cured by a drug containing antimony; the same drug is still in use today, but the Indian child's mother could not afford it, even though it is cheap.

In the early 1950s the great DDT campaign against malaria eradicated kala azar from India until it became a forgotten disease, but in the Seventies the ending of the DDT campaign caused it to revive. It is now more endemic than before because improved means of transport have favored its spread. Desowitz charges that it has become a neglected disease and that too little money is being spent on kala azar research because it does not afflict Americans.

By contrast, from early colonial times until about 1940, malaria was one of the most visible American diseases. During the American Civil War half the white and four-fifths of the black soldiers got malaria every year. Desowitz's history describes the "confusion and sometimes chaos" that for seventy years handicapped malaria research, before the life cycle of the malaria parasite and its transmission were understood.

More than a hundred years ago a French army doctor in Algeria, Alphonse Laveran, discovered the parasite that causes malaria. This is how his notebook described it:

> D. aged 24, soldier in the eighth squadron of artillery, had been in Algeria since December 5th 1879 and entered the Constantine Hospital on November 4th 1880. The patient had grown thin; was markedly anemic, the skin had the earthy tint characteristic of the wasting of malaria. Temperature 39.5°, 5th November. Temperature 38.5° in the morning. Examination of blood… numerous crescent-shaped bodies. I prescribed sulphate of quinine 0.60 grams. Examination of blood on 6th November; crescent-shaped bodies still numerous; spherical bodies with mobile flagella whose existence I noticed for the first time.

Later, Laveran added:

> These parasites evidently live at the expense of the red blood corpuscles which become pale as the parasites grow.[2]

Like the organisms that were found to cause kala azar, they were protozoa, single-celled organisms of the kind first noticed in stagnant water by Antony van Leeuwenhoek more than two hundred years earlier, but never before associated with disease. The individual protozoon became known as *Plasmodium falciparum*. It eats hemoglobin, the protein of the red blood cells, which is why they turn pale. In 1990, scientists at the Rockefeller University in New York finally discovered how *Plasmodium* digests it, which may open new approaches to the design of antimalarial drugs.[3]

Desowitz writes:

> [Laveran's] was a remarkable discernment; modern malariologists claim that Laveran's microscope was so optically defective it had only one-half the magnification power necessary to reveal what he so accurately saw and described.

The thirty-five-year-old colonial army doctor's discovery is all the more impressive, because he made it only a few years after Louis Pasteur's discovery of the bacterial origin of a silkworm disease, and Robert Koch's of a disease in sheep, and before anyone had found a microorganism that causes a human disease.

As often happens when an outsider makes a discovery, nobody believed Laveran. Italian doctors rejected his thesis even when he showed them the parasites in the red blood cells of malaria patients. According to Desowitz, Koch went so far as to dismiss as an idiot anyone who believed Laveran. All the same, in 1907, twenty-seven years after his discovery, Laveran was to receive the Nobel Prize for Physiology or Medicine, only two years after Koch got his prize for discovering the cause of tuberculosis.

It took eighteen years of detective work from Laveran's discovery of the malaria parasite to the discovery of the creature that transmitted it. Unnecessarily so because many inhabitants of Africa and Asia apparently realized that mosquitoes were the culprits, but educated Westerners dismissed such primitive tales. One of them was Ronald Ross, a young physician in the Indian Army Medical Service who even refused to believe Laveran. As late as 1893, he published a paper in the *Indian Medical Gazette* claiming malaria to be an intestinal infection by bacteria that should be treated with calomel, i.e., mercurous chloride, a toxic laxative and cure-all that must have aggravated the already severe debilitating effects of malaria. It was one of the many futile and counterproductive drugs with which doctors used to torture patients.

Ross might have continued on his mistaken track but for a visit to London in 1894, where he met Patrick Manson, a physician described as the father of tropical medicine. Manson had discovered that filariasis, a parasitic worm infection, is transmitted by mosquitoes.[4] Ross wrote in his memoirs:

> In November I called upon [Manson] again and found him just starting for the Seamen's Hospital. I went with him, and remember distinctly that as we were walking along Oxford Street at about 2:30 p.m. he said to me: "Do you know, I have formed the theory that mosquitoes carry malaria just as they carry Filariae." I replied at once that I had seen the same conjecture in one of Laveran's books.[5]

Manson believed malaria to be transmitted by drinking water from wells contaminated with infected mosquitoes, on the model of typhoid fever that was known to be spread by infected water. Ross bought a microscope to take back to India and began his search for the carriers of malaria.

In June 1885 Manson wrote to him, "Mosquito water or mosquito dust should be taken or inhaled first thing in the morning and on an empty stomach, so that there would be no danger of its included germs being destroyed by the stomach juices." In August Ross replied: "All the facts can be explained *best* on the supposition that the poison is conveyed by mosquitoes into isolated pots of drinking water in the houses of men," but when he drank about 1,500 to 2,000 malaria "sperms" on two occasions, he felt nothing. In April 1897, Ross suffered a severe attack of malaria and still attributed it to contaminated air or water. One of Ross' great problems was to find which mosquitoes carried the disease. Desowitz suggests that he could have saved himself years of labor if he had only bothered to study the characteristics of the known varieties of Indian mosquitoes. This is unjustified. Ross did ask an expert, A. Alcock, who replied that Indian mosquitoes had not been subjected to scientific study. In 1895, the first entomology of European mosquitoes had only just been published in Italy.

On August 16th, 1897, a servant brought Ross ten mosquitoes of a kind he had never seen before. They were of the type called *Anopheles*. Ross let them feed on a malaria patient in a ward of the local hospital. Desowitz writes:

> Twenty-five minutes after all the mosquitoes had their blood meal... two were killed and dissected. Nothing; and then there were eight. Twenty-four hours later, two were found dead in the cage and two more were dissected. Nothing; and then there were four. On the fourth day, August 20, 1897, one had died a natural death and Ross dissected one of the three remaining mosquitoes. There it was. A minute, round cyst on the exterior of the stomach wall. "The Angel of Fate fortunately laid his hand upon

> my head" was Ross's dramatic description of that moment. The last mosquito was sacrificed the next day. Not only were there cysts on the stomach wall but they were larger than those of the day before and had malarial pigment granules within them. They were alive and growing.

Ross' observation proved that the *Plasmodium falciparum* protozoon moves from the human host into the *Anopheles* mosquito and multiplies there, but it did not reveal how *Plasmodium* gets from the mosquito into the next human host. Ross had just started to work on this problem when the army ordered his transfer to a semidesert station that was almost free from malaria. He wrote "a humble letter" to the surgeon-major general asking to be transferred to a station where he could finish his malaria work, but received the blunt reply: "I don't know what this officer means. He was sent to his command by H.E. (His Excellency) the Commander-in-Chief and there he will remain until H.E. orders him away."

After five months in the wilderness Ross wrote to Manson: "I have just received orders to be off to Calcutta at once, thanks to your action," but at first Calcutta was a disappointment because Ross found no malaria patients on whom to feed his mosquitoes. Having discovered malaria parasites in two sparrows at his desert station, he set to work on bird malaria instead. On May 9, 1898, he complained to Manson: "I am nearly mad. Sparrows all round and I can't catch them." On July 6 he wrote, "I am nearly dead and blind with exhaustion—but triumphant." Ross had made two decisive observations. He had let the *Anopheles* mosquitoes feed on three healthy sparrows and later found the sparrows to be swarming with the parasites; he had also dissected mosquitoes that had fed on infected sparrows and found the parasites first in the mosquitoes' stomachs, later in their thorax, and finally in their salivary glands, ready for injection into another victim.

At last Ross had incontrovertible proof that malaria is transmitted by mosquitoes, not as Manson and he had thought, through drinking water, but simply by their bite. He wrote: "Such moments come only to one or two persons in a generation. The pleasure is greater than that given by any triumph of the orator, the statesman or the conqueror." Four years later, in 1902, Ross received the Nobel Prize for Physiology or Medicine, only the second one ever to be awarded; the first had gone to Emil Adolf von Behring for his discovery that diphtheria could be cured by injection of the serum of immunized horses.

Desowitz describes Ross as arrogant and opinionated, a poet manqué who had only just scraped through his medical exams, a determined and ambitious but ignorant romantic. Reading Ross's regular reports to Manson shows this view to be unfair. Ross emerges as exceedingly able, resourceful, and energetic, a scientist who goes for what is most important with superb skill and acute powers of observation. Who else would have discovered malaria in sparrows and gone out into the streets of Calcutta to catch them when human patients failed him, or stopped the fan that made only just bearable the hours peering down his microscope in stifling heat, because its draft blew away the delicate organs that he had teased out of his mosquitoes in search of parasites? It is hard to imagine the strength of character Ross needed to carry on with his work for years in isolation, with primitive means, in an appalling climate, and without encouragement from his superiors! After his great discovery, Ross devoted the remainder of his life to the eradication of malaria and yellow fever from British colonies. His methods of mosquito control proved decisive in the building of the Panama Canal. In his spare time, he wrote creditable poems about India.[6]

The Forgotten Plague[*]

What did Cardinal Richelieu, Heinrich Heine, Frédéric Chopin, Anton Chekhov, Franz Kafka, George Orwell, and Eleanor Roosevelt have in common? They all died of tuberculosis because the treatments available until about fifty years ago did little to prolong sufferers' lives. Ryan's story of the discovery of the antibiotics and other agents used against tuberculosis makes as exciting reading as Paul de Kruif's *Microbe Hunters*, which must have lured more idealistic young people into medical research than any other book ever written. In earlier times, a clean, mild climate was often prescribed, but Chopin wrote ruefully from his villa in Mallorca:

> I have been sick as a dog the last two weeks; I caught cold in spite of 18 degrees C, of heat, roses, oranges, palms, figs and three most famous doctors on the island. One sniffed at what I spat up, the second tapped where I spat it from, the third poked about and listened how I spat it. One said I had died, the second that I am dying, the third that I shall die... All this has affected the "Preludes" and God knows when you will get them.

Perhaps the common fear of consumption was the source of the German Romantic poets' preoccupation with early death, which Schubert set to music in his *Winterreise* with tears streaming down his cheeks. Later in

*A review of the books: *The Forgotten Plague: How the Battle Against Tuberculosis Was Won—and Lost* by Frank Ryan (Little, Brown); *Living in the Shadow of Death: Tuberculosis and the Social Experience of Illness in American History* by Sheila M. Rothman (Basic Books); and *Silent Travelers: Germs, Genes, and the 'Immigrant Menace'* by Alan M. Kraut (Basic Books).

the century, both Verdi's opera *La Traviata* and Puccini's *La Bohème* end with the heroine's death from consumption. In her recent book *Living in the Shadow of Death*, Sheila Rothman has reconstructed the tragic lives of consumptives in nineteenth-century New England from collections of their letters. The disease gradually drained all energy from Deborah Fiske, an enterprising and intelligent young woman, until she could barely manage even to give directions to the servants looking after her household. She died in 1844, aged thirty-eight. Since consumption was not regarded as contagious, Deborah Fiske never seems to have worried that her husband might catch it, which of course he did.

Some patients sought better health in California, Arizona, or Florida, where they were often exploited by ruthless employers or cheated by the people who ran bogus sanatoriums. Jeffries Wyman, a New Englander whose forebears had settled in Massachusetts before the arrival of the Mayflower, contracted tuberculosis in 1833 as a medical student at Harvard, but he did not succumb to the disease until he had reached his sixties. He combined his duties of professor of anatomy at Harvard with winter expeditions of biological exploration in Florida's Everglades which restored his health. In 1871 he wrote to his brother: "I have not been sick a day since leaving Hibernia, have slept with and without the tent, have gained strength, have taken long rows in the boat once of nine and twice of ten miles, each without fatigue. My cough has not gone but is greatly diminished and my appetite is always good." However, in 1872 he reported that he had had several hemorrhages and that "the trouble is there where it has been and I see no reason why I should ever be free from it again."

Except for peaks during the Crimean War and the First and the Second World Wars, mortality from tuberculosis in Britain declined throughout the second half of the nineteenth and the first half of this century; by 1947 it had a fallen to about one-eighth of what it is estimated to have been in 1800, even in the absence of any effective treatment. We do not know exactly when the decline started; it is generally attributed to improving standards of living, but these actually deteriorated during much of the nineteenth century as a result of the movement of large numbers of people from the country to overcrowded slums in the new industrial towns.

Natural selection may have played its part in the decline: the disease may have killed those genetically least resistant before they reached reproductive age, so that a gradually increasing fraction of the population inherited enough resistance to survive infection. It is sometimes alleged that vaccination and the newly introduced antibiotics and drugs had little effect on the decline of deaths from tuberculosis, which was merely continuing its historic downward trend. In fact the pace of that trend became over fifty times faster and the annual number of new infections declined twice as fast as they had done before they were introduced. In 1935, nearly 70,000 people died from tuberculosis in England and Wales; in 1947, just after antibiotics were introduced, 55,000 died; in 1990 that number was reduced to 330. Similar reductions have been achieved in the United States and other industrialized countries.

This spectacular success inspired ambitious plans to eradicate the disease in underdeveloped countries, where its incidence remains many times larger than in developed ones. With the assistance of the International Union against Tuberculosis and help from the U.S. and other richer countries, some very poor countries in Africa, including Mozambique in the middle of its civil war, developed highly effective national programs for the control of tuberculosis. On the other hand, in 1990 the annual mortality rate for tuberculosis in sub-Saharan Africa was still about 1,500 times greater than that in England. In many African countries the incidence of tuberculosis looked as if it was just beginning to drop when the AIDS epidemic struck. This has been ruinous for all health services, including those for the control of tuberculosis, because AIDS suppresses the natural immune system. In many Asian countries that lack effective programs for controlling their already very high tuberculosis rates, the explosion of the AIDS epidemic could lead to disaster; in Western industrialized countries resistance to antibiotics has recently caused a resurgence of tuberculosis.

Ryan's book begins on March 24, 1882, with Robert Koch's announcement to the German Physiological Society in Berlin of his discovery of the rod-shaped tubercle bacilli. Koch detected these tiny, elusive bacteria partly thanks to his invention of a new method of staining, but he might not have spotted them if Carl Zeiss, the founder of the great optical firm in Jena, had not presented him with the first of the newly developed oil immersion lenses, which enormously improved the power of Koch's microscope. Articles in the *New York Tribune* and the London *Times* hailed Koch's discovery and looked forward to an early vaccine against tuberculosis, but this did not emerge until many years later. The immediate benefit of Koch's discovery was proof that tuberculosis is contagious.[1] According to René and Jean Dubos, American doctors still doubted this even ten years afterward,[2] but before the end of the century many countries were reducing the spread of the disease by isolating tuberculosis patients in sanatoriums and forbidding spitting in public places.

Early in the century, Albert Calmette, a pupil of Louis Pasteur, and Camille Guérin at Lille discovered that virulent tubercle bacilli from cows lost their virulence after being cultured for eleven years in 231 successive solutions that contained ox bile. Animals infected with these attenuated bacilli became immune to tuberculosis. Calmette and Guérin made these bacilli into the vaccine now known as BCG (for Bacilli Calmette-Guérin). They first used it in 1921 on a child whose mother had died of tuberculosis in childbirth, but it came into disrepute in 1930 after 73 out of 249 children vaccinated with it in Lübeck had died. Their deaths were later proved to have been caused by contamination of the vaccine with virulent bacilli. Scandinavian countries therefore resumed BCG vaccination in 1938, but in England doubts lingered on because most of the evidence for the effectiveness of BCG remained anecdotal until 1950, when the Medical Research Council's Tuberculosis Research Unit began a countrywide trial of its safety and effectiveness.

Of over 16,000 children between fourteen and fifteen-and-a-half years old, half, chosen at random, were vaccinated with BCG and half were left unvaccinated. During the next twenty years 248 of the unvaccinated children and only 62 of the vaccinated ones developed the disease.[3] Two trials in the United States, one on twenty-year-old American Indians and the other on infants in Chicago, and another trial in Haiti also proved to be highly effective, but others in the United States, in Puerto Rico, and in Southern India showed little or no evidence that BCG offered protection against tuberculosis. Studies in several other countries showed that between 53 and 74 percent of the people vaccinated did not contract tuberculosis, and a much higher proportion were protected against tubercular meningitis. There is no satisfactory explanation for the contradictory results of the trials,[4] but some researchers now believe that previous infection with common and harmless bacteria related to *Mycobacterium tuberculosis* rendered many children in Puerto Rico and Southern India immune. In consequence, there was only a small difference in the incidence of tuberculosis between those children who had been vaccinated and those who had not.

Early in this century, German chemists were prominent in the fight against infectious diseases. Paul Ehrlich discovered Salvarsan, his "magic bullet" against syphilis in 1910. The next great discovery in Germany, which led to the isolation of sulfanilamide, was made by Gerhard Domagk at Elberfeld in 1935.

In his search for antibacterial agents, Domagk was driven by his memories of World War I. As a medical student he worked at a hospital on the Russian front and he could never forget the terrible suffering of the young soldiers infected with gas gangrene. After the war, he took a medical degree, and when he was only thirty-two he was appointed director of research in experimental pathology and bacteriology at the laboratories of the giant chemical firm I.G. Farben-industrie at Elberfeld, a post which he held for the rest of his working life. The firm asked Domagk to make a wide-ranging survey of chemical agents that could be used against bacterial infections. An Englishman who visited Domagk was shown "enormous laboratories in which they did nothing but take compound after compound and test its ability to deal with infections in animals caused by a variety of organisms." For several years this work yielded no clear results.

According to Leonard Colebrook's authoritative memoir of Domagk,[5] two chemists, Fritz Mietsch and Klarer, gave him in 1932 a red compound which they had synthesized several years earlier as a "fast" dye for leather. This was Prontosil, the forerunner of the sulfanilamides. Even though the dye failed to stop the growth of bacteria in cultures, it killed lethal streptococci in animals and brought about miraculous recoveries in some mortally ill human patients.

Domagk did not publish his findings until 1935, when Prontosil had been patented. Within a few weeks of his publication Jacques Tréfouël and his colleagues at the Pasteur Institute in Paris found that the active component in Prontosil was sulfanilamide, a colorless molecule split from the larger molecule of Prontosil, which made the patent worthless.

I had been under the impression that the Hungarian physician Ignaz Semmelweiss had done away with puerperal fever in 1861, when he wrote that midwives and doctors must wash their hands before attending childbirths, but I learned from Ryan's book that maternal mortality at childbirth remained around 46 deaths per 10,000 births from 1850 to 1930 and had dropped only slightly, to 36 deaths, by 1940. Colebrook recalls the dramatic impact of Domagk's discovery on mortality from puerperal fever and other infectious diseases. It opened up an immense field for therapeutic advance.

> The haemolytic streptococcus [which Prontosil killed] played a major role in the heavy and tragic mortality of puerperal and acute rheumatic fevers, of countless septicaemic conditions resulting from war wounds [and civilian injuries—including burns], as well as of erysipelas (a skin infection) and the numerous acute inflammatory conditions of the respiratory tract and the ear—the latter being chiefly responsible for the immeasurable miseries of deafness.
>
> In England it was quickly found that cerebro-spinal meningitis (which had usually proved fatal) could readily be arrested and cured. Pneumonia, too,—"the captain of the men of death"—was similarly brought under control. And the dreaded venereal disease, gonorrhoea, usually cleared up in a few days.[6]

The original sulfanilamide was not very effective in curing pneumococcal pneumonia, but a modification, sulphapyridine, introduced in the early Forties, proved highly effective. It cured Winston Churchill's pneumonia in North Africa in 1943.

On October 26, 1939, some weeks after Germany's attack on Poland, Domagk received a telegram that he had been awarded the Nobel Prize for Physiology or Medicine. Hitler had forbidden Germans to accept Nobel prizes after the Norwegian Parliament had awarded the peace prize to the imprisoned German pacifist Carl von Ossietzky. I don't suppose he knew that the prize for medicine is awarded by the Medical Karolinska Institute in Stockholm. Perhaps because Domagk had not indignantly refused the prize as an insult to the Führer, the Gestapo arrived to arrest him, and armed soldiers took him to the town prison for interrogation. He was kept there for a week, fearing for his life. After his release the Gestapo arrested him once more and ordered him to sign a letter declining the prize. In 1947, Domagk was finally able to travel to Stockholm to receive the Nobel medal, but without the accompanying check; the money had been returned to the general funds.

The sulfanilamides failed to kill tubercle bacilli. They also had unpleasant and dangerous side-effects, including vomiting, skin rashes, and damage to the liver, kidneys, and bone marrow. But their discovery stimulated many chemists and microbiologists to search for other antibacterial drugs. One of them was Selman Waksman, who had arrived in the United States in 1910 from a small Jewish town in the Ukraine. By 1939 he was head of a leading laboratory for soil microbiology at Rutgers Agricultural College in New Jersey. His research covered subjects ranging from nitrogen-fixing bacteria and the taxonomy of fungi to the use of human feces for compost. He was aware of reports that some virulent bacteria, including tubercle bacilli, do not survive for long in soil.

According to his memoirs, the outbreak of the Second World War in September 1939 made him switch the work of his laboratory to a search for antibacterial agents produced by micro-organisms in soil. This work bore fruit five years later when Waksman's graduate student Albert Schatz extracted a compound from the fungus *Actinomyces griseus* which stopped the growth of many virulent bacteria, including tubercle bacilli cultured in test tubes. This compound was the new antibiotic, streptomycin.

I could not help being struck by the contrast between Alexander Fleming, the reticent and inhibited Scotsman who discovered penicillin, and the dynamic go-getter Selman Waksman. Fleming found in 1929 that the broth in which he had cultured his mold *Penicillium notatum* inhibited the growth of the streptococci and staphylococci that infected wounds and of the organisms responsible for gonorrhea, meningitis, and diphtheria. The broth was harmless to white blood cells; it could be injected with impunity into mice and rabbits, and the mold itself could be eaten without ill effects. Having carried out these experiments, Fleming unaccountably failed to take the obvious next step—the step that Ernst Chain and Howard Florey were to take eleven years later when they decided to find out whether an injection of Fleming's broth would protect mice from lethal infection. The spectacular success of their experiment encouraged them to turn their Oxford laboratory into a factory to make enough penicillin for human trials. Fleming could have done all this eleven years earlier.

By contrast, Waksman maintained regular contact with two physicians from the Mayo Clinic, Dr. William H. Feldman and Dr. H. Corwin Hinshaw, throughout his search for antibacterial agents. Within a few weeks of Schatz's discovery, Waksman sent them ten grams of streptomycin for animal tests. In Feldman's words:

> The results suggested that, despite its many impurities, this new substance was well tolerated at a therapeutic level sufficient to exert a marked suppressive effect on otherwise irreversible tuberculous infection in guinea pigs. However, despite the marked suppressive effect on the pathogenesis of the infection, it was recorded that streptomycin had not destroyed all of the tubercle bacilli in the tissues of the treated animals and that, therefore, under the conditions imposed, the action of the drug was essentially bacteriostatic rather than bacteriocidal [that is, it prevented bacteria from multiplying, but didn't destroy them].[7]

On the strength of Feldman and Hinshaw's experiments with only four guinea pigs, Waksman persuaded the nearby pharmaceutical firm Merck to call a board meeting. At first the board members hesitated to support efforts to isolate and manufacture streptomycin, but then the company's founder, George Merck, joined the meeting, and took the courageous and farsighted decision to put fifty people to work on streptomycin immediately.

Although streptomycin stopped the growth of tubercle bacilli without killing them, it still seemed to bring about spectacular cures of terminally ill patients. In 1945 news of these results prompted the British Medical Research Council to use the limited supplies of streptomycin then available in England for a controlled double-blind clinical trial. Philip d'Arcy Hart, who set up the trial, chose 107 patients with acute pulmonary tuberculosis; in each case doctors had decided that the only possible treatment was bed rest. Fifty-five patients, chosen at random, were treated with the available streptomycin and prescribed bed rest. The other fifty-two simply stayed in bed. At the end of six months four of the treated and fourteen of the untreated had died; the lungs of twenty-eight of the treated and of only four of the untreated had improved.

However, five years later the hopes raised by these results were shattered; by that time thirty-two of the treated and thirty-five of the untreated patients had died. Clearly streptomycin alone was not going to cure tuberculosis; Feldman and Hinshaw had already discovered that tubercle bacilli resistant to streptomycin began to emerge after only four weeks of treatment. Streptomycin also damaged the inner ear, and prolonged treatment with large doses of it could cause deafness. In 1948, a course of streptomycin injections gave George Orwell a remission that enabled him to finish *Nineteen Eighty-Four*, but he developed such severe allergic reactions against the drug that treatment had to be discontinued after only fifty days. He suffered a relapse and died in January 1950. All the same, Waksman received the Nobel Prize in 1952, "for his discovery of streptomycin, the first antibiotic effective against tuberculosis," a discovery which stimulated the immensely fruitful search for other micro-organisms in soil that secrete antibiotics.

But Albert Schatz, the real discoverer of streptomycin, and the first author listed on the original publication, got no prize. He was the son of poor Jewish farmers in Connecticut and had studied soil microbiology to find ways of increasing the yields on his father's unproductive farm. He embarked on the search for antibiotics only because Waksman made it a condition of his meager offer of $40 a month to work in his laboratory; but then Schatz threw himself into the research, testing hundreds of different soil micro-organisms for antibacterial activity. After slaving away for three and a half months, Schatz found in the throat of an infected chicken and in a compost heap fungi which stopped the growth of several virulent bacteria, including some known to resist penicillin.

Against the advice of his colleagues, and apparently not on Waksman's instructions, he then tested the effects of the fungus on tubercle bacilli and was elated to discover that it inhibited their growth in cultures. He next devised a way of extracting the active compound from cultures of the fungus. The discovery was published unglamorously in the *Proceedings of the Society for Experimental Biology and Medicine* of 1944 under the joint authorship of Schatz and Waksman, but in the subsequent publicity Schatz's contribution was soon forgotten.

In his rather pedestrian and humorless autobiography,[8] Waksman reproduced the text of a lecture he delivered at the Mayo Clinic in October 1944. Announcing the great discovery he declared: "In September 1943, my assistants and I succeeded in isolating in our laboratory an organism that produced an antibiotic...," when Waksman himself had sat in his office while Schatz toiled away in a basement laboratory. He said not a word in the lecture about Schatz, or about René Dubos or H. Boyd Woodruff, whose earlier discoveries of soil antibiotics that proved too toxic for clinical use had prepared the way for the discovery of streptomycin. Instead, Waksman used throughout the lecture the *plural majestatis* "We." The ethics of scientific and medical research have now been elevated to an academic subject, to which an entire new journal is to be devoted, but no one seems to care about what I regard as the First Commandment: "Thou shalt not take the credit for thy junior collaborators' work," a commandment whose observance would avoid much injustice and bitterness.

Isolated from one another by the war, but stimulated by Domagk's discovery of Prontosil, Domagk and other scientists in Europe and the U.S. also searched for a drug that would work against tuberculosis. One was the biochemist Jorgen Lehmann, the son of a Danish professor of theology who headed the pathology department of Sahlgren's Hospital in Gothenburg, Sweden. Ryan describes him as an eccentric, imaginative genius who had already made his name showing that dicoumarol—a chemical compound extracted from spoiled sweet clover, a derivative of which is marketed under the name warfarin—was an effective anticlotting agent. In 1940, he read a report that salicylic acid, a compound related to aspirin, makes tubercle bacilli respire—i.e., take in oxygen—faster than normally. This showed that the bacilli ate the compound and that a chemically modified version of it might poison them.

According to Lehmann's report in the *Lancet*, chemists at the Swedish pharmaceutical firm Ferrosan synthesized about fifty different derivatives of salicylic acid, one of which, PAS, inhibited the growth of tubercle bacilli in cultures. Guinea pigs tolerated the compound well and two tuberculosis patients treated with it showed marked improvement.

Swedish medical colleagues at first received Lehmann's discovery with disbelief, but this soon proved unjustified. In 1952, a study by the British Medical Research Council showed that PAS, when taken alone, was indeed less effective than streptomycin taken alone, but treatment with both almost completely prevented the emergence of streptomycin-resistant strains after prolonged treatment. For the first time, tuberculosis could be cured. As the news of the efficacy of PAS against tuberculosis spread, the Ferrosan company vastly increased its production to meet the huge, worldwide demand. Lehmann's discovery of PAS was just as important as Schatz and Waksman's of streptomycin, and was made at the same time, even though it was published two years later. Lehmann was yet another scientist unjustly denied the Nobel Prize, for reasons that we shall not know until 2002, when the Nobel files for 1952 will finally be open to inspection.

At Elberfeld, Domagk continued his search for an anti-tuberculosis drug throughout the Second World War, regardless of the town's destruction by Allied bombers. By 1945 he had found an effective compound called Conteben and traveled around war-ravaged Germany under terrible hardships to convince his skeptical colleagues of its therapeutic value against tuberculosis. The efforts of Domagk and others to improve on Conteben led to the discovery of one of the most potent antitubercular drugs yet found: isoniazid.

By the early 1950s there were several powerful anti-TB drugs on the market—conteben, isoniazid, PAS, and streptomycin, but no single one cured all patients because it took many months of treatment to clear the body of the bacilli, and drug-resistant mutants were liable to emerge much sooner. A combination of drugs offered better hopes. Suppose the probability of the emergence of a bacillus with resistance against any one of the drugs were one in a million, then the probability of the emergence of a bacterium with resistance against two drugs in combination would be one in a million million, and against three in combination, one in a million million million. The British Medical Research Council therefore began a long series of trials designed to discover the right combination, and dose, of drugs that would prevent the emergence of resistant strains.[9] Another trial, in India, showed no difference between the time needed to cure patients who took the drugs while resting in sanatoriums and those who carried on relatively normal lives at home. These findings led to the gradual closing of the world's tuberculosis sanatoriums, including those in Davos, the setting of Thomas Mann's *Magic Mountain*. They have long been converted into hotels.

The Medical Research Council's trials were designed according to statistical analyses made by Bradford Hill and, later, by Ian Sutherland. They have been crucial to the near-eradication of tuberculosis in developed and many underdeveloped countries throughout the world. They were the first to evaluate the efficacy of treatments free from human bias and according to rigorous mathematical criteria, and they have helped to transform clinical practice from an art into a science.

By 1980 the annual incidence of new cases of tuberculosis—not deaths from it—in the Western industrial countries had dropped to about one in ten thousand. Yet only ten years later the Centers for Disease Control in Atlanta, Georgia, warned that tuberculosis was not only reappearing but was out of control in the U.S. By 1991 there were 4000 new cases of tuberculosis in New York alone. AIDS is believed to be the chief culprit. Many of us, myself included, contracted tuberculosis in childhood, recovered, and have since been immune; but apparently tubercle bacilli survive in our scar tissues for years or even decades, and we are safe only as long as they are kept in check by our immune systems. When AIDS destroys this, the dormant bacilli revive.

Immigration from underdeveloped countries where tuberculosis has never been suppressed is another source of infection. In 1990 the Centers for Disease Control reported the incidence of tuberculosis among foreign-born persons in the United States to be thirteen times higher than among native Americans. In a book called *Silent Travelers*, Alan M. Kraut writes that the number of active tuberculosis cases in New York City rose from 2,545 in 1989 to 3,673 in 1991. The incidence rose by 56 percent among

African-Americans, 52.3 percent among Hispanic residents, and 46.9 percent among Asians, compared to 13.2 percent among native whites.

In the United States as a whole, the incidence rose by nearly 10 percent between 1989 and 1990, but the accompanying graph (see Figure 1) shows no evidence that new immigrants are largely responsible for these increases. In London, one in fifty of the homeless has tuberculosis, but this is mainly a consequence of poverty rather than of AIDS. Malnutrition and exposure make the homeless susceptible to infection; sleeping in crowded shelters promotes infection, and the movement of infected people from shelter to shelter causes it to spread.

The bacilli of some recent patients have proved resistant to all commonly used drugs and antibiotics, often because patients failed to comply with the prolonged combined drug regimen[10]—isoniazid, rifampicin, and pyrazinamide for six months—that is essential for a cure. Others foolishly treat themselves with single antituberculosis drugs that pharmacies in some countries sell over the counter, or they are prescribed inadequate drug treatments by incompetent doctors. In the United States, about half the cases of tuberculosis that resist the commonly used drugs can still be cured by special combinations of drugs, but at a cost of about $200,000 for each patient. Stephen E. Weis and others report that drug resistance and relapse among tuberculosis patients are mostly owing to non-compliance with the prescribed drug regimens, and therefore can be drastically reduced by treatment under direct observation by local health departments. In a trial in Tarrant County, Texas, directly observed therapy reduced the rate of relapse from 20.9 to 5.5 percent and the rate of multidrug-resistant relapse from 6.1 to 0.9 percent even though many of the patients were alcoholics, drug users, homeless, or psychiatric cases.[11] Kraut recommends rigorous public health measures to combat the spread of the disease, not just in New York, but throughout the United States. Others plead for worldwide campaigns.

K.M. Citron at the Brompton Hospital in London has written that "BCG vaccination of the newborn usually protects against serious forms of tuberculosis, is safe and cheap and should be used in developing countries where tuberculosis is most prevalent."[4] Many reports confirm that in children it prevents both tuberculosis and meningitis. Others believe the most important task to be an intensive search for new vaccines that would be effective worldwide for both adults and children. Molecular biologists are trying to make better vaccines by using recombinant DNA technology.

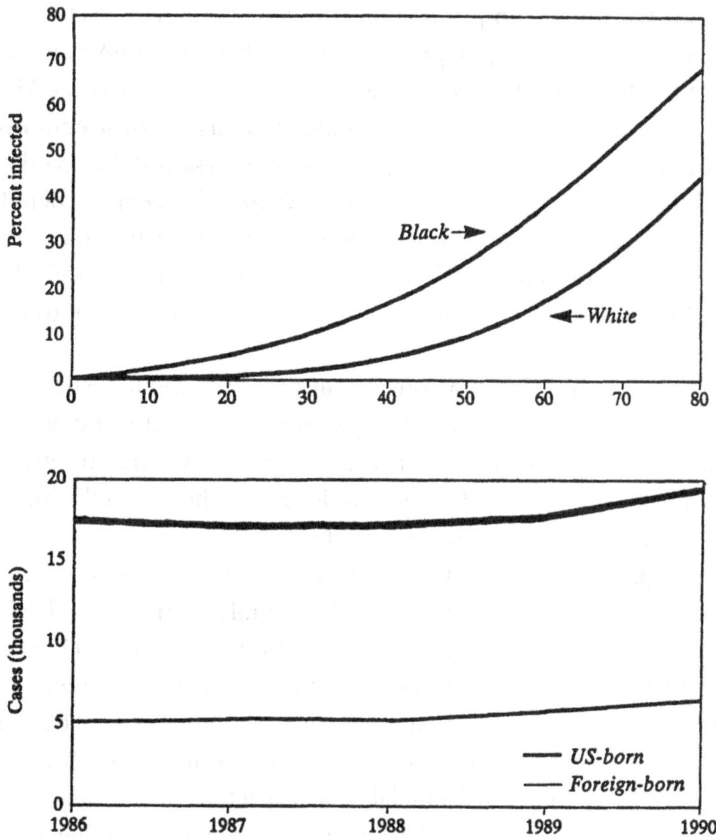

Figure 1. Infection with tuberculosis in the United States. Percentage of people who have been infected with tuberculosis at one stage in their lives and have become immune (top) and incidence of tuberculosis in U.S.-born and in foreign-born cases from 1986 to 1990 (bottom). Note that the percentage of foreign-born cases has not increased dramatically. This figure was reproduced from B.R. Bloom and C.J.L. Murray, 1992, "Tuberculosis: Commentary on the Re-emergent killer," *Science* **257:** 1055-1064.

Alexander Pope advised:

Be not the first by whom the new are tried,
Nor yet the last to lay the old aside.

Most doctors in Ryan's story seem to have paid more attention to the first rather than the second line. The often maligned drug industries come out well. Domagk's search for active compounds remained fruitless for many years, yet the managers of I.G. Farben continued to put huge resources at his disposal. They made no profit from Prontosil because the active principle turned out to be a split-product of the dye, nor from Conteben, because the victorious Allies confiscated their patent rights. George Merck turned his international patent rights on the manufacture of streptomycin over to Rutgers University, so that any firm could make it if they paid royalties to Rutgers. Ferrosan employed a mere seventy-five people when they took up the synthesis of PAS, which proved cumbersome and uneconomical until one of the company's chemists invented a quick and cheap method of making it. Its risk paid off. By 1964 it was selling 3000 tons per year.

Ryan's book reminds us how precarious our lives used to be before the discovery of antibiotics, when a cut finger or a sore throat could be deadly, and he emphasizes that our battle against deadly bacteria can rarely be finally won, because natural selection leads to the multiplication of mutants which can overcome our best defenses.

Micro-organisms that resist antibiotics have become a common problem of many hospitals. Among them are the hemolytic streptococci, which give rise to puerperal fever and wound infections, staphylococci causing boils, and klebsiella, which are responsible for a particularly virulent form of pneumonia. Penicillin-resistant gonorrhea is common in Africa and Southeast Asia and affects about one case in ten in the United Kingdom. In the U.S., complacency and financial cuts have let the once excellent state and local systems of communicable-disease surveillance crumble, so that new and antibiotic-resistant infections may spread undiagnosed and unreported. The Centers for Disease Control in Atlanta plan to reverse this trend by strengthening surveillance, by opening centers for the discovery of new infections, and by starting a global warning system for disease migration.[12] We also need to discover new antibiotics and vaccines all the time if Ryan's heroes are not to have battled in vain.

Toward the end of Domagk's life, the biochemist Otto Warburg told him that he deserved monuments in each valley and on every mountain. Domagk replied that no one is interested any longer in diseases that can be cured. Tuberculosis must now be removed from the list of those diseases; it presents a renewed deadly menace to mankind,[13] not only in the Third World, where it is the leading cause of death, but everywhere.

What Holds Molecules Together?

Like Michael Faraday, Linus Pauling was born poor; he struggled at menial jobs to support himself at high school and while studying for a degree in chemical engineering at the Oregon Agricultural College. All the same, he became interested in the electron theory of valence when he was only eighteen, and after this he "continued to hope that the empirical information about the properties of substances could eventually be encompassed in a theory of the structure of molecules." Fulfillment of his hope began to take shape in 1926 when he spent a postdoctoral year at Munich with Germany's greatest teacher of theoretical physics, Arnold Sommerfeld, just as Erwin Schrödinger published his first papers on wave mechanics. Sommerfeld immediately recognized their importance and gave a course of lectures on them. Having absorbed what knowledge he could in Munich, Pauling next went to Zürich to work with Schrödinger himself, but found his stay disappointing because Schrödinger liked to work all by himself and took little notice of him. Pauling then returned to the California Institute of Technology at Pasadena where he had taken his doctoral degree. Part of Pauling's PhD thesis was on the structure of the mineral molybdenite (MoS_2), which he determined by X-ray analysis. After this, minerals became his first interest; he became intrigued especially by the silicates whose complex structures formed the pride of W.L. Bragg's school at Manchester in the 1920s. In 1929, Pauling formulated a set of simple rules of coordination, based on ionic charges and radii, and on the postulate of local neutralisation of charges, which allowed the structures of silicates and of many other minerals to be understood and often predicted. With these rules, he put an entire subject on a rational chemical basis.

One of his first papers on the chemical bond was revolutionary because it explained the tetrahedral coordination of carbon and the square or octahedral coordination of transition metals on wave mechanical principles. The introduction conveys something of the flavour of his paper:

> During the last four years the problem of the nature of the chemical bond has been attacked by theoretical physicists, especially Heitler and London, by the application of quantum mechanics. This work has led to an approximate theoretical calculation of the energy of formation and other properties of very simple molecules, such as H_2, and has also provided a formal justification for the rules set up in part by G.N. Lewis for the electron pair bond. In the following paper it will be shown that many more results of chemical significance can be obtained from the quantum mechanics equation, permitting the formulation of an extensive set of rules for the electron-pair bond supplementing those of Lewis. Those rules provide information regarding the relative strengths of bonds formed by different atoms, the angles between bonds, free rotation, or lack of free rotation about bond axes, the relation between the quantum numbers of bonding electrons and the number and spatial arrangement of bonds, and so on. A complete theory of the magnetic moments of molecules and complex ions is also developed, and it is shown that for many compounds involving elements of the transition group this theory together with the rules of electron pair bonds leads to a unique assignment of electron structures as well as a definite determination of the type of bonds involved.

Nothing like this had been done before.

In a later paper, Pauling applied to chemistry the concept of resonance originally introduced by Heisenberg into quantum mechanics.

> It is found that there are many substances whose properties cannot be accounted for by means of a single electronic structure of the valence bond type, but which can be fitted into the scheme of classical valence theory by the consideration of resonance among two or more structures.

Benzene was the prime example.

Pauling's valence bond theory established the relationship between interatomic distances, mostly derived from accurate crystallographic data, and bond energies, on which much of his successful interpretation of the chemical properties of organic compounds was based. With that in mind, he founded at Pasadena an outstanding research school of structural chemistry. Pauling's Munich experience also led him to write an *Introduction to Quantum Mechanics* for chemists, together with E. Bright Wilson, which was first published in 1935 and remains a useful classic.

As a chemistry student in Vienna in the 1930s, I had been made to memorize the 759 pages of Karl Hoffman's *Inorganic Chemistry* and the 866 pages of Paul Karrer's *Organic Chemistry*. I looked upon such tasks as feats of endurance which gave me a certain sporting satisfaction—like walking from Land's End to John O'Groat's—but they gave me little intellectual satisfaction because the books did not explain the properties of matter. Why does water freeze at 0°C and methane at –184°C? Why does one form of selenium melt at a temperature 76°C higher than the other? Why is sulphur soft and diamond hard? Why is one form of silica, quartz, optically active, while the two others, tridymite and crystoballite, are not? Why is salicylic acid stronger than benzoic acid? No such questions were answered for me.

In 1936 I became a research student in X-ray crystallography at Cambridge. For Christmas, 1939, a girl friend gave me a book token which I used to buy Linus Pauling's recently published *Nature of the Chemical Bond*. His book transformed the chemical flatland of my earlier textbooks into a world of three-dimensional structures. It stated that "the properties of a substance depend in part upon the type of bonds between its atoms and in part on the atomic arrangement and the distribution of bonds," and it proceeded to illustrate this theme with many striking examples. For instance, Pauling discusses the cause of a discontinuity in the melting points of the fluorides of the second row elements thus:

> An abrupt change in properties in a series of compounds, such as in the melting points or boiling points of metal halides has sometimes been considered to indicate an abrupt change in bond type. Thus of the fluorides of the second-row elements, those of high melting points, have been described as salts, and the others as covalent compounds; and the drop in melting point of 1100°C in going from aluminium fluoride to silicon fluoride has been interpreted as showing that the bonds change sharply from the extreme ionic type to the extreme covalent type[1,2]. I consider the bonds in aluminium fluoride to be only slightly different in character from those in silicon fluoride and attribute the abrupt changes in properties to a change in the nature of the atomic arrangements. In NaF, MgF_2 and AlF_3 each of the metal atoms is surrounded by an octahedron of fluorine atoms, and the stoichiometric relations then require that each fluorine atom be held jointly by several metal atoms. In each of these crystals the molecules are thus combined into giant polymers, and

> the processes of fusion and vaporization can take place only by breaking the strong chemical bonds between metal and nonmetal atoms: as a consequence the substances have high melting points and boiling points. The stable coordination number of silicon relative to fluorine is, on the other hand, four, so that the SiF_4 molecule has little tendency to form polymers. The crystals of silicon fluoride consist of SiF_4 molecules piled together only by weak van der Waals forces".

Characteristically for Pauling's showmanship, the first reference is to N.V. Sidgwick's classic *The Electronic Theory of Valency*, and the second reference to one of his own papers. By such examples Pauling's book fortified my belief, already inspired by J.D. Bernal, that knowledge of three-dimensional structure is all important and that the functions of living cells will never be understood without knowing the structures of the large molecules composing them.

In my physical chemistry practical at Vienna I had had to prove to myself that acetic acid in solution forms dimers, but it needed Pauling to drive home the importance of the hydrogen bonds that are responsible for their formation:

> Although the hydrogen bond is not strong it has great significance in determining the properties of substances. Because of its small bond energy and the small activation energy involved in its formation and rupture, the hydrogen bond is especially suited to play a part in reactions occurring at normal temperatures. It has been recognized that hydrogen bonds restrain protein molecules to their native configurations, and I believe that as the methods of structural chemistry are further applied to physiological problems it will be found that the significance of the hydrogen bond for physiology is greater than that of any other single structural feature.

This was a remarkable prophecy made when nothing was known as yet about protein structures.

Pauling's imaginative approach, his synthesis of structural, theoretical, and practical chemistry, his capacity for drawing on a wide variety of observations to prove his generalizations, and his vivid writing drew the dry facts of chemistry together into a coherent intellectual fabric for me and thousands of other students for the first time.

I first met Pauling ten years after I had read his book and was intrigued by his lectures, where he would reel off the top of his head atomic radii, interatomic distances, and bond energies with the gusto of an organist playing a Bach fugue. Pauling's lectures reinforced the chief message of his book: to understand the properties of molecules, not only must you know their structures, but you must know them accurately.

Besides directing many people's research, Pauling used to deliver a freshmen's introductory course of lectures which he first published as a textbook of General Chemistry in 1947. Its 1970 edition contains over 900 pages; it begins with an introduction to the atomic and molecular structure of matter, covers most important aspects of physical and inorganic chemistry, touches on the elements of organic and biochemistry, and ends with nuclear chemistry. The lectures were spectacular and often dramatic. Jack Dunitz described one to me: A large beaker filled with what looked like water stood on the bench. Pauling entered, picked a cube of sodium metal from a bottle, tossed it from hand to hand (done safely if your hands are dry) and warned of its violently explosive reaction with water. He then threw it into the beaker. As students cowered in fear of an explosion, he said nonchalantly "but its reaction with alcohol is much milder."

Pauling also made several important excursions into biological subjects; the first was into immunology. He conceived the idea that antibodies adapt their structures to those of antigens by a refolding of their polypeptide chains in solution: he was wrong, but it was a reasonable theory before anything was known about the genetic basis of protein structure. He next became interested in haemoglobin. Having read that Faraday found blood to be diamagnetic despite iron and oxygen being paramagnetic, he told his student Charles Coryell to measure the magnetic susceptibility of haemoglobin in the presence and absence of oxygen. In 1936, Coryell found oxyhaemoglobin to be diamagnetic, and deoxyhaemoglobin to be paramagnetic with a spin of $S = 2$.

Forty years later, I asked Pauling what made him think of this experiment which later turned out to have been crucial for an understanding of haemoglobin's function. Pauling replied that it had not been clear whether oxygen formed a chemical bond with the iron in haemoglobin or whether it was merely being adsorbed, and he thought that formation of a chemical bond might be accompanied by a magnetic change. In 1970 I found that the spin transition of the iron is the trigger for the allosteric change that accompanies the reaction of haemoglobin with oxygen.

In 1940, Linus Pauling and Max Delbrück anticipated molecular complementarity that later proved to be the basis of DNA structure and replication. They attacked the German theoretician Pascual Jordan, who had advanced the idea that there exists a quantum-mechanical stabilizing interaction, operating preferentially between identical or near-identical molecules, which is important in biological processes such as the reproduction of genes. Pauling and Delbrück pointed out that interactions between molecules were now rather well understood and gave stability to two molecules of *complementary* structure in juxtaposition, rather than to two molecules with necessarily *identical* structures. Complementariness should be given primary consideration in the discussion of the specific attraction between molecules and their enzymatic synthesis.

In 1948, Pauling followed this up with the prediction: "I think that enzymes are molecules that are complementary in structure to the activated complexes of the reactions that they catalyze, that is, the molecular configuration that is intermediate between the reacting substances and the products of the reaction." Seventeen years later, lysozyme, the first enzyme structure solved, was indeed found to have an active site complementary to the transition state of its substrate, and the same had been true of all enzyme structures solved since.

In 1949, someone got Pauling interested in sickle-cell disease—a genetic disorder affecting mainly black people—which causes the red cells to be distorted to various sickled shapes on loss of oxygen. Pauling suggested to his young collaborators Itano, Singer, and Wells that they should examine the electrophoretic mobilities of normal and sickle-cell haemoglobin. They found them to be different because sickle-cell haemoglobin carries two fewer negative charges than normal haemoglobin. Pauling published this result in *Science* under the dramatic title: "Sickle cell haemoglobin, a molecular disease." This paper led Vernon Ingram and John Hunt in my Medical Research Council Unit in Cambridge to show that sickle-cell haemoglobin differed from the normal protein merely by the replacement of a single pair of glutamic acid residues by valines. Their discovery of the effect of a genetic mutation on the amino acid sequence of a protein posed poignantly the question of the genetic code, which Crick, Brenner, and their colleagues answered a few years later with their demonstration that triplets of nucleotides determine the amino acid sequence of proteins.

Among biochemists, Pauling is most famous for his discovery of the α-helix that came as the culmination of X-ray analyses of the structures of amino acids pioneered in the 1930s and 40s by his collaborators Robert Corey and Edward Hughes. At that time, glycine, alanine, and diketopiperazine presented a degree of complexity that carried X-ray analysis to the very limits of the possible. The results provided Pauling with the stereochemical data he needed for the interpretation of the meagre X-ray diffraction patterns that Bill Astbury at Leeds had obtained from protein fibres such as hair, nails, and muscle. Pauling argued that in a long-chain polymer made of chemically equivalent units, all units must occupy geometrically equivalent positions, which was possible only in a helix. Further, the structure of diketopiperazine had shown that the peptide bond had partial double bond character so that the atoms

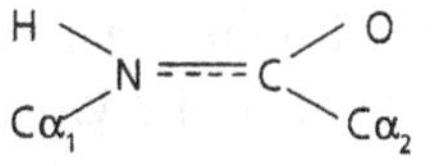

H O
N ==== C
$C\alpha_1$ $C\alpha_2$

must all lie in a plane. Finally, all amido (NH) groups should form hydrogen bonds with carbonyl (CO) groups.

Lying in bed with 'flu in Oxford in 1948, Pauling amused himself by building a paper chain of planar peptides and found a satisfactory structure by folding them into a helix with 3.6 residues per turn. Shortly after this he visited Kendrew and myself in Cambridge. Ignorant of his Oxford experiment, I proudly showed him my three-dimensional Patterson of haemoglobin which indicated that its polypeptide chain was folded in the same way as Astbury's fibres, but to my disappointment Pauling made no comment; he did not announce his discovery until the delivery of a dramatic lecture at Pasadena in the following year. It helped to earn him the Nobel Prize in Chemistry in 1954, but he had really deserved it for his many other outstanding contributions to chemistry much earlier.

After the α-helix, Pauling produced one more fundamental paper. When the amino acid sequences of the haemoglobins of different animals were beginning to be known, it became clear that the number of amino acid substitutions increased with the distance between species on the evolutionary tree. This inspired Pauling and his young collaborator Emile Zuckerkandl to propose the existence of an evolutionary clock which ticks at the rate of about one amino acid substitution per hundred residues per 5 million years. Like many of Pauling's papers this one initiated an entire new field of research which has since occupied many other scientists' lives. Pauling continued publishing papers almost to the end of his life, but nothing as fundamental as his earlier work appeared, possibly because he became too preoccupied, first with the threat of nuclear war and later with vitamin C.

In the McCarthy era, Pauling's anti-nuclear stance earned him the reputation of a 'red'. In 1952, when the Royal Society organized a Discussion on the Structure of Proteins at which Pauling should have been the principal contributor, he could not come because the State Department had withdrawn his passport. In 1954, the English philosopher Bertrand Russell gave Christmas lectures on the radio warning of the dangers of nuclear war. In the following year he drew up a manifesto against nuclear arms which Albert Einstein signed a few days before his death. This concluded with the words: "There lies before us if we choose, continual progress in happiness, knowledge and wisdom. Shall we, instead, choose death, because we cannot forget our quarrels? We appeal, as human beings, to human beings: remember your humanity and forget the rest. If you can do so, the way lies open to a new paradise; if you cannot, there lies before you the risk of universal death." Pauling signed this manifesto together with seven prominent physicists and the geneticist Hermann Muller; it led to the convening of the first Pugwash Conference at which Soviet and Western scientists discussed measures to reduce the dangers of nuclear war.

In 1958, Pauling published a book: *No More War* and handed to Dag Hammarskjöld, the Secretary General of the United Nations, a petition signed by 9235 scientists

> urging that an international agreement to stop the testing of nuclear weapons be made now....inasmuch as it is the scientists who have some measure of the complex factors involved in the problem, such as the magnitude of the genetic and somatic effects of the released radioactive materials.

In May 1961, Pauling organized a conference of forty scientists on nuclear disarmament in Oslo, and afterwards led hundreds of people in a torchlight procession against nuclear war through the streets of that city. His campaign, conducted with the same panache as his lectures on chemistry, made a vital contribution to the conclusion of the atmospheric test ban in 1963 and won him the Nobel Prize for Peace in December of that year. He also campaigned against the war in Vietnam, undeterred by being called a traitor by some of his fellow countrymen.

Pauling used to suffer from severe colds several times a year. In 1966 he followed a friendly doctor's suggestion and began to take a daily dose of 3 grams of vitamin C. After this his colds became rarer and milder. This experience may have sparked off his idea that huge doses of vitamin C were vital for health, and that they cured the common cold and even cancer. Chemists shops in Britain still sell vitamin C as "Linus Powder," of which he swallowed about eighteen grams every day; probably about 100 milligrams would have been absorbed and the rest excreted. Ascorbic acid is a scavenger of free radicals and a deficiency of it may therefore increase the likelihood of cancer, but there is no solid evidence that such massive doses have any prophylactic effect. It seems tragic that this should have become one of Pauling's major preoccupations for the last 25 years of his life and spoilt his great reputation as a chemist. Perhaps it was related to his greatest failing, his vanity. When anybody contradicted Einstein, he thought it over, and if he found he was wrong, he was delighted, because he felt he had escaped from an error, and that he now knew better than before, but Pauling would never admit that he might have been wrong. When, after reading Pauling and Corey's paper on the α-helix, I discovered an X-ray reflection at 1.5 Å spacing from planes perpendicular to the axis of protein fibres which excluded all conformations other than the α-helix. I thought he would be pleased, but no, he attacked me furiously because he could not bear the idea that someone else had thought of a test for the α-helix of which he had not thought himself. I was glad when he forgot his anger later and became a good friend.

Pauling's fundamental contributions to chemistry cover a tremendous range, and their influence on generations of young chemists was enormous. In the years between 1930 and 1940 he helped to transform chemistry from a largely phenomenological subject to one based firmly on structural and quantum mechanical principles. In later years the valence bond and resonance theories which formed the theoretical backbone of Pauling's work were supplemented by R.S. Mulliken's molecular orbital theory, which provided a deeper understanding of chemical bonding. For instance, it allowed C. Longuet-Higgins and W. Lipscomb to predict and explain the structures of the boranes, which would not have been possible on the basis of Pauling's concepts. Nevertheless, resonance and hybridization have remained part of the everyday vocabulary of chemists and are still used, for example, to explain the planarity of the peptide bond. Many of us regard Pauling as the greatest chemist of the century.[3]

I Wish I'd Made You Angry Earlier

Fifty years ago, the great unsolved problem of biology seemed to be the structure of proteins.

Bill Astbury, a physicist and X-ray crystallographer working for the Wool Research Association in Leeds (United Kingdom), discovered that the fibrous protein keratin, found in wool, horn, nails, and muscle, gave a common X-ray diffraction pattern consisting of just two reflections, a meridional one at 5.1 Å and an equatorial one at 9.8 Å.

Astbury called this the α-keratin pattern. When these fibers were stretched under steam, a new pattern appeared with meridional reflection at 3.4 Å and two equatorial ones at 4.5 and 9.7 Å. Astbury called this the β-keratin pattern. He concluded that it arose from the regular repeat of amino acid residues along straight polypeptide chains, whereas the chains in α-keratin must be folded or coiled such that several amino acid residues form a pattern repeating every 5.1 Å along the fiber axis.

It appeared that the key to understanding the structure of proteins lay in the elucidation of this ubiquitous fold, but the meager information contained in the X-ray diffraction patterns did not provide enough clues to uncover it.

By 1950, J.C. Kendrew and I had evidence that the same fold of the polypeptide chain occurred also in the two globular proteins myoglobin and hemoglobin. W.L. Bragg, the pioneer of X-ray crystallography who was our professor at the Cavendish Laboratory in Cambridge, stimulated us to attack the problem by building molecular models.

To start us off, he hammered nails representing amino acid residues into a broomstick in a helical pattern with an axial distance between successive turns (or pitch) of 5.1 centimeters. Kendrew and I had great difficulty in building real models of helical polypeptide chains with the right pitch; no matter whether we built them with two, three or four amino acid residues per turn, their bond angles were always strained. After some months we published our work jointly with Bragg in the *Proceedings of the Royal Society*, but without any firm conclusions about the correct fold.

One Saturday morning shortly after our paper had appeared, I went to the Cavendish library, and in the latest issue of the *Proceedings of the National Academy of Sciences of the United States of America* found a series of papers by Linus Pauling together with the crystallographer R.B. Corey. In their first paper they proposed an answer to the long-standing riddle of the structure of α-keratin, suggesting that it consisted of helical polypeptide chains with a non-integral repeat of 3.6 amino acid residues per turn. Their helix had a pitch not of 5.1 Å, as Astbury's X-ray picture seemed to demand, but of 5.4 Å, which was consistent with the repeat found in fibers of certain synthetic polypeptides by C.H. Bamford and A. Elliott and their colleagues at the Courtaulds Research Laboratories.

I was thunderstruck by Pauling and Corey's paper. In contrast to Kendrew's and my helices, theirs was free of strain; all the amide groups were planar and every carbonyl group formed a perfect hydrogen bond with an imino group four residues further along the chain.

The structure looked dead right. How could I have missed it? Why had I not kept the amide groups planar? Why had I stuck blindly to Astbury's 5.1 Å repeat? On the other hand, how could Pauling and Corey's helix be right, however nice it looked, if it had the wrong repeat? My mind was in a turmoil. I cycled home to lunch and ate it oblivious of my children's chatter and unresponsive to my wife's inquiries as to what the matter was with me today.

Suddenly I had an idea. Pauling and Corey's α-helix was like a spiral staircase in which the amino acid residues formed the steps and the height of each step was 1.5 Å. According to diffraction theory, this regular repeat should give rise to a strong X-ray reflection of 1.5 Å spacing from planes perpendicular to the fiber axis. As far as I knew, such a reflection had never been reported, either from "natural" proteins like hair and muscle or from synthetic polypeptides. Hence, I concluded, the α-helix must be wrong.

But wait! Suddenly I remembered a visit to Astbury's laboratory and realized that the geometry of his X-ray setup would have precluded observation of the 1.5 Å reflection because he oriented his fibers with their long axes perpendicular to the X-ray beam, while observation of the 1.5 Å reflection would have required inclining them at the Bragg angle of 31°. Furthermore, Astbury used a flat plate camera that was too narrow to record a reflection deflected from the incident X-ray beam by 2 x 31°.

In mad excitement, I cycled back to the lab and looked for a horse hair that I had kept tucked away in a drawer. I stuck it on a goniometer head at an angle of 31° to the incident X-ray beam; instead of Astbury's flat plate camera I put a cylindrical film around it that would catch all reflection with Bragg angles of up to 85°.

After a couple of hours, I developed the film, my heart in my mouth. As soon as I put the light on I found a strong reflection at 1.5 Å spacing, exactly as demanded by Pauling and Corey's α-helix. The reflection did not by itself prove anything, but it excluded all alternative models that had been put forward by ourselves and others and was consistent only with the α-helix.

On Monday morning I stormed into Bragg's office to show him my X-ray diffraction picture. When he asked me what made me think of this crucial experiment, I told him that the idea was sparked off by my fury over having missed building that beautiful structure myself. Bragg's prompt reply was, "I wish I had made you angry earlier!" because discovery of the 1.5 Å reflection would have led us straight to the α-helix.

The inconsistency between the 5.4 and 5.1 Å continued to worry me until one morning about two years later when Francis Crick arrived at the lab with two rubber tubes around which he had pinned corks with a helical repeat of 3.6 corks per turn and a pitch of 5.4 centimeters. He showed me that the two tubes could be wound around each other to make a double helix such that the corks neatly interlocked. This shortened the pitch of the individual chains, when projected onto the fiber axis, from 5.4 to 5.1 centimeters, as required by the X-ray pattern of α-keratin.

Such a double helix was eventually found by my colleagues A.D. McLachlan and J. Karn in the muscle protein myosin.

*Big Fleas Have Little Fleas...**

The early conceptual development of molecular biology was dominated by two physicists: Max Delbrück and Francis Crick. Both owe their fame to a small number of seminal papers and their influence to their formidable powers of imagination and argument.

Delbrück began his scientific career with Otto Hahn and Lise Meitner, who engaged him as a theoretician to help them interpret their bombardment of uranium with neutrons. Delbrück failed to grasp that they had split the uranium nucleus in two, but he applied the target theory learnt from them to the Russian geneticist Timoféeff-Ressovsky's work on the irradiation of fruitflies with X-rays. He calculated the number of X-ray quanta needed to produce one genetic mutation in the flies. From that he could calculate that the size of the target on the flies' chromosomes corresponded to the size of a single large molecule containing no more than a few hundred atoms. This was the first experimental evidence about the size of a gene. Even though his estimate was wrong because he neglected the effects caused indirectly by the generation of free radicals in the surrounding medium, the paper secured him a Rockefeller Fellowship to go and study in Pasadena; it also stimulated Schrödinger to write his influential book *What is Life?* in which he predicted the gene to be a molecule with an aperiodic structure; and it made a young medical graduate learn-

*A review of the book *Licht und Leben: Ein Bericht über Max Delbrück, den Wegbereiter der Molekularbiologie* (*Light and Life: An Account of Max Delbrück, Pioneer of Molecular Biology*) by Peter Fischer (Konstanz, FRG; Universitätsverlag Konstanz).

ing physics in Enrico Fermi's laboratory in Rome decide to work with Delbrück on the nature of the gene. His name was Salvatore Luria.

Delbrück's entry into biology was inspired by a lecture, "Light and Life", by Niels Bohr. Bohr predicted that the study of life at the atomic level would lead to a paradox similar to that posed earlier by atomic spectra, a paradox that was resolved only by the new quantum mechanics:

> The existence of life must be considered as an elementary fact that cannot be explained, but must be taken as a starting point in biology, in a similar way as the quantum of action, which appears as an irrational element from the point of view of classical mechanical physics, taken together with the existence of elementary particles, forms the foundation of atomic physics. The asserted impossibility of a physical or chemical explanation of the function peculiar to life would be... analogous to the insufficiency of the mechanical analysis for the understanding of the stability of atoms [Niels Bohr, *Nature* **131**, 458 (1933)].

Delbrück himself has described the search for this "Elementary Fact of Life" as the sole motive behind all of his work. He should have listened instead to Linus Pauling, with whom he published the joint paper mentioned in the preceding chapter. Pauling realized that the hydrogen bond accounts for most biochemical reactions without having to invoke any new "Elementary Facts."

In 1937, for scientific rather than ideological or racial reasons, Delbrück left Berlin for T.H. Morgan's laboratory at the California Institute of Technology. He hoped to reduce the genetics of the fruitfly *Drosophila* to simple physical principles, but was disappointed to find no quantitative data susceptible to theoretical interpretation. He was about to give up when he discovered that in the basement of the same building another biologist, E.L. Ellis, was working on bacteriophages in *Escherichia coli*. A glance at Ellis' plaques convinced Delbrück that the bacteriophage was the hydrogen atom of biology for which he had been looking, and that study of it might lead him to the "Great Paradox of Life." He and Ellis soon discovered that a single phage adsorbed to a single bacterium multiplies "upon or within" that bacterium until it bursts with the release of an average of 60 progeny phages; such a mechanism had been proposed by d'Herelle, but never proved, while others thought they had found evidence in favour of continuous release of phage by infected bacteria. By clear thinking and application of the simple theory of exponential growth, Delbrück and Ellis opened the way to the analysis of phage genetics. Delbrück soon attracted an enthusiastic band of disciples who formed the flourishing Phage School that met each summer at Cold Spring Harbor in New York state.

In 1938, Salvatore Luria obtained an Italian government fellowhip to work with Delbrück in Pasadena, but Mussolini's racial laws annulled that because he was a Jew, and it was not until September 1940 that Luria reached New York, now as a refugee. He sought out Delbrück, who had become an instructor at Vanderbilt University in Nashville, Tennessee. They first worked together on interference between two bacterial viruses acting on the same host, where Delbrück hoped to find something analogous to Pauli's exclusion principle in physics.

The paper which was to earn them the Nobel Prize 26 years later was conceived in 1942, after Luria had found a job at Bloomington, Indiana. Luria tried to discover whether bacterial resistance to phage infection was caused by an adaptive change, as many believed, or whether it arose from mutations. He was perplexed by the extreme variability of the numbers of resistant bacteria present in different cultures of the same organisms, until the correct explanation dawned upon him one night at a dance while watching a game machine. If the change from susceptibility to resistance was a random event due to mutations, then a mutation occurring early in the life of a culture would give rise to a large clone of resistant bacteria, while several mutations arising later would each produce only small clones. Luria wrote to Delbrück, telling him of his idea; Delbrück put it into mathematical form and proved rigorously that the distribution of resistant bacteria in Luria's different cultures was consistent only with their being due to random mutations, and that these mutations occurred with a constant frequency of 2.45×10^{-8} per bacterial division. These results opened up the field of bacterial genetics. Just like Delbrück and Ellis' earlier results on phage, they involved nothing that could not have been found out years earlier; the only new ingredient was clear thought.

In the ecology of science the opening of a new habitat immediately attracts a crowd. Delbrück escaped from the multitude congregated around phage genetics by switching to phototropism of the fungus *Phycomyces*, hoping again that simple experiments and clear thinking would lead him to a breakthrough. Twenty years of work, however, failed to bring the solution of this very difficult problem any nearer.

Delbrück was a German Romantic searching for the Holy Grail, Bohr's "Elementary Fact of Life." To those who have tried to understand the workings of large biological molecules in terms of simple chemical laws, Bohr's and Delbrück's belief in some mystical principle looks like vitalism. Delbrück had a proverbial and often misplaced scepticism of new work. For example, he objected to Beadle and Tatum's one-gene-one-enzyme hypothesis on the ground that it could not be falsified by experiment; he dismissed Lwoff's lysogeny of phage as a non-phenomenon; and he disbelieved Meselson and Stahl's demonstration of the semi-conservative replication of DNA. Delbrück wanted to model himself on his two great teachers by combining Bohr's insights with Pauling's mordant criticism, or as he put it, by becoming God and Mephisto all in one, but I have the impression that Delbrück really wanted to disbelieve any advance that removed the elusive "Elementary Fact" further from his grasp.

*How the Secret of Life Was Discovered**

Of course, now that we know the answer, it seems so completely obvious that no-one nowadays remembers just how puzzling the problem seemed then.[1]

FRANCIS CRICK

In 1936, I left my hometown of Vienna, Austria, for Cambridge, England, to seek the Great Sage. I asked him, "How can I solve the secret of life?" He replied, "The secret of life lies in the structure of proteins, and X-ray crystallography is the only way to solve it." The "Sage" was John Desmond Bernal, a flamboyant Irishman who headed the Crystallographic Department of the Cavendish Laboratory and who had been the first to discover that protein crystals give detailed X-ray diffraction patterns extending to spacings of the order of interatomic distances.[2] We really called him Sage, because he knew everything. During a discourse at the Royal Institution in London in 1939, Bernal said:

> The structure of proteins is the major unsolved problem at the boundary of chemistry and biology today. It is difficult to exaggerate the importance of this study to many branches of science. The protein is the key unit in biochemistry and physiology. ...All protein molecules that we

*Presented at the COGENE/Symposium, "From the Double Helix to the Human Genome: 40 Years of Molecular Genetics." UNESCO, Paris. 21–23 April 1993.

> know now have been made by other protein molecules, and these in turn by others.[3]

When Bernal advanced this argument in a later BBC discussion, the physicist W.H. Bragg asked him where the first protein had come from. Instead of replying "I don't know," Bernal skilfully sidestepped Bragg's awkward question. Inspired by Bernal's enthusiasm, I became a crystallographer and began to work on the structure of haemoglobin, because it was the protein that was most abundant and easiest to crystallise.

What had attracted me to Cambridge as an undergraduate in Vienna was Gowland Hopkins' work on vitamins and enzymes. He was the founder of Cambridge biochemistry. In the nineteen-thirties he still had had to battle against vitalism to get acceptance for his then heretical view that

> The living cell, at one definite level of its organisation, admitting that higher levels may be superimposed, is to be pictured as the seat of diverse but organised chemical reactions, in which substances identifiable by chemical methods undergo changes which can be followed by chemical methods. The molecules of these substances are activated and their reactions directed in space and time by the catalytic agencies which are commonly known as intracellular enzymes. The influence of these differs in no essentials from that of catalysts in non-living systems save that it is displayed in relations which are exceptionally specific.[4]

Even today some philosophers would reject Hopkins' views as reductionist.

Hopkins was the professor of biochemistry. The reader was J.B.S. Haldane, a devoted communist in the guise of an English squire, and one of the most imaginative scientists of his generation. He pointed out in 1937 that many enzymatic reactions, as well as blood groups, are genetically controlled.

> Two possibilities are now open. The gene is a catalyst making a particular antigen, or the antigen is simply the gene or part of it let loose from its connection with the chromosome. The gene has two properties. It intervenes in metabolism, sometimes at least by making a definite substance. And it reproduces itself. The gene, considered as a molecule, must be spread out in a layer one building block deep. Otherwise it could not be copied. The most likely method of copying is by a process analogous to crystallization, a second similar layer of building block being laid down on the first.[3] But we could conceive of a process analogous to the

copying of a gramophone record by the intermediation of a "negative" perhaps related to the original as an antibody to an antigen.

Haldane dismissed the idea that genes might be made of nucleic acids and asserted that the most likely substances are the histones, which are proteins attached to DNA in the chromosomes.

Three years later, Pauling and Delbrück published a seminal paper. It was an attack on the German theoretician Pascual Jordan, who had advanced the idea that there exists a quantum-mechanical stabilizing interaction, operating preferentially between identical and near-identical molecules that governs biological processes such as the reproduction of genes.[5] Pauling and Delbrück pointed out that interactions between molecules were now rather well understood to give stability to two molecules of *complementary* structure in juxtaposition, rather than two molecules with necessarily identical structures.[6] Haldane's and Pauling and Delbrück's ideas were to prove prophetic.

In 1944, the almost universally held view that genes consist of proteins was overturned by Avery, MacLeod, and McCarty's discovery that the transforming principle of pneumococci is made of DNA. Oswald Avery, who took the first step as a young man and persevered until his retirement, was an even more reluctant revolutionary in biology than Max Planck had been in physics earlier in the century. He was born in Halifax, Nova Scotia in 1877, the son of a Baptist Minister who had emigrated there from England and later became pastor of a Baptist church in a poor district of New York.

Motivated by his family's Christian missionary background, Oswald Avery became a doctor, but soon abandoned practice for research. This remained humdrum until, aged 36, he was appointed bacteriologist to the hospital of the Rockefeller Institute for Medical Research in New York. At that time, lumbar pneumonia killed 50,000 people a year in the United States. Avery's mother had died of it. Avery wondered why it killed some people in his hospital ward, while others recovered.

The first pointers came in 1916. Avery's friend A.R. Dochez discovered in the filtrate of a pneumococcus culture a specific soluble substance that was precipitated by an antiserum against that pneumococcus. Dochez and Avery then found that same substance in the blood and urine of their patients and concluded, wrongly as it turned out, that it was a protein. In succeeding years, Avery convinced himself that it was the same substance as that which formed the slimy bacterial capsule originally described by Neufeld and Handel in Robert Koch's laboratory in Berlin in 1910.

Avery sensed that this specific soluble substance played a vital part in the disease and wanted to find out more about it. When the organic chemist Michael Heidelberger joined the Institute, Avery used to agitate a tube of it in front of him saying, "Michael, the whole secret of bacterial specificity is in this tube. When can you work on it?" When Heidelberger finally did take it up, he found it to consist, not of protein, but of polysaccharide, long chains of sugar-like molecules.[7]

The previous year Fred Griffith at the laboratory of the Ministry of Health in London had discovered the nature of the difference between virulent and non-virulent pneumococci that Avery had sought for so long. He found that the virulent cocci were encapsulated by the slimy shell and the nonvirulent ones were not. Griffith called the encapsulated ones smooth and the others rough. So the specific soluble substance consisted of polysaccharide and was associated with virulence.

Heidelberger and Avery tried to produce an antiserum against it. They failed, but Avery did make an effective antiserum against pneumonia by immunizing horses with virulent pneumococci, and this saved many lives.

The next great advance came in 1928 with a startling discovery by Griffith. He tried an experiment which every sane person could have told him was a waste of time. He injected mice with a mixture of killed smooth bacteria and live rough ones, after he had made doubly sure that the smooth ones were really dead. Two days later he was surprised to find all the mice dead and full of smooth virulent pneumococci. This meant that the dead smooth pneumococci had transformed the live rough ones into smooth ones. A bizarre and unbelievable result. Griffith interpretated the transformation as a Lamarckian kind of adaptation.

> When the R form of either type is furnished under suitable experimental conditions with a mass of the S form of the other types, it appears to utilize that antigen as a pabulum (food) from which to build up a similar antigen and thus it develops into an S strain of that type.[8]

He thought that the recipient rough cocci had retained the power to synthesize the capsular polysaccharides of several types and needed only the specific stimulus of the killed encapsulated cells to adapt themselves and make the polysaccharide capsule again. Apparently he never thought of genetic mutations, perhaps because bacteria were then believed not to exhibit genetics.

According to his colleague René Dubos, Avery refused to believe Griffith's results until another member of the Institute, Henry Dawson, repeated them while Avery was away sick.[9] Dawson and Sia did away with the mice and transformed rough pneumococci into smooth ones in cultures in beef broth. After this, Avery was convinced and used to ask persistently, day after day, year after year: "What is the substance responsible for the transformation?", but he was handicapped by being a bacteriologist rather than a biochemist.

After Dawson had left the Institute in 1930, Avery encouraged his successor, J.L. Alloway, to pursue the problem. He dissolved the virulent cocci in deoxycholate, a natural detergent extracted from bile juice, filtered off the cellular debris, and found the solution to be active. The transforming activity came down as a thick, syrupy precipitate on addition of alcohol and could be redissolved in water without loss of activity but, unbelievably, it took another 13 years to find out what this precipitate was made of. Not even Colin MacLeod's pregnant observation, made in 1936, that the transforming activity was destroyed by ultraviolet light, which was known to attack nucleic acids, gave the clue.

The slow progress was due partly to the very inconsistent yields of transforming activity extracted from the pneumococci. It was not until 1941 that Avery surmounted this trouble by heating the transformed cocci to 65°C for 30 min *before* lysing them with deoxycholate. This inactivated the bacterial enzymes that had destroyed the activity, and it left the transforming principle intact.

The final attack started in September 1941 with the arrival of Maclyn McCarty, a young medical doctor with good biochemical training. It seemed most likely that the transforming principle was a protein, whence enzymes that digest proteins should have destroyed it. Nowadays such enzymes can be bought off the shelf, but in those days McCarty was lucky to be given two such enzymes, trypsin and chymotrypsin, by Northrop at the Rockefeller Institute in Princeton, who had been the first to isolate and crystallise them in pure form.

Northrop's enzymes left the transforming activity intact, which meant that it could not be a protein. Could it reside in ribonucleic acid, RNA for short? Northrop's colleague Moses Kunitz had just crystallised ribonuclease, an enzyme which splits RNA, and gave Avery some of his crystals. Ribonuclease also left the transforming activity intact. The same held for enzymes that split polysaccharides. Avery and McCarty therefore set out to rid an extract of the virulent bacteria of protein, RNA, and polysaccharides without destroying the transforming activity, and then isolating that activity in pure form.

After a lot of hard work McCarty obtained a gooey precipitate of white fibres which took up a stain characteristic of DNA and showed properties similar to the DNA that another member of the Rockefeller Institute, Alfred Mirsky, had isolated from calf thymus. Now came the crucial test. Would the transforming activity be destroyed by deoxyribonuclease, the enzyme that splits DNA? No one could provide this. McCarty had to isolate it laboriously from dog intestinal mucosa, or swine kidneys, or rabbit blood. He made sure that it degraded neither protein, nor RNA, nor polysaccharide, but it did destroy the transforming activity, showing that it must consist of DNA.[10] Avery and McCarty now piled proof upon proof to convince themselves of their finding, yet it was so revolutionary that it took Avery a long time before he finally summoned the courage to publish it. The paper leaves no shadow of doubt that the transforming factor consists of DNA and nothing but DNA.[11]

Did Avery comprehend the full significance of his discovery? Rollin Hotchkiss, who worked with him at the Rockefeller Institute, has testified that Avery "was well aware of the implication of DNA transforming agents for genetics and infections." The Australian virologist Macfarlane Burnet, who visited Avery in 1943, wrote home to his wife that Avery "had just made an extremely exciting discovery which, put rather crudely, is nothing less than the isolation of a pure gene in the form of DNA."[12] It was a revolutionary finding, but Avery was no revolutionary. He was a small, delicate monkish bachelor who lived only for his science, and for his life's aim to find the cause and cure of virulent pneumonia. He wore pince-nez, was fastidious with his words, ever cautious in his public utterances, never went on lecture tours, wrote no books, and never travelled. He never co-authored any paper on research to which he had not actively contributed, did not patent his discovery, and never became rich.

Aaron Levene, a chemist at the Rockefeller Institute, had proposed that all DNAs are made up of regular sequences of the same four nucleotides, whence they could not carry information. Alfred Mirsky, who had spent years working on DNA without realizing its significance, was sure that the transforming activity was carried by protein impurities in Avery's DNA. As late as 1947, Mirsky said at Cold Spring Harbor, "In the present state of knowledge, it would be going beyond the experimental facts to assert that the specific agent in transforming bacterial types is DNA."[10] Seeing that as little as 3 billionths of a gram of DNA had been sufficient to transform the cocci in half an assembly of infected test tubes, the idea that transformation could have been effected by a protein impurity seems far fetched. Perhaps because of Mirsky's continued smear campaign, Avery, MacLeod, and McCarty never received the Nobel Prize for one of the century's greatest discoveries. Yet to its credit, the Royal Society of London recognised the discovery by making Avery a Foreign Member and awarding him their highest honour, the Copley Medal. Sir Henry Dale, the president, said in his citation. "Here surely is a change to which, if we were dealing with higher organisms, we should accord the status of a genetic variation and the substance inducing it—the gene in solution, one is tempted to call it—appears to be a nucleic acid of the deoxyribose type."[13]

Robert Olby concludes his authoritative account of the history of the transforming factor with the words, "With the passage of time the work of Avery, MacLeod, and McCarty looks, if anything, more significant than in 1953; perhaps it was the most important discovery in the path to the double helix."[12] But most people remained sceptical for many years afterwards. Some were convinced when Hotchkiss[14] transferred penicillin resistance to a non-resistant strain of pneumococci with DNA from a resistant strain. Others dropped their doubts when Alfred Hershey and Martha Chase[15] demonstrated that on infection of coli bacteria with a virus, called phage, only the phage DNA and not the phage protein enters the bacteria. References to the transforming principle in successive editions of J.N. Davidson's textbook on nucleic acids illustrate the scepticism that persisted for a long time even after that. He wrote in 1950:

> If the active DNA is in fact protein-free, we have here an example of a specific biological property, the ability to induce the synthesis of a characteristic immunological polysaccharide, which is peculiar to one form of DNA and no other. Not only is this the first good example of biological activity attributable to a nucleic acid per se; it indicates that there may be important differences between one specimen of DNA and another which may not be detectable by chemical means.[16]

And even in 1960, seven years after Watson and Crick's discovery of the double helix:

> It seems reasonable therefore to interpret bacterial transformation as indicating that DNA is the active material of the gene;... It has proved a matter of some difficulty to find evidence in confirmation of that hypothesis.[17]

Davidson did not specify the nature of that difficulty.

One day in September 1951 a strange young head with a crew-cut and bulging eyes popped through my door and asked, without saying as much as hello, "Can I come and work here?" He was Jim Watson, who wanted to join the small team of enthusiasts for molecular biology which I led at the Physics Laboratory in Cambridge, England.

My colleagues were John Kendrew, a chemist like myself, and Francis Crick and Hugh Huxley, both physicists. We shared the belief that the nature of life could be understood only by getting to know the atomic structure of living matter, and that physics and chemistry would open the way, if only we could find it.

In his best-selling book, *The Double Helix*, Watson mirrors himself as a brash western cowboy entering our genteel circle, but this is a caricature. Watson's arrival had an electrifying effect on us because he made us look at our problems from the genetic point of view. He asked not just, "what is the atomic structure of living matter?" but, foremost, "what is the structure of the gene that determines it?" Watson found an echo in Crick who had begun to think along similar lines. Crick was 34, a more than mature graduate student due to years lost by the war; Watson was 22, a whizz-kid from Chicago who had entered university aged 15, and got his Ph.D. in genetics at 20.

They shared the sublime arrogance of men who had rarely met their intellectual equals. Crick was tall, fair, dandyishly dressed, and talked volubly, each phrase in his King's English strongly accented and punctuated by eruptions of jovial laughter that reverberated through the laboratory. To emphasise the contrast, Watson went around like a tramp, making a show of not cleaning his one pair of shoes for an entire term (an eccentricity in those days), and dropped his sporadic nasal utterances in a low monotone that faded before the end of each sentence and was followed by a snort.

To say that they did not suffer fools gladly would be an understatement—Crick's comments would hit out like daggers at *non sequiturs* and Watson demonstratively unfolded his newspaper at seminars that bored him. Watson had put Crick's mind to the structure of DNA, yet their relationship was something of teacher and pupil because there was little that Watson could teach Crick, but much that Crick could teach Watson. Crick has a profound understanding of that hardest of the sciences, physics, without which the structure of DNA would never have been solved. This crucial fact is obscured in Watson's *Double Helix*. Yet Watson had an intuitive knowledge of the features that DNA ought to have if it were to make genetic sense.

At some stage there was much argument as to whether genes consist of two or three chains of DNA wound around each other. Watson lodged with a lady retired from the stage who kept a boarding house for young girls. One day she noticed him pacing restlessly and muttering to himself, "There must be two... there must be two..." She guessed that this referred to matters of the heart. But we knew better. He reasoned on genetic grounds that genes must be made of two chains of DNA, and he was right.

Like Leonardo, Crick and Watson often achieved most when they seemed to be working least. They did an immense amount of hard work, studying while hidden away, often at night, but when you saw them they were more likely engaged in argument and apparently idle. This was their way of attacking a problem that could be solved only by a tremendous leap of the imagination, supported by profound knowledge. Imagination comes first in both artistic and scientific creations. But in science Nature always looks over your shoulder. To paraphrase Winston Churchill, "In science you don't need to be polite, you only have to be right."

Dangerous Misprints[*]

Watson and Crick's famous double helix showed how the genetic information is written on DNA and how it is copied every time a cell divides. Some years after this, scientists also deciphered the genetic code. Occasionally, very rarely in fact, an error occurs in the copying of the message that makes up a gene. Even more rarely, such an error manifests itself in a malfunction or absence of the protein specified by that gene, and gives rise to a congenital disease. Such an error occurred in Queen Victoria who carried the gene for haemophilia which surfaced in some of her male descendants as a malfunctioning of a protein needed for the clotting of blood; consequently several of them bled to death.

Thanks to discoveries made over a span of many years, the genes responsible for haemophilia and other congenital diseases have recently been mapped. The scene was set by an apparently irrelevant observation made in 1952 by Jean Weigle, a Swiss biologist in California. Weigle was puzzled by a virus that thrived in one strain of coli bacteria, languished at first in another closely-related strain, and then mysteriously regained its former vigour. Years later, another Swiss biologist, Werner Arber, decided to follow up Weigle's seemingly trivial observation. Arber found that Weigle's second strain restricted the virus's growth, because it contained a scissor-like enzyme that cut its DNA. After a while the virus mobilised its defences against the bacterial scissors and thrived as before. The enzyme did not cut the DNA randomly, but at a specific word which, like 'madam',

*A review of the book *Genome* by Jerry Bishop and Michael Waldholz (Touchstone).

read the same either way and which the enzyme evidently recognised.

Until 1973, Arber's discovery did not look like one that could be put to practical use. In that year, Daniel Nathans, a biologist in Baltimore, wanted to find out which of the genes carried by a tumour virus actually made tumours grow. He happened to work in the same laboratory as Hamilton Smith, who had succeeded in isolating and purifying an enzyme similar to Arber's. Nathans therefore borrowed some of it to cut his long viral DNA chain into shorter pieces. After that he put the fragments on a strip of wet blotting-paper and passed an electric current through it. Being negatively charged, the DNA fragments migrated to the positive pole; the smallest fragments went fastest and the largest lagged behind. When the current was switched off, the fragments formed a set of bands according to their size. Each band contained one or several genes; one of these contained the tumour gene that Nathans had been looking for. In deference to Weigle's original observation, Smith called his enzyme a restriction enzyme and Nathans called the pattern of bands a restriction map. It offered a simple way of isolating genes and of characterising DNA, and it was to mark the birth of recombinant DNA technology.

Since that pioneering work, scientists have isolated hundreds of different restriction enzymes from micro-organisms, each cutting DNA at different words and into fragments of different lengths which line up in an electric current to give a different restriction map. The maps are characteristic for each species—maps from dog DNA differ from those of human DNA. In addition, not all human maps turn out to be exactly the same. Scientists found slight differences in the positions of single bands on the maps derived from different individuals, which implies that at some time in the past a misprint in the genetic message has shifted the position of the word at which a restriction enzyme cut the DNA. These harmless misprints are inherited. Since a child inherits half of its DNA from each parent, half of the bands on any one restriction map are inherited from the father and half from the mother.

Very occasionally, a harmless misprint that causes the DNA to be cut in a position different from that found in the bulk of the population occurs in an individual who also suffers from a genetic disease, due to an unknown misprint that you are trying to detect. If the disease and the harmless misprint are inherited together, there is a likelihood that the latter occurs on the same chromosome as the misprint responsible for the disease. This can be a clue to guide the investigator to the position of the diseased gene, but only if he can prove that the harmless misprint and the disease have been inherited together in many affected families over several generations.

Leonore Weber was a healthy, active, middle-aged woman who suddenly went down with a terrible disease that rots and finally kills the brain—Huntington's Chorea. Unlike cystic fibrosis, Huntington's Chorea kills even if it is inherited from only one parent. Carriers do not realise that they harbour it until it suddenly erupts in middle age, perhaps only after they have passed the fatal gene on to some of their own children. Mrs. Weber's daughter Nancy Wexler was nagged by the fear that she, too, harboured the deadly gene, but she soon learned that there was no way of finding out. To remedy this, her father set up a foundation for the study of inherited diseases, and she mobilised scientists to search for the responsible gene.

She heard of a community in a remote Venezuelan fishing village where Huntington's Chorea was common and dreaded as "El Mal." Nancy Wexler went there, befriended the affected families and organised expeditions to establish its incidence among them. She took blood samples from them back to America for analysis of their DNA and found scientists eager to search for the gene, even though the chances seemed remote.

Eventually 49 scientists spread over four American and two British laboratories joined in the search; in 1993 they found the gene and discovered the nature of the misprints responsible for the disease. It was a triumph, but it raised heart-rending dilemmas, because genetic tests can now tell if an individual has inherited the fatal gene, yet there is as yet no mitigating treatment or cure in sight. As Nancy Wexler said: "Knowledge alone does not provide the support you need for your life; you must know there is hope." And as yet there is none. A woman with a family history of the disease finally did decide to take the test, because she found that "it's the waiting and wondering that kills. It kills from within." When the result was negative, a crushing weight was lifted from her mind, but later, when her brother came down with it, she felt guilty at having escaped it herself. Some people have committed suicide after being told that they harbour the gene. On the other hand, there is the immeasurable joy of affected couples whom pre-natal diagnosis has assured that their child will be free from it.

A succession of exciting scientific adventures led to the discovery of the genes of other congenital diseases: muscular dystrophy, cystic fibrosis, retinoblastoma (an inherited cancer of the eye), and Tay-Sachs Disease, an inherited degeneration of the brain in small infants, frequent among Sephardic Jews. Scientists have also located genes that can predispose people to colorectal, lung, or breast cancer, and to cardiovascular disease, manic depression, schizophrenia, or alcoholism, but genetic disposition accounts for only a small fraction of affected individuals. Some of these diseases manifest themselves if inherited from only one parent; others only if inherited from both. Some are caused by defects in single genes, while others surface only when two or more genes are affected, or if habits like smoking or unhealthy diets collude with the predisposition.

Techniques are now available for detecting the presence of any of these genes in a single human cell. It might be a cell that has been detached from a fertilised egg after it has divided a few times, or from the membrane surrounding an eight-week-old embryo, or it might be an adult white blood cell. The information derived from such a genetic screen may bear on a person's future health, life-expectancy, and mental stability. Science has presented us with these far-reaching new possibilities before their implications have been thought through.

In the United States, doctors and scientists are worried that commercial pressures and fear of damage suits may precipitate routine genetic screening before it has been made sufficiently reliable, before counsellors have been trained to explain the often complicated meaning of the results, and before the public has been taught the elements of human genetics and statistical probability. Screening could lead to unjust discrimination. Employers may turn down men with a predisposition to alcoholism because they fail to understand that four-fifths of the people with such a genetic predisposition never become alcoholics and that only a quarter of the men who do become alcoholics are genetically predisposed to the habit. Parents may be persuaded to abort foetuses with a genetic predispostion to alcoholism or manic depression, forgetting that Ernest Hemingway was an alcoholic, Virginia Woolf a manic depressive, Dostoevsky an epileptic, and Lincoln may have suffered from a hereditary connective tissue disorder. Great achievement does not always go with good health.

Genetic screening could be beneficial if it were to be restricted to the most common life-threatening diseases, or to cases where there is a family history of congenital disease. Otherwise, we might finish with a society of genetic hypochondriacs. α_1-Antitrypsin deficiency is an inherited defect that predisposes people to emphysema, particularly if they become smokers. Some years ago, the Swedish Government decided that much ill health could be avoided if new-born babies were screened for the deficiency, and parents of affected babies were told to warn them later on to avoid smoking. In fact, this screening led to so much morbidity as a result of feelings of guilt, recrimination between couples, and quarrels with in-laws for having brought a deleterious gene into the family, that the Swedish Government had to abandon it.

Will the benefits of genetic screening outweigh its drawbacks? I believe that detailed knowledge of the human genome will reduce the sum total of human suffering, partly by its potential for diagnosis, and partly because it will deepen our understanding of cancer and other non-infectious diseases. In time this will also lead to better treatments. The dangers to society are real, but they could be mitigated by enlightened legislation. The billions of dollars to be spent on the Genome Project are likely to benefit only the people in the rich world who are already very healthy, but will do nothing to rid the vast majority of people in the poor world of the host of crippling and deadly parasitic diseases that are the real scourge of mankind. On these, little money is being spent and there has been no clarion call to stir scientists to action.

A Deadly Inheritance*

This is the story of a Jewish paediatrician at the Harvard Medical School in Boston who saved the life of Dayem, a small Iranian Arab boy afflicted with a deadly inherited anaemia. In the course of 26 years of treatment David Nathan transformed the boy from a dwarfed, grotesquely hideous cripple to a good-looking, active, independent businessman and *bon vivant*.

Dayem suffers from thalassaemia, a disease due to defective synthesis of haemoglobin, the protein of the red blood cells that transports oxygen from the lungs to the tissues and helps the return transport of carbon dioxide from the tissues back to the lungs. His red cells contained too little haemoglobin, and even that little haemoglobin was liable to precipitate and lead to the cells' premature destruction. As a result, his body was chronically short of oxygen. Red cells have a lifetime of about 120 days and are continuously replaced by fresh ones growing in the bone marrow. In response to his shortage of oxygen, Dayem's bone marrow produced red blood cells at the double and expanded at the expense of the bony mass surrounding it. When Dayem, aged six, arrived at Nathan's clinic, he had the stature of a two-year-old and had already suffered multiple fractures of his fragile bones. His haemoglobin level was little more than one tenth of normal, which is viewed as below the minimum needed to sustain life, his pulse was racing and he was out of breath, yet Dayem walked in vigorous-

*A review of the book *Genes, Blood, and Courage: A Boy Called Immortal Sword* by David G. Nathan (Harvard University Press).

ly and confidently, accompanied by his parents and two healthy younger brothers, and kept everyone around him laughing.

They had come from Lisbon where Dayem's father had his business; before that they had consulted Europe's leading specialist in Zurich who had told Dayem's parents that he could be kept alive by regular blood transfusions, but that the iron overload caused by them would poison his liver and kidneys and would eventually kill him. There was a faint hope that his body would respond to the shortage of adult haemoglobin by making another haemoglobin that is normally present only in the foetus, and this foetal haemoglobin might keep Dayem alive without transfusions. This was the option Dayem's parents had adopted before they took him to Nathan. It had kept him alive, but only just.

Nathan set himself the double task of rescuing Dayem from certain death and of relieving his parents of their feelings of guilt for having inflicted this terrible disease on their child. He persuaded them that Dayem must have regular blood transfusions if he was to grow and have no more fractures, and told them of his hope that research would find a way of dealing with the iron overload. To assuage their guilt, he told them how the genetic lesions responsible for thalassaemia have accumulated among Arabs and other peoples living in malarial zones of the world.

Thalassaemia is a recessively inherited disease; carriers with one defective haemoglobin gene are healthy. When two carriers marry, one quarter of their children are liable to become anaemic because they have inherited defective haemoglobin genes from both parents. For reasons that are still not fully understood, babies who have inherited one defective haemoglobin gene are less likely to die from malaria than babies with normal haemoglobin genes. Their greater chances of survival more than compensates for the early deaths of children with two defective haemoglobin genes. Natural selection has therefore led to an accumulation of the thalassaemia gene where malaria is common. In former times, when there was no blood transfusion and malaria was rampant in the Middle East, the couple's possession of defective haemoglobin genes would have given them more healthy children than if their haemoglobin genes had been normal. Therefore, they need not have felt guilty about Dayem.

In 1968, when Dayem arrived in Boston, bone marrow transplants from a compatible donor offered a chance of curing thalassaemia, but at great risk. The risk of death has now shrunk very much, but it still presents parents with an agonising dilemma. The author writes:

> In pediatrics a decision to accept or reject a particular therapy is made by parents on behalf of a beloved child. If that decision has a bad outcome, the parents will be completely devastated by guilt. For that reason, the physician must make the parents believe that the decision, once made, is the one that he or she completely endorses. He must lift the burden of responsibility from the shoulders of the parents and place it where it belongs, on his or her own back. But all this must be done without taking the choice away from the parents in the first place.

For Dayem, this dilemma did not arise because the marrow of his brothers was incompatible with his own; at first he did well on the transfusion régime, but after a few years 'flu brought a crisis that could be overcome only by removing his grossly enlarged spleen. Even so, Dayem developed symptoms of severe iron overload that Nathan decided to correct with a then new drug, Desferal, which, when injected, binds iron and eliminates it through the kidneys. Daily injections of Desferal from the age of 12 cured Dayem's symptoms, but they were painful and he hated them. As a teenager removed from his mother's control, he revolted, first about his repulsive face, which plastic surgeons were able to correct, but as soon as his new face made him acceptable to girls, he abandoned himself to a carefree life, ignored all warnings and forgot about Desferal until he was twice brought back to the Children's Hospital with heart failure and told that a third incidence was bound to be fatal.

Dayem's carelessness threw Nathan into rages of frustration, but to no avail. As Dayem told him years later:

> It was really cramping my style... I couldn't swim when others wanted to swim. I couldn't do this when the others wanted to do that. It was annoying. It was painful. It itched... so I thought whether I skipped a year or two wasn't going to make much difference in the long run. So I told myself everything was fine... on top of all that, I was being told that I was going to die anyway... at one point, even though you thought that I had suicidal tendencies, and you even wanted my to see a psychologist, I stopped seeing him because... I told him whatever I thought he wanted to hear, and he wasn't doing anything for me though he was a very nice man.

Meditation finally made Dayem start a new life, got him over the shame of being different, and of having been written off by his successful businessman father. He realised that the disease saved him from having to live somewhere around the Persian Gulf and following in his father's footsteps and enabled him to live the independent life he wanted.

When Nathan was a young physician, one of his teachers told him that the greatest joy of paediatrics was to watch sick children shake off the fetters of chronic illness and become productive adults. On the other hand, Nathan is often asked whether it is right to go to the limits of technology in an attempt to save the life of a chronically sick child when the same health dollars, $30,000 a year for a thalassaemic patient, could be spent on vaccinations that would save thousands of lives. Nathan refutes that argument. As a physician, he regards himself as the servant and advocate of the individual child. If this view flies in the face of official cost control and resource utilisation, he cannot help it. For him, the child's needs come first.

Darwin Was Right

Three hundred years ago Isaac Newton claimed that scientists work from the particular to the general, first observing phenomena and only later drawing generalisations from them. In his *Principia*, he stated that "In experimental philosophy particular propositions are inferred from the phenomena and afterwards rendered general by induction." Karl Popper has disputed this, arguing that imagination comes first: scientists begin by formulating hypotheses and then proceed to test them by observation. Only hypotheses that can be falsified experimentally are scientific. If the hypothesis turns out to be inadequate, scientists formulate a new and improved one that can be again subjected to experimental tests. In this way, science has evolved by an interplay of imaginative conjections and experimental refutation. Popper elaborated these ideas in *Conjectures and Refutations* (Routledge and Kegan Paul, 1972). In his other great work, *The Open Society and its Enemies*, Popper turned against all those philosophies that claim the future of mankind to be determined by its past. He condemns Karl Marx's assertion that the contradictions of capitalism must lead to class war and the dictatorship of the proletariat. Popper disputes the existence of historical laws and holds that our future is in our own hands. He detests determinism in all its guises.

The same philosophical outlook permeates Popper's ideas on the evolution of species. He accepts Darwinism and defines it by the law that "organisms better adapted than others are more likely to leave offspring." But he argues that it is always good for theories to have competitors.

Because Darwinism has no competitor, Popper creates one by splitting Darwinism into a passive and an active form.

By passive Darwinism he apparently means the generally accepted theory that random mutation and natural selection lead inexorably to the evolution of higher forms of life. Popper condemns that theory as deterministic, as merely another version of the philosophical historicism that he demolished in *The Open Society*. He argues that "the idiosyncracies of the individual have a greater influence on evolution than natural selection" and that "the only creative activity in evolution is the activity of the organism." According to him, organisms sought better environments from the very beginning of life because adaptation includes the power of actively searching for food. The environment is passive, only organisms are active, as they seek better niches for themselves, and Popper regards this activity as the primary driving force of evolution.

According to Popper, passive Darwinism is a mistaken idea of adaptation, a consequence of faulty deterministic ideologies that have ruled biology and which today find their expression in sociobiology. We should instead think of evolution as a huge learning process, as an active preference of species for better niches.

Let us assume, he said, that we have managed to create life in the test tube. The organism will, by definition, not be adapted to the test tube, and it cannot seek a more favourable niche. We therefore have to adapt conditions in the test tube to the needs of the organism, which requires much knowledge. In Nature, on the other hand, life on Earth may have arisen not just once, but many times unsuccessfully, until an organism arose that *knew* how to adapt itself by actively seeking a better environment. Popper thus equates adaptation with knowledge, but knowledge in the form of function—such as the ability of a microorganism to detect and move toward a particular chemical (chemotaxis)—rather than of structure. He admits that this is an anthropomorphism, but he asserts that we cannot do biology without thinking in anthropomorphic terms. He justifies such thinking as a way of devising hypotheses that are founded on the common evolutionary origins of many biological functions.

Popper also pointed out that natural selection is not comparable to artificial selection by breeders; that notion was a mere teleological metaphor of Darwin's. Selective pressure, rather than natural selection, is a better term. Even that phrase carried teleological overtones, but this is unavoidable because organisms are problem solvers seeking better conditions—even the lowest organism performs trial and error measurements with a distinct aim. This image brought to my mind Howard Berg's striking film of chemotactic bacteria. He showed how a bacterium's flagellar motor makes it run and tumble randomly until the bacterium senses a gradient of nutrient. The bacterium then reduces the frequency of tumbling and lengthens the runs towards a greater concentration of nutrient. Yet the bias of the flagellar motor arises not from mystical knowledge, but from the activities of protein receptors that measure differences in the concentration of food at opposite ends of the bacterium. It is pure chemistry.

Yet at this stage Popper asserted that biology is irreducible to physics and chemistry. He maintained that biochemistry must include biological purpose, and is therefore irreducible to chemistry. In the 18th century, the philosopher Immanuel Kant argued that we possess an inborn, *a priori* sense of space and time that precedes our knowledge gained from observations. According to Popper, biological evolution involves a similar *a priori* knowledge by organisms. It is this *a priori* knowledge, foreseen by Kant, that led to adaptation in the long term. Darwin was a determinist because he regarded evolution as a passive process, while Lamarck was not.

When I asked Popper on what grounds he thought biochemistry to be irreducible to chemistry, I received the magisterial reply that I would see the answer if I thought about it for an evening. Having since done so, I still cannot think of a single biochemical function that is different *in vitro* from *in vivo* because, as Popper told another questioner, *in vivo* it works with a purpose—unless he meant it merely in the sense that a battery acquires a purpose when it is put into a torch. Nor can I think of a biochemical reaction that cannot be reduced to chemistry.

Popper's assertion reopens battles that were fought early in this century. Then, biochemists tried to convince the scientific world that the dynamics of living cells are not due to the purposeful action of protoplasm, but can be dissected into chemical reactions, each catalysed by a specific enzyme. In 1933, Gowland Hopkins, who in Cambridge pioneered the study of enzyme chemistry, complained that "justification for any such claim has been challenged in advance from a certain philosophic standpoint"—for instance by the Cambridge philosopher A.N. Whitehead's axiom that each whole is more than the sum of its parts.

Hopkins proved that biochemical reactions in living cells are nothing more than the sum of reactions that can be performed in the laboratory and interpreted in chemical terms. Since then his views have been vindicated by the demonstration that such fundamental processes as the replication of DNA, the transcription of DNA into messenger RNA, the translation of RNA into protein, the transduction of light into chemical energy, and a host of metabolic reactions can all be reproduced *in vitro*, without ever a hint of their activities in the cell being anything more than the sum of the chemical reactions of their parts in the test tube. It might be argued that it is the organisation that gives the cell purpose and thus makes the sum be more than its parts. This is true, but the organisation is intrinsic and chemical. The living cell is like an orchestra without a conductor that has its score laid down in its DNA.

Let me now examine some of the evidence relating to Popper's two forms of Darwinism—active, purposeful evolution as opposed to passive, deterministic evolution. I shall take my examples from haemoglobin, because they are the ones with which I am most familiar.

The camel and the llama are closely related species with different habitats. Camels live in the plains and llamas high up in the Andes. The camel possesses a haemoglobin molecule with an affinity for oxygen that is normal for an animal of its size. But because of a single mutation in the gene coding for one of the two globin chains that make up a haemoglobin molecule, the llama has a haemoglobin with an unusually high oxygen affinity. The variant haemoglobin helps the llama to breathe in the rarified mountain air. The geneticist Richard Lewontin, of Harvard University, pointed out to me that this mutation is likely to have occurred *before* llamas discovered that they could graze at altitudes barred to competing species. In other words, a mutation that adapts a species to a new environment probably happens before the species occupy that environment. While the mutation is an event whose happening is determined purely by the laws of chance, and in that sense is deterministic, the exploitation by the animals of that chance event needed a purposeful search for a better environment of the kind Popper seems to have in mind.

An even more striking example is provided by two species of geese: the greylag goose that lives in the plains of India all the year round, and its relative, the bar-headed goose that migrates across the Himalaya at 9000 metres to find better feeding grounds in summer. The bar-headed goose can reach these heights thanks to a haemoglobin with high oxygen affinity that has been generated by a chance mutation different from that in the llama. Before the bar-headed geese acquired that haemoglobin, the birds might have flown north by a longer, more roundabout route. The mutation allowed them to explore the shortcut across the high mountains. Alternatively, the geese might have started migrating across the Himalayas before they rose to such great heights. The mutation might have adapted them to the recent rise which is believed to amount to at least 1300 metres in the past 1,500,000 years.

Let me now turn to an example where adaptation may have been both active and passive. The deer mouse, *Peromyscus maniculatus*, is spread over the plains and mountains of North America. Its haemoglobin is polymorphic, which means that each individual's blood may contain either one of two haemoglobins that differ in their oxygen affinity, or an equal mixture of both. M.A. Chappell and L.R.H. Snyder from the University of California at Riverside have discovered a relationship between the altitudes of the habitats occupied by the deer mice and the oxygen affinities of their blood: the higher the habitat, the higher the oxygen affinity (Figure 1). To ensure that this correlation reflects an adaptive mechanism, the researchers allowed the mice to acclimatise for two months at altitudes of 340 metres and 3800 metres, and then measured the amount of oxygen they consumed during exercise. Chappell and Snyder found that at 340 metres the

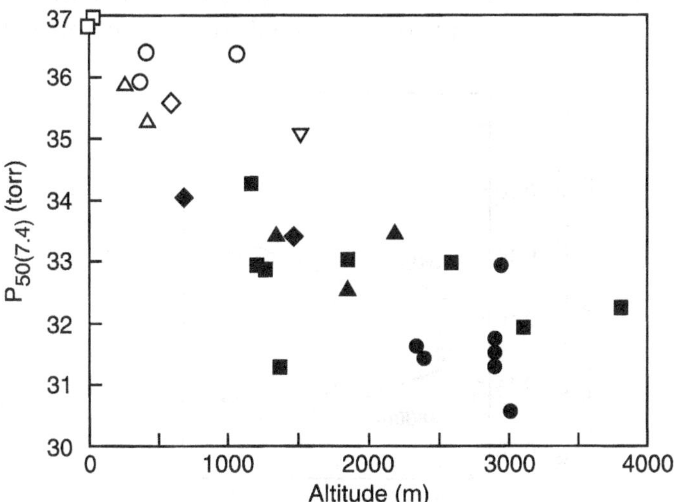

Figure 1. The higher up deer mice live, the greater the oxygen affinity of their blood. Such an adaptation can be regarded as both active and passive. P50 (7.4) is the partial pressure of oxygen at which half the hemoglobin molecules in a solution of pH 7.4 are saturated with oxygen.

mice with the haemoglobin with the lowest affinity for oxygen consumed oxygen at the highest rate and could therefore exercise longest. At 3800 metres the reverse was true, which proved that the differences in oxygen affinity really did adapt the mice to life at different altitudes (Figure 2).

Slight variations in the structure of a given protein—so-called protein polymorphisms—are widespread in nature, and many researchers have speculated about their survival value. Chappell and Snyder were the first to show that a polymorphism can affect the physiology of an animal by altering biochemical reactions that can be measured *in vitro* and related directly to Darwinian fitness. Their results strongly suggest that the polymorphism is maintained by selective pressure—that is, in one habitat one form of the protein is at a premium, but in another habitat, a different structure is more advantageous to the organism. Is this an example of active or passive Darwinism? Mice whose habitat lies on a mountain slope are likely to migrate to altitudes that best fit the oxygen affinity of their haemoglobin. On the other hand, mice living on a high mountain plateau or in a low lying plain are likely to stay put, and those with the haemoglobin best suited to that environment are likely to produce the most offspring. Hence active and passive Darwinism will work side by side.

Let me now discuss two inherited haemoglobin diseases in humans; sickle-cell anaemia and thalassaemia. Each is due to different mutations in the haemoglobin genes. If inherited from only one parent, the mutations

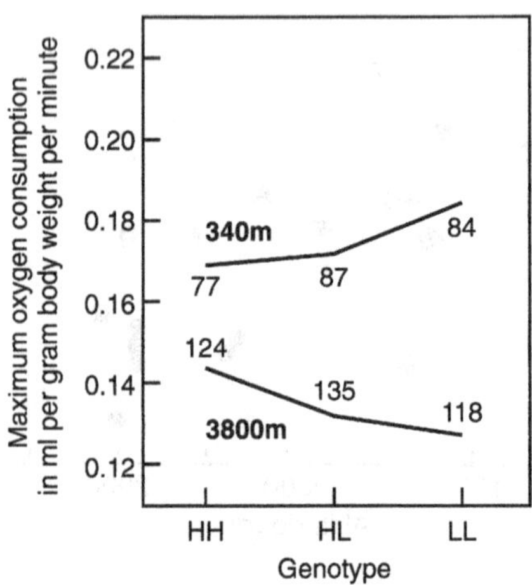

Figure 2. Differing genes for haemoglobin really do fit deer mice to their particular environment. HH = mice with hemoglobins of high oxygen affinity. HL = mice with hemoglobins of intermediate oxygen affinity. LL = mice with hemoglobins of low oxygen affinity.

are generally harmless; if inherited from both parents, their effects tend to be crippling. Sickle-cell anaemia is most common in Africa, while thalassaemia is most widely spread in Mediterranean countries, East Asia, and certain Pacific islands. In 1949, the Scottish geneticist J.B.S. Haldane first spotted an association between these diseases and malaria. This has now been confirmed by extensive studies in several parts of the world.

In Papua, New Guinea, thalassaemia is prevalent near sea level where malaria is common, and rare among mountain tribes not exposed to it. It is also a common disease in malarial islands of Melanesia and rare in islands free from malaria. In regions of Africa where malaria kills a large proportion of infants, up to 40 percent of the native population carry the sickle-cell gene. What made it accumulate? When two carriers of the sickle-cell gene marry, half of their offspring are likely to be carriers, a quarter to have normal haemoglobin and a quarter to have the disease. For reasons that we do not really understand, infants carrying the sickle-cell gene are more resistant to malaria than normal infants and therefore stand a better chance of surviving to adult age. Was the sickle-cell mutation a unique chance event in one individual from which all the carriers are descended? Analyses of the fine structure of the gene in different African populations reveal that modern carriers of the gene have descended from three to four individuals, proving that the sickle-cell mutation must have occurred three or four times.

Thalassaemia also arises from a variety of different mutations. It looks as though the mutations causing either sickle-cell anaemia or thalassaemia arise spontaneously in human populations. In the absence of malaria, selective pressure penalises the carriers of the variant haemoglobins and they die out, but in its presence selective pressure favours them and they multiply. It would be absurd to suggest that the carriers of these diseases actively sought a malarial environment where their infants have a selective advantage. They represent a form of adaptation by natural selection which is wholly passive and deterministic, because there can be no escape from the laws of chance. G. Pontecorvo and John Maynard Smith point out that plants have evolved well even though the dispersal of seed and pollen is passive.

Popper has done a useful service to the theory of Darwinian evolution by drawing attention to the importance of the individual actively seeking better environments. But my examples convince me that this is only one facet of the theory. Darwinian evolution may be either active or passive or a mixture of both. I have also found that scientific advances are not made by any one single method. Some arise from following Popper's hypothetico-deductive one; others are the result of induction from observation that Newton prescribed. In practice, scientific advances often originate from observations, made either by accident or by design, without any hypothesis or paradigm in mind. The discovery of pulsars by Tony Hewish and his colleagues was accidental and came as a surprise. The idea that radio pulses might be emitted by rotating neutron stars arose afterwards.

A Passion for Crystals

In October 1964, the *Daily Mail* carried a headline "Grandmother wins Nobel Prize." Dorothy Hodgkin won it "for her determination by X-ray techniques of the structures of biologically important molecules."

She used a physical method first developed by W.L. Bragg, X-ray crystallography, to find the arrangements of the atoms in simple salts and minerals. She had the courage, skill, and sheer willpower to extend the method to compounds that were far more complex than anything attempted before. The most important of these were cholesterol, vitamin D, penicillin, and vitamin B_{12}. Later, she was most famous for her work on insulin, but this reached its climax only five years after she had won the prize.

In the early 1940s, when Howard Florey and Ernest Chain had isolated penicillin from Alexander Fleming's mould, some of the best chemists in Britain and the United States tried to find its chemical constitution. They were taken aback when a handsome young woman, using not chemistry but X-ray analysis, then still mistrusted as an upstart physical technique, had the face to tell them what it was. When Dorothy Hodgkin insisted that its core was a ring of three carbon atoms and a nitrogen which was believed to be too unstable to exist, one of the chemists, John Cornforth, exclaimed angrily, "If that's the formula of penicillin, I'll give up chemistry and grow mushrooms." Fortunately he swallowed his words and won the Chemistry Prize himself 30 years later. Hodgkin's formula proved right and was the starting-point for the synthesis of chemically modified penicillins that have saved many lives.

Pernicious anaemia used to be deadly until the early 1930s when it was discovered that it could be kept in check by liver extracts. In 1948, the active principle, vitamin B_{12}, was isolated from liver in crystalline form, and chemists began to wonder what its formula was. The first X-ray diffraction pictures showed that the vitamin contained over a thousand atoms, compared to penicillin's 39; it took Hodgkin and an army of helpers eight years to solve its structure. Like penicillin, vitamin B_{12} showed chemical features not encountered before, such as a strange ring of nitrogens and carbon atoms surrounding its central cobalt atom and a novel kind of bond from the cobalt atom to the carbon atoms of a sugar ring that provided the clue to the vitamin's biological function. The Nobel Prize was awarded to Hodgkin not just for determining the structures of several vitally important compounds, but also for extending the bounds of chemistry itself.

In 1935 Dorothy Crowfoot, as she then was, put a crystal of insulin in front of an X-ray beam and placed a photographic film behind it. That night, when she developed the film, she saw minute, regularly arranged spots forming a diffraction pattern that held out the prospect of solving insulin's structure. Later that night she wandered around the streets of Oxford, madly excited that she might be the first to determine the structure of a protein, but next morning she woke with a start: could she be sure that her crystals really were insulin rather than some trivial salt? She rushed back to the lab before breakfast. A simple spot test on a microscope slide showed that her crystals took up a stain characteristic for protein, which revived her hopes. She never imagined that it would take her 34 years to solve that complex structure, nor that once solved it would have practical application. It has recently enabled genetic engineers to change the chemistry of insulin in order to improve its benefits for diabetics.

Dorothy Crowfoot was born in Cairo in 1910. Her father, J.W. Crowfoot, was Education Officer In Khartoum and an archaeologist; her mother too was an archaeologist, with a particular interest in the history of weaving. When Dorothy was a child, they lived next door to the Sudan Government Chemist, Dr. A.F. Joseph. It was "Uncle Joseph's" early encouragement that excited her interest in science. Later he introduced her to the Cambridge Professor of Physical Chemistry, T. Martin Lowry, who advised her to work with J.D. Bernal.

When Dorothy Crowfoot was 24 and working in Cambridge with Bernal on crystals of another protein, the digestive enzyme pepsin, Bernal made his crucial discovery of their rich X-ray diffraction patterns. But, on the day that he did, her parents had taken her to London to consult a specialist about persistent pains in her hands. He diagnosed the onset of the rheumatoid arthritis that was to cripple her hands and feet, but never slowed her determined pursuit of science.

At Oxford, Dorothy Hodgkin used to labour on the structure of life in a crypt-like room tucked away in a corner of Ruskin's Cathedral of Science, the Oxford Museum. Her Gothic window was high above, as in a monk's cell, and beneath it was a gallery reachable only by a ladder. Up there she would mount her crystals for X-ray analysis, and descend precariously, clutching her treasure with one hand and balancing herself on the ladder with the other. For all its gloomy setting, Hodgkin's lab was a jolly place. As Chemistry Tutor at Somerville College, she always had girls doing crystal structures for their fourth year and two or three research students of either sex working for their Ph.D.s. They were a cheerful lot, not just because they were young, but because her gentle and affectionate guidance led most of them on to interesting results. Her best-known pupil, however, made her name in a career other than chemistry: Margaret Roberts, later Margaret Thatcher, worked as a fourth-year student on X-ray crystallography in Dorothy Hodgkin's laboratory.

In 1937, Dorothy had married the historian Thomas Hodgkin. Some women intellectuals regard their children as distracting impediments to their careers, but Dorothy radiated motherly warmth even while doing scientific work. Concentration came to her so easily that she could give all her attention to a child's chatter at one moment and switch to complex calculation the next.

She pursued her crystallographic studies, not for the sake of honours, but because this was what she liked to do. There was magic about her person. She had no enemies, not even among those whose scientific theories she demolished or whose political views she opposed. Just as her X-ray cameras bared the intrinsic beauty beneath the rough surface of things, so the warmth and gentleness of her approach to people uncovered in everyone, even the most hardened scientific crook, some hidden kernel of goodness. She was once asked in a BBC radio interview whether she felt handicapped in her career by being a woman. "As a matter fact," she replied gently, "men were always particularly nice and helpful to me *because* I was a woman." At scientific meetings she would seem lost in a dream, until she suddenly came out with some penetrating remark, usually made in a diffident tone of voice, and followed by a little laugh, as if wanting to excuse herself for having put everyone else to shame.

Dorothy Hodgkin's uncanny knack of solving difficult structures came from a combination of manual skill, mathematical ability, and profound knowledge of crystallography and chemistry. It often led her and her alone to recognise what the initially blurred maps emerging from X-ray analysis were trying to tell. She was a great chemist; a saintly, gentle, and tolerant lover of people; and a devoted protagonist of peace.[1]

The Top Designer[*]

We owe chemists and physicists our knowledge of the composition of living matter, of the conversion of the sun's heat into chemical energy, and of the myriad molecular interactions that sustain life. Steven Vogel's *Cats' Paws and Catapults* is the first book that has made me look at biology through the eyes of an engineer and compare the mechanics of animals and plants with the objects produced by man. At first sight this project looks unpromising. How can you compare organisms that consist predominantly of carbon with machines that are made of metals? Many of nature's engines are on the molecular scale, which means that they are about 100 million times smaller than a car engine. Man's artifacts are deliberately designed, while nature's structures have evolved blindly, haphazardly, over millions of years, by the reshuffling of genes, by mutation and natural selection of features that have led to more successful offspring. On the other hand, this very success requires living organisms to be constructed on sound engineering principles, which Vogel tries to explain.

He describes how nature has sometimes inspired man's engineering designs. Otto Lilienthal, the German engineer, who was the first man to have lifted himself off the ground, modeled his gliders' wings on a careful

*A review of the books *Cats' Paws and Catapults: Mechanical Worlds of Nature and People* by Steven Vogel (Norton); *Of Flies, Mice, and Men* by François Jacob, translated by Giselle Weiss (Harvard University Press).

study of storks' wings. The streamlined shapes of dolphins suggested the shapes of airplane bodies. Alexander Graham Bell, the inventor of the telephone, was not an electrical engineer, but a professor of vocal physiology at Boston University, where he taught deaf people how to speak. The anatomy of the ear led him to construct the first microphone. Our eardrums are thin membranes which transmit sound vibrations to tiny bones in the middle ear; those bones in turn set up vibrations in the liquid-filled canals of the inner ear which our auditory nerves sense by a mechanism still unknown. In Bell's own words:

> It occurred to me that if a membrane as thin as tissue paper could control the vibrations of bones that were, compared to it, of immense size and weight, why should not a larger and thicker membrane be able to vibrate a piece of iron in front of an electromagnet... and a simple piece of iron attached to a membrane be placed at the other end of the telegraphic circuit?

Bell replaced our eardrum by a thin metal plate which he attached to a magnetized iron rod surrounded by an independently fixed coil of copper wire. The metal plate transmitted sound vibrations to the iron rod, and its vibrations in turn induced vibrating electric currents in the copper coil. A wire transmitted these to a receiving coil which set up vibrations in an iron rod attached to a metal plate. This converted the electrical vibrations back into sound vibrations. So Bell made the analogy with the ear work both ways, as transmitter and receiver. (See Figure 1.)

The ancient Egyptians made paper from papyrus; later, paper was made first from linen and then from cotton rags. In 1719, when the supply of rags threatened to fall behind demand, the French entomologist

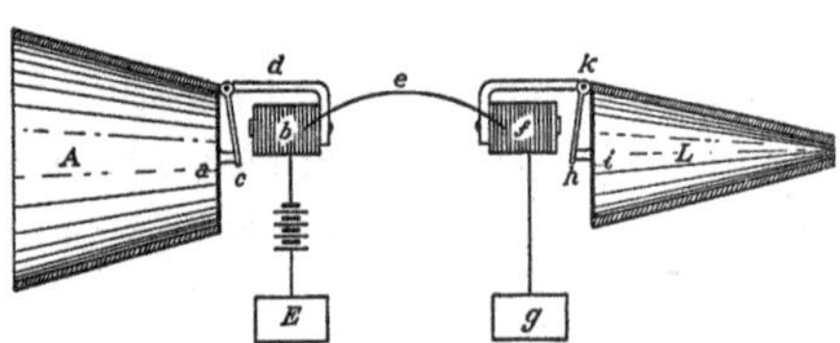

Figure 1. Bell's own diagram of his transmitter and receiver. The drum *A* is sealed by a thin metal plate at *a*; *b* and *f* are coils of thin wire surrounding a magnetized iron rod; *E* and *g* may be batteries; *i* is a thin metal plate sealed to the drum *L*; *c*, *d*, *k*, and *h* are the mechanical connections from the metal plates to the iron rods; *e* is a wire that transmits the signal from the transmitting coil to the receiving coil.

René-Antoine Réaumur noted that the American wasp (*poliste*) makes fine paper for its nests from wood, and he published an essay suggesting that this might be copied. In 1800 Réaumur's article inspired one Matthias Koops in London to make paper from straw and wood. He incorporated this recipe in a book on the history of papermaking, but he went bankrupt after building a large mill, and it took a few more decades before wood paper became cheaper than rag paper.

Velcro, Vogel writes, is another example of man copying from nature. A Swiss engineer, Georges de Mestral, saw that the burs that clung to his socks and his dog after a walk in the hills had tiny hooks at the tips of their bristles, which he managed to reproduce in nylon felt. Such examples recall Descartes's view:

> The only difference I can see between machines and natural objects is that the workings of machines are mostly carried out by apparatus large enough to be readily perceptible to the senses (as is required to make their manual activity humanly possible), whereas natural processes almost always depend on parts so small that they virtually elude our senses.

Science has since sharpened our senses so that we can now see how these small parts are constructed and how they work. Nature's engines are protein molecules, complex assemblies of thousands of atoms of carbon, oxygen, nitrogen, hydrogen, and sulphur; they are a thousand times smaller than the smallest object distinguishable in a light microscope, but electron microscopes make them visible, and another physical technique, X-ray crystallography, allows us to find the arrangement of the individual atoms inside them. These methods have shown that muscles contract by the interaction of two kinds of straight, long, rigid protein fibers called myosin and actin. Myosin has extensions that resemble oars all around and along it. After each forward stroke, the oars hook on to the nearest actin fibers; at the backward stroke they propel the actin fiber relative to the myosin; so muscles work like long galleys propelled by oars all around them. (See Figure 2.) The power output of our muscles is weight for weight as good as that of electric motors and half as good as that of automobile engines, but aircraft turbines do thirty times better. In efficiency, muscles are comparable to gas turbines, piston engines, or electric motors. They all convert about 20 percent of the consumed energy into mechanical work. In all living organisms that energy comes from the universal biological fuel adenosine triphosphate, ATP for short, which is made by a remarkable engine described after Vogel had written his book.[1]

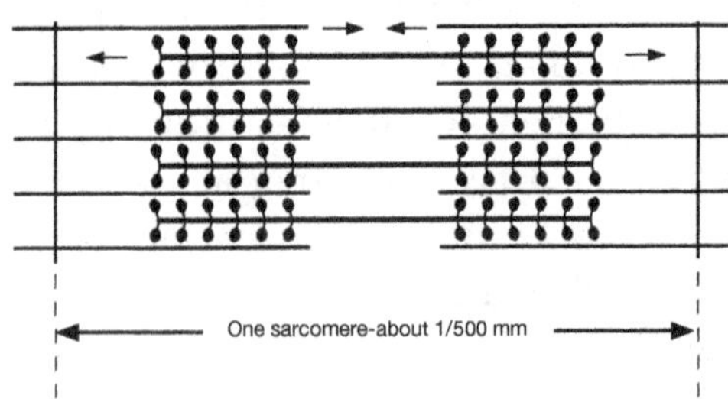

Figure 2. Diagrammatic sketch of a thin section of one segment of a muscle fiber. The thin horizontal lines are the actin filaments and the thick lines with the oars attached are the myosin filaments. The vertical lines are membranes dividing one segment from the next. The segments are called sarcomeres, and about ten thousand of them are arranged end-to-end in an inch of muscle. When a muscle contracts, the rowing of the oars makes the thin filaments move to the middle so that the gap between them closes up and the sarcomere shortens.

In 1997 my colleague John Walker received the Nobel Prize for chemistry for unraveling the atomic structure and mechanism of that engine, which turned out to be a molecular turbine. A central protein shaft forms its axis; the shaft is surrounded by six protein molecules that catalyze the synthesis of ATP in three chemical steps. (See Figure 3.) The turbine spins

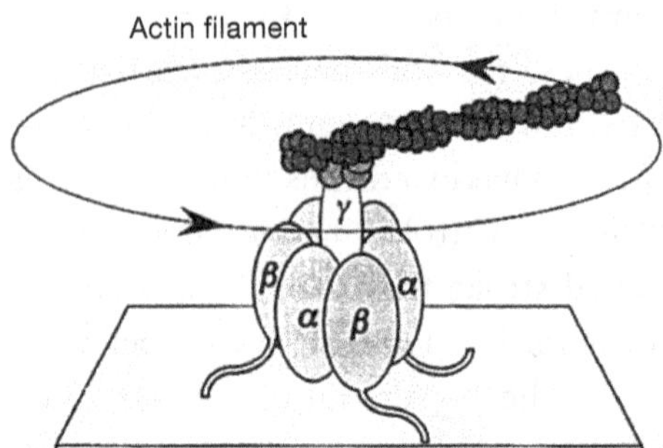

Figure 3. Molecular turbine which manufactures ATP in three successive chemical steps; α, β, and γ are protein molecules, each containing over fifty thousand atoms; γ is the central shaft around which α and β turn. Its width is about a two and a half millionth of an inch. M. Yoshida at the University of Tokyo attached an actin filament to the shaft, and linked a fluorescent dye to it which made the rotation visible under a microscope. (Reproduced, by permission, from H. Nojic, R. Yasuda, M. Yoshida, and K. Kinosita, Jr., "Direct observation of the rotation of F1-ATPase," *Nature*, Vol. 386 (1997), pp. 299–302.

at the rate of about 150 revolutions a second, churning out one molecule of ATP at each turn. It is driven by a protein dynamo which is fueled by the burning of sugars and whose structure has recently been elucidated by the research of Daniella Stock at the Medical Research Council Laboratory in Cambridge. As far as we know, the same engine is to be found in all living organisms from bacteria to man. It must therefore have arisen very early in evolution.

In the 1970s, biologists discovered that some bacteria also have turbines driving lashlike appendages called "flagella" that act as propellers. (See Figure 4.) Close to the turbines are sensors, molecular noses that smell food. In the absence of food the propellers idle, but when the sensors smell food, they drive the bacteria toward it. We still don't understand how the turbines work or how they are switched on and off. They may be driven like the ones that drive the ATP factory.

Nature has evolved molecular turbines, but no wheels; nor did primitive societies have them, because wheels need roads. Cart wheels cannot go over bumps higher than a quarter of their diameter; they also sink in sand.

Vogel writes that the ancient Egyptians transported the stone blocks for their pyramids, each weighing 2.5 tons, by rolling them along on logs, and that there are Egyptian and Assyrian illustrations showing huge statues mounted on sleds being dragged along by large numbers of men. But the Australian-born Cambridge engineer R.H.G. Parry has found evidence that they used a more efficient method. The great pyramid of Cheops contains some 2.3 million limestone blocks weighing on average 2.5 tons; some weigh up to 6.7 tons. It is believed that the pyramid was built in only twenty years; Parry calculates that the stones would have to have been placed at an average rate of one every two minutes during all the daylight hours of that period. He argues that all methods hitherto proposed would have been too slow to account for the rapid transport of the stones from the distant

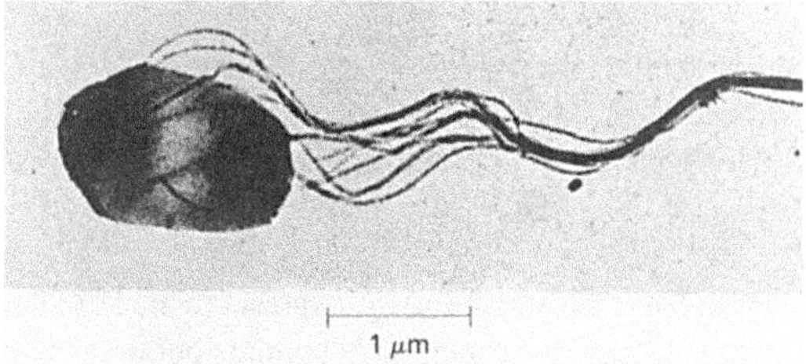

Figure 4. Bacterium with wavy flagella. When the bacterium senses food, the flagella spin around their long axes and drive it forward.

quarries to the site, and it would have been too hazardous to lift them into position, especially the massive granite blocks of the King's chamber.

Small wooden cradlelike objects found in the New Kingdom Foundation Deposits gave Parry the idea that these resembled the huge cradles used to envelop the large blocks of stone and so turn the cradled stones into wheels that could be rolled along the ground. One such cradle would have been fitted to each of the four sides of a stone block to form circular runners. Parry worked out that the sliding of a 2.5-ton stone along a flat surface would take about twenty-five men, while pushing it along on circular runners would take no more than two men, and that rolling it up the slope of the pyramid needed a force of only one quarter of that required to slide it up. (See Figure 5.) Such a method, Parry suggests, is more likely to have been used by a highly intelligent and sophisticated people than the primitive ones proposed earlier.

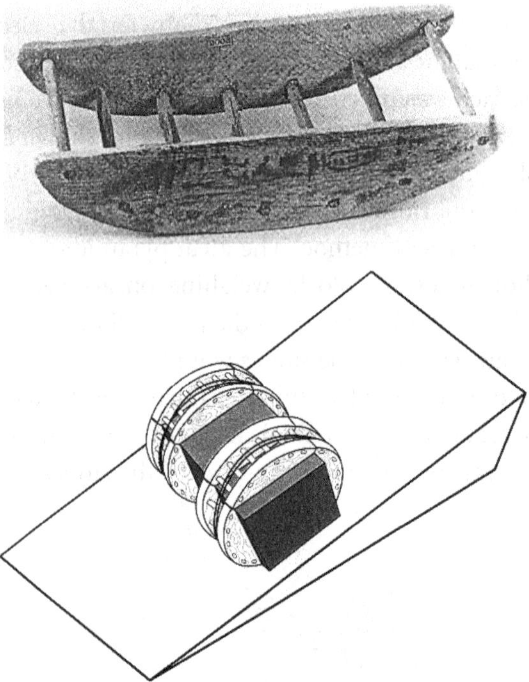

Figure 5. At top, small wooden cradle which Parry interprets as a model of large ones fitted to the blocks of stone for wheeling them to the pyramids. At bottom, four wooden cradles tied to a large block of stone to turn it into a wheel.

Nature makes no wheels, but it does make bearings which are as smooth as those of the best engines. At our hips, knees, and elbows, bones are covered with cartilage and form ball and socket joints; they are lubricated by synovial fluid which allows them to slide over each other with almost no friction.

Nature has also evolved molecular syringes. Just as small fleas proverbially are said to have even smaller fleas upon their backs to bite them, *coli* bacteria are bitten by viruses, which were discovered early in this century by the French physician F. d'Herelle, who called them bacteriophages (*phage* is Greek for "glutton"). D'Herelle's phage consists of a protein cylinder that acts as a container for the long chain of DNA that carries the genes, about 250 of them, needed to specify this simplest of organisms. Attached to the cylinder is a tail with long fibers at its tip. The phage attaches itself to its victim with these fibers; the tail then contracts and injects the DNA into the bacterium. The DNA replicates itself, and its replicas direct the synthesis and assembly of more phage proteins. Twenty minutes later the dead bacterium bursts with the release of up to two hundred progeny phages. (See Figure 6.) Such phages are among the fastest reproducing organisms we know of. D'Herelle hoped that his phages would become effective weapons against bacterial infections, but his hopes were disappointed.

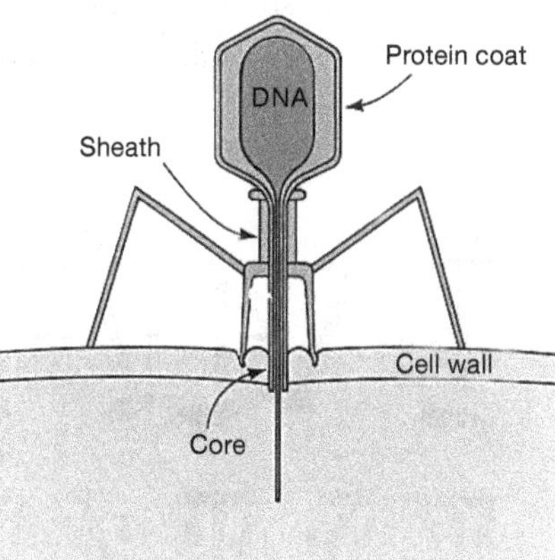

Figure 6. Diagrammatic sketch of a bacterial virus, or bacteriophage, about 200,000 times enlarged, attacking a bacterium. The head of the phage harbors its DNA, which contains the genes. It is encapsulated in the protein shell. When the DNA is injected into the bacterium, the shell remains behind, dead like the bee after its sting.

Living creatures walk, run, jump, swim, or fly; much of Vogel's book turns on the engineering principles of their motions. Muscle has little elasticity; once contracted, it does not reexpand unless stretched by another muscle or by an elastic element coupled to it, or by gravity. Vogel explains that

> we cheapen our walking gait with a little gravitational, pendulum-like, storage [of energy] between strides. When we walk faster than the pace set by the period of our legs as pendulums, they are switched to jogging, for a middle-sized human at about five miles an hour. [See Figure 7.]

When a creature runs or jumps, energy is stored in stretched tendons, which consist of protein chains; they stretch elastically by no more than 10 percent of their length, but that is enough to assist a kangaroo's jump. On landing from one jump, its tendons are stretched and help lift it on to the next one. About 40 percent of the stored energy assists its lift and the rest is lost as heat. This seems wasteful, but most man-made engines waste more energy. Fleas store elastic energy by compressing an elastic pad made of a highly resilient protein. Release of the pressure launches them on a leap of several hundred times their own length. They may be the catapults referred to in the title of Vogel's book.

The Russian physicist Peter Kapitsa set his students the examination question "How fast would Jesus have had to walk over the waves in order not to sink?" Using the soles of his feet as surfboards, he would have needed prodigious speed to make their small area provide enough lift for his weight. An aquatic bug weighing ten million times less can do it easily

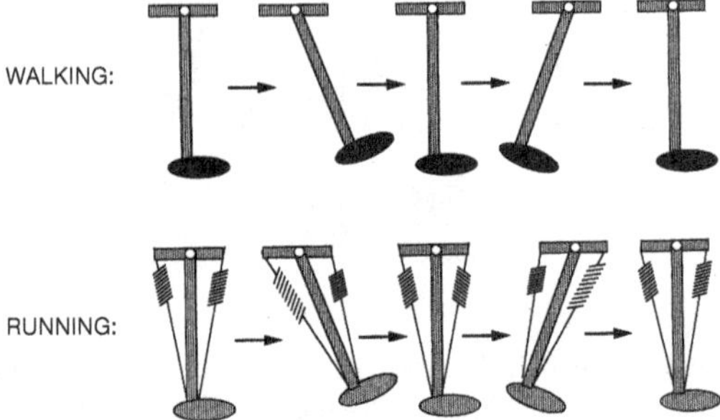

Figure 7. The essential difference between walking and running. In walking, what matters is the weight of the legs, since energy is stored between strides as in the swinging of a pendulum clock. In running, the elastic stretch of the tendons first absorbs and then releases energy.

because water is covered by a skin of water molecules which stick to each other firmly enough to support it on its six wax-coated feet.

Marine animals and birds, penguins for example, swim under water rather than on it, surfacing in order to breathe. Vogel explains that the movement of an underwater swimmer is opposed only by the drag of the water adhering to its body; drag can be minimized by streamlining the body as in a whale or a dolphin. On the other hand, a surface swimmer's movement is opposed also by the waves he generates. Their wavelength equals the length of his body, while their speed rises with the square root of the wavelength, doubling with a fourfold increase in length. When a swimmer or ship tries to go faster than its waves, it has either to cut through them or to go perpetually uphill. A ship a hundred feet long reaches that critical speed at about fifteen miles an hour, but for a duck it's only one and a half miles an hour, twenty times slower than its speed in air. It can swim much faster under water.

The famous geneticist J.B.S. Haldane was a large, burly man with a small moustache. Educated at Eton and dressed in tweeds, he might have been taken for a pillar of the establishment, but he was an active member of the Communist Party and wrote a weekly science column for the *Daily Worker*. In 1927, Vogel recalls, he wrote "On Being the Right Size," a playful essay on the ratio between an animal's height and its weight. Gulliver's giants of Brobdignag, who were a hundred feet tall, would have weighed 280 tons, or 4,600 times more than Gulliver, but since their bones would have been only three hundred times thicker they would have crumbled under their own weight. Gulliver's Lilliputians, on the other hand, were only six inches tall. They would have profited from being able to fall a long way without hurting themselves, because the air's drag would have slowed their fall more than that of a full-sized man. The drag that slows a falling object is proportional to its surface area. The ratio of a Lilliputian's surface area to his weight would have been sixteen times greater than that of a full-sized man, and this would have slowed his fall. An even greater ratio of surface area to weight lets mice fall down mine shafts and run away unharmed.

Vogel points also to a handicap that would have afflicted Lilliputians, but had escaped Haldane's notice: the heat lost by animals is also proportional to the ratio of their surface area to their weight, so that they would have lost heat sixteen times faster than Gulliver and might have frozen to death in cold weather. This is why babies have to be kept warm. The smallest warm-blooded animals, shrews and hummingbirds, have to eat almost continuously just to keep warm. To reduce their loss of heat during the night, they lower their body temperature as if they were hibernating.

Vogel extols the advantages of being curved. Bookshelves sag because they are flat, but sagging can be reduced by making them curved, as well as by beams or struts. Many kinds of leaves are stiffened by being just a little curved, but that makes them catch the wind, so that a gale can fell a tree. Vogel found a defense. To measure air drag on single leaves, he suspended them in a highly turbulent wind tunnel at speeds they might meet in a storm. As this storm caught the leaves from a maple or a tulip poplar, they coiled up into cones that minimized air drag. (See Figure 8.)

Vogel's chapter on the stress put on objects—through compression and tension—dwells on cathedrals and bridges, but says too little about the impressive construction of bones, including teeth. Gothic cathedrals were designed to resist compression: "Vaulted roofs press outward on the walls, and exterior buttresses press inward, with the two in very close balance." On the other hand, Filippo Brunelleschi encircled the great dome of the cathedral of Santa Maria del Fiore in Florence with massive iron chains to balance the tension on its periphery. For material that resists both compression and tension, engineers use composites, like concrete reinforced with steel rods. In nature, wood is a composite of cellulose fibers and the glue lignin, and bones are composites of protein fibers made of collagen and a concretelike mineral, calcium phosphate. A rat's incisor teeth are made up of that hard mineral arranged in layers of mutually perpendicular rods, like crossed beams in a ceiling, that give the teeth great strength

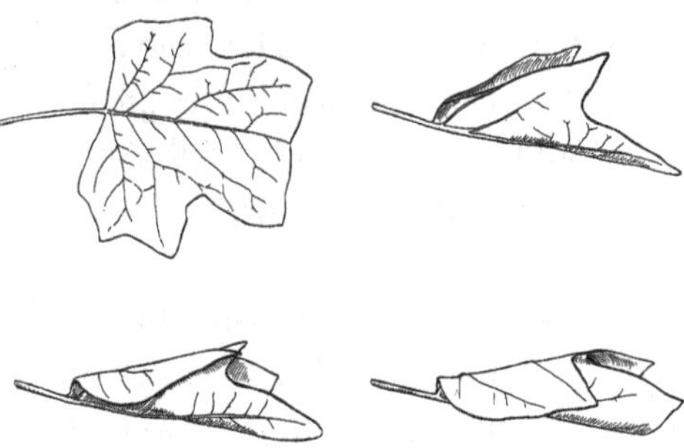

Figure 8. The leaf of the tulip poplar, *Liriodendron*, in still air and in winds of eleven, thirty-three, and forty-four miles per hour.

and resistance to wear. (See Figure 9.) Vogel, I was disappointed to find, gives no statistics to compare the strength of wooden beams and iron girders, or of bones and steel tubes, or of tendons and nylon strings, matters that engineers must think about when designing artificial limbs.

Vogel mentions two kinds of levers: amplifiers of force used by humans and amplifiers of distance used in nature. Nutcrackers increase the force of one's hand by having a long arm from the hand to its pivot, and a short one from the pivot to the nut, while insect wings are anchored so as to transmit force over a long distance from their pivots. Some insect wings are coupled to elastic elements whose vibrations make them swing up and down more than a thousand times per second. Vogel's book is filled with many other examples of nature's engineering designs. Its only continuous theme is nature's ingenuity, which he demonstrates with unbounded enthusiasm and without any scientific jargon.

Maintenance of mechanical devices includes the replacement of parts. Replacements of hips and knees have now become routine, but most of us live with the organs and joints we were born with, because the cells that make them up are being replaced continuously by freshly synthesized ones. This is vital, because many of our proteins are unstable, i.e., they decompose into small compounds which may be either recycled or excret-

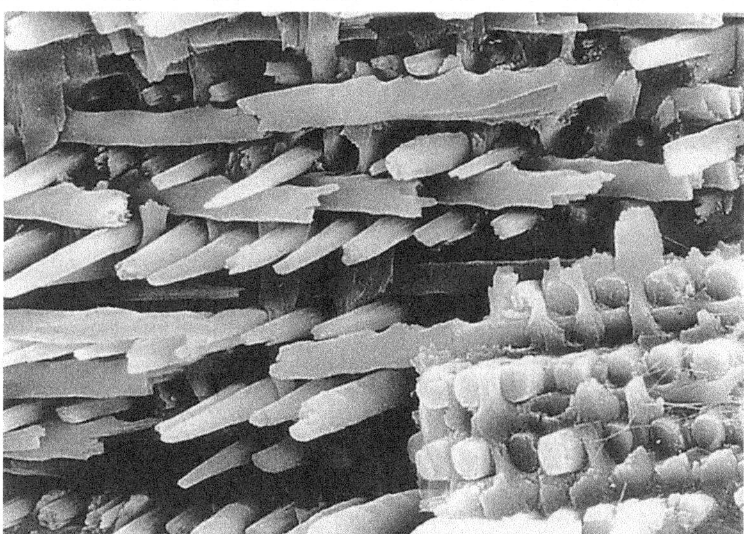

Figure 9. A rat's incisor tooth viewed under an electron microscope enlarged 360,000 times.

ed. Then why are we not immortal? Partly because our maintenance system omits our hearts and brains; partly because that system is error-prone and the errors may accumulate with time. Our life span may also be genetically programmed by structures of DNA attached to the ends of our chromosomes and other devices yet unknown.

François Jacob made his scientific name by discovering how genes are switched on and off in bacteria, and became known as a writer for his moving autobiography *The Statue Within*. In his latest book, *Of Flies, Mice, and Men*, he asks what decides whether a fertilized egg develops into a mouse, a fly, or a human. He writes:

> What is...wonderful about the appearance of a new human being is not the nature of the receptacle in which the first stage takes place. It would not even be the accomplishment of making the entire development take place in a test tube. The incredible thing is the process itself. It is that the meeting of the sperm with the egg initiates a gigantic set of chemical reactions, hundreds of thousands of which follow each other, overlap and cross each other in an orderly network of unbelievable complexity. All this to result... in the appearance of a human baby and never a little duck, a little giraffe or a little butterfly.[2]

How can we penetrate that complexity and discover what decides the differences between the development of different species? The bodies of all organisms are made largely of proteins. Protein molecules form the machinery that makes them alive, and the blueprints for them reside in the genes, of which humans possess about 80,000. One might think that the differences among species would be reflected in differences among the structures of these proteins, but this is not so. The structures of protein molecules with similar functions in distantly related species are so closely alike that they could not possibly determine the macroscopic differences among these species. Some species contain proteins that others don't have, and higher organisms have more genes than lower ones and therefore a greater variety of proteins with different functions. Even so, the variety of species conceals an astonishing unity in the makeup of their molecules.

The first clue to the factors that determine the development of an organism came from studies of the geneticists' pet, the humble fruit fly. There are flies that grow legs on their heads in the place where they should have antennae. The mutant gene responsible for this monstrosity lies on one of the fly's chromosomes, and it belongs to a family of genes that determine the fly's body plan. Are these genes unique to the fly, or do similar genes determine the development of a human embryo? The answer to that question has had to wait for the invention of recombinant DNA technology, which has made it possible to isolate, copy, and amplify genes from a fly and to introduce them into a mouse or vice versa. François Jacob describes a group of genes that determine the sequence of embryonic developments along an axis from the front to the rear of the fly's larva. He writes:

> There was hardly a chance of finding these genes in organisms other than insects, seeing how different their embryonic developments are. But people looked for them all the same. Just to see. They were stunned. They found them. Everywhere. First in a frog, then a mouse, then in man, in a leech, in a worm.... In short, one finds a group of genes very similar to those of the fly in all animals. Everywhere, their role seems to be the same: to define the identity of different cells along the axis from the front to the rear of the animal. If one takes a mutant fly which lacks one of these genes and inserts in its place the homologous gene from a mouse, it works, and it fulfils the same function as the normal fly gene.

Insect eyes differ fundamentally from animal eyes: the fly focuses light over a wide angle through hundreds of separate facets, while animals focus light over a narrower angle through a single lens. One would therefore have expected the development of their eyes to be controlled by different genes. To everyone's astonishment this has proved untrue. Certain fruit flies fail to develop eyes, a defect that has been traced to mutations in either of only two genes. When geneticists added the homologous mouse gene to the fly, the fly developed an additional fly's eye and not a mouse eye. (See Figure 10.)

Similarly, a mouse which had its own gene controlling development of the eye replaced by the homologous fly gene developed a normal mouse eye. Some stillborn human babies have no eyes. Their defect was found to be due to mutations of the same genes that govern the growth of the flies' eyes.

This discovery raises a question that vexed Charles Darwin. How could an organ as subtle and complex as the eye have arisen by evolution and natural selection, rather than been designed and brought into being by an omniscient creator? The biologist Ernst Meyr believed he had at least a partial answer when it seemed as though eyes had evolved independently in about forty different species; but if the same gene initiates the development of eyes in all species, then this suggests that they have all developed from a single light-sensitive cell which arose early in evolution.

This discovery has not answered Darwin's question, but only deepened the mystery. If the development of eyes is initiated by the same gene

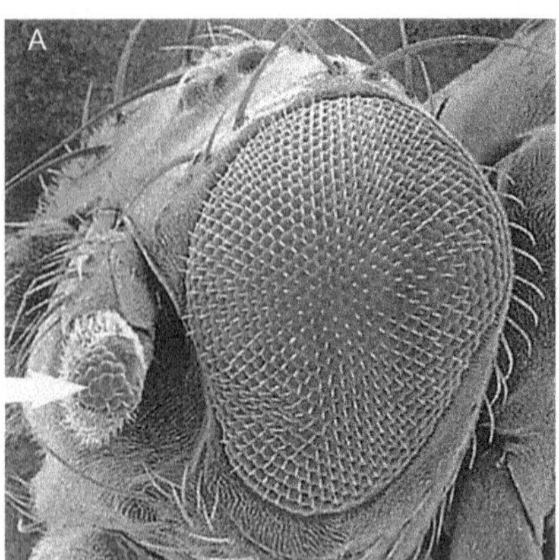

Figure 10. (*A*) The large eye on the right is the natural one of the fruit fly Drosophila. The growth of the small eye at left (magnified in *B*) was induced by addition of the gene for a mouse eye, but its structure is that of a fruit fly's eye. (Reproduced, by permission, from Walter Gehring, Genes to Cells 1, 11–15, Blackwell Science Ltd., 1996.)

in humans and flies, then why are they so different? The Swiss biologist Walter Gehring suggests that in human beings the single gene switches on a cascade of as many as 2,500 other genes. They would code for 2,500 different proteins whose complex interplay would then govern the growth of the eye. Some of these proteins might be common to human beings and flies and others different. We know as yet next to nothing about them.

Will we ever be able to unravel these genes' labyrinthine workings? In the concluding chapter of his book, Jacob asks whether there are limits to scientific knowledge. For instance, research may slow down because it has become unmanageably elaborate and extensive, "like a building that cannot rise to infinite heights." Or else there might be a limit to our understanding, which could be "like a net that can catch only fish larger than its holes, or a microscope that cannot resolve details smaller than the resolving power of its lenses." There might indeed be a limit to the degree of complexity that we can comprehend, such as the interactions between thousands of genes or between the billions of neurons in our brains. Jacob fears that "the human brain may be incapable of understanding the human brain." I share his fears.

Jacob's thesis, repeated in this book, is that much of evolution has arisen from Nature's tinkering. Just as mechanics put new cars together by tinkering with bits and pieces from several old ones, the processes of Nature make new genes which code for proteins with new functions by putting together bits and pieces from several existing genes in new ways, or simply by replacing bits and pieces in existing genes. How are these processes initiated in one case and not in another? Here Jacob writes:

> The whole of the living world looks like some kind of giant Erector set. Pieces can be taken apart and put together again in different ways to produce different forms. But fundamentally the same pieces are always retained.

For example, the "expression of genes" —i.e., their production of proteins—is switched on by proteins known as transcription factors. There is a specific combination of transcription factors for each gene. It combines with the gene on receipt of a specific signal, say a hormone like insulin. There are hundreds, if not thousands, of different transcription factors, all similarly constructed with just a few bits and pieces in different places to make them respond to different chemical signals and then combine with different genes.

New combinations of parental genes form the basis of inherited individuality. "Each of us," Jacob writes, "is different from all other human beings who ever lived, now live, or will live on earth." Unless people are mad enough to have themselves cloned.

Jacob has scathing criticisms of eugenic proposals to use "frozen sperm from carefully selected donors." He writes that some people are even excited by the idea of fertilizing human eggs with the sperm of Nobel laureates (since they probably don't know any Nobel laureates). But how should we select for complex traits directed by many genes about which we know nothing? Which genes are we to consider the best? Or which genes would we wish to eliminate? Each of us contains a mosaic of good and bad ones. Abraham Lincoln is thought to have suffered from an inherited bone disease, Dostoyevsky was an epileptic, Virginia Woolf a manic depressive. Had eugenics been used to eliminate human fetuses with deleterious genes, none of the three would have lived.

Knowledge of the human genome, and mapping of the genes that make us susceptible to a variety of diseases, will mean, Jacob predicts, that "people will become patients before their time. Their condition, their future will be discussed in medical terms, even though they feel fine and will remain in good health for years...." And "whether means of treatment exist or not, potential disorders will announce themselves in future as never before." I would add that they will do so in rich societies that can afford to spend large sums on genetic screening—that is, among a small minority of mankind. The health of most people on earth is still threatened primarily by parasitic and infectious diseases and by malnutrition, whose conquest presents a greater challenge to medicine and society.

Part of Jacob's book is about his life as a scientist and about science in general. "The history of science,"he writes, "is the history of the battle of reason against revealed truth" (in which I would include political ideologies such as Marxism). "Pandora introduced a fundamental ambiguity into the world. From now on every good would be twinned with its evil counterpart, every light with its shadows...."

Discussing these shadows, Jacob writes that "no one would have expected that the speed and growth" of medicine and public health since the end of the last century "would lead to overpopulation, which poses one of the greatest threats to humanity." It now looks as though AIDS may become the greatest threat instead. It is well on the way to decimating India; in parts of Africa as many as 25 percent of the people are already affected, and eventually it may almost wipe out entire populations.

Like its predecessors, Jacob's book is masterly in combining erudition, wit, and wisdom. It is marvelously clear in describing what we know about the fundamental questions of life and the laws that determine the growth of each species—and what we don't know.

The Great Sage[*]

In 1936, after four years of chemistry at Vienna University, I took the train to Cambridge to seek out the Great Sage, and asked him: 'How can I solve the riddle of life?' 'The riddle of life is in the structure of proteins,' he replied, 'and it can be solved only by X-ray crystallography.' The Great Sage was John Desmond Bernal, a flamboyant Irishman with a mane of fair hair, crumpled flannel trousers and a tweed jacket. We called him Sage, because he knew everything, from physics to the history of art. Knowledge poured from him as from a fountain, unselfconsciously, vividly, without showing off, on any subject under the sun. His enthusiasm for science was unbounded.

He became my PhD supervisor. I could not have wished for a more inspiring one. Yet he was erratic, his desk was chaotic, he was always travelling hither and thither, which left him no time to finish things, and in the middle of one of his discourses he would suddenly look at his watch, remember that he had something more important to do and rush off. His lectures were spontaneous, fluent, wide-ranging, inspiring for the cognoscenti but difficult for undergraduates. When one of them left his notebook behind, we found it empty, with the single heading: 'Bernal's Bloody Business.'

*A review of the book *J.D. Bernal: A Life in Science and Politics*, edited by Brenda Swann and Francis Aprahamian (Verso).

Now 28 years after his death, a book about him has appeared. It is made up of 12 essays by different authors plus his D-Day diaries. Bernal never talked about himself, only about the world around him. Much of what I have now read about the early experiences that shaped his character and outlook made me wish that I had known it when he was still alive. It would have helped me understand him better.

Bernal's father was an Irish Catholic gentleman farmer married to an American woman of wide education and interests. She brought Desmond up bilingually in English and French, 'the language of gentleness,' he later remembered. In his early teens, the contrast between wealthy English Protestant landowners and the abjectly poor Irish Catholic labourers turned Desmond into a revolutionary who vowed to study science in order to make war on the English and drive them out of Ireland.

The young Bernal's suspicious expression on the dustcover puzzled me until I read of his experiences as a boarder at Bedford School. 'There I could be tortured, humiliated, waste my time and my interests on dully repetitive games and military drill...I lived like a hostage in an enemy land. My companions were cheerful thieves and liars, and furtive sexual perverts. I merely thought they were English and kept my hatred of the race to myself.' But he did make two friends, one of whom wrote home: 'He is the cleverest chap in the school...not a bit conceited...a simply topping chap.' He left school 'amazingly ignorant of the world.' He wrote: 'Cambridge was a liberation. All the richness of thoughts was open to me...I could meet for the first time intelligent people and be accepted by them...I read and talked violently, discursively.'

Halfway through his first term an exhilarating all-night discussion with a friend transformed Bernal's life; it was like the conversion of St Paul. 'This socialism was a marvellous thing...the theory of Marxism, the great Russian experiment, what we could do here and now, it was all so clear, so compelling, so universal...It would bring the Scientific World State.' That state's creation remained Bernal's Holy Grail for the rest of his life. His Catholic faith peeled away gradually: 'First God, then Jesus, then the Virgin Mary and lastly the rites of the Church, not for any scientific reason, because I had long before reconciled the phenomenal world of science with the transcendental, symbolic world of revelation.'

Instead of Catholicism, he now fell for Freudianism. Within two days of reading Freud's *Interpretation of Dreams* he was expounding the new religion. I read that book at about the same age 13 years later; it struck me as a far-fetched piece of fiction masquerading as a work of science. But Bernal was, and remained, uncritical.

Soon afterwards he joined the still embryonic Communist Party and remained faithful to it ever after. He often travelled to the Soviet Union, where he was received with honours and was later awarded the Order of Lenin. I was shocked to find him photographed here with Lysenko, that pseudoscientific crook who gained the ear of Stalin for his Lamarckian belief in the inheritance of acquired characters and was responsible for the death in prison of Vavilov, the greatest of Russian geneticists, for the persecution of many other geneticists and for the decline of Russian biology over an entire generation.

In 1923, after taking a degree in physics, Bernal went to the Royal Institution in London to join W.H. Bragg, who had founded the first research school in X-ray crystallography, a method of determining the atomic positions in crystals which his son, W.L. Bragg, had introduced in Cambridge in 1913. The next year, Bernal published a paper on the positions of the carbon atoms in graphite, and later another on a simple method of interpreting X-ray diffraction pictures that was widely adopted. In 1927, he applied for a lectureship in structural crystallography at Cambridge, to run the sub-department in crystallography at the Cavendish Laboratory, which was then headed by Lord Rutherford, the discoverer of the atomic nucleus. C.P. Snow described his job interview:

> Bernal had come into the room and sat down. His head was sunk into his chest...He sat there, looking sullen...They couldn't get anything else out of him. Finally, in despair, Professor Hutchinson, who was in the chair, asked him what he would do...if he were given the job. At which Bernal threw his head back, hair streaming, like an oriflamme...and gave an address, eloquent, passionate, masterly, prophetic, which lasted 45 minutes. There was nothing left for it but to elect him.

At Cambridge, Bernal determined the shape of the molecule of cholesterol, showed that the chemists' formula must be wrong and helped them to correct it, and discovered the rich X-ray diffraction patterns of crystals of proteins and viruses, which opened up a new field of research. He also wrote his monumental book on *The Social Function of Science*, and co-authored a fundamental paper on the structure of water. In 1939, the outbreak of war found him as professor of physics at Birkbeck College in London.

Bernal had faith in the efficacy of committees. When I was his graduate student, he was reputed to be on 60 of them, mostly devoted to the socialist cause. In the first year of the war he worked on committees persuading the authorities to make better use of science for the war effort. Then in 1942, Sir John Anderson, the minister responsible for civil defence, commissioned Bernal and the anatomist Solly Zuckerman to report on the physical damage, casualties, effects on absenteeism and morale caused by the German bombing of two typical cities, with a view to assessing the likely effects of British bombing raids on Germany. Anderson did this at the request of Lord Cherwell, Churchill's scientific advisor. Bernal and Zuckerman chose the cities of Birmingham and Hull. They found no breakdown of morale; any loss of production was almost certainly due to damage done to factories rather than private houses; indirect effects on labour turnover or the health and efficiency of workers were insignificant. Cherwell jumped the gun, however. A week before he received Bernal and Zuckerman's report, he sent a minute to Churchill claiming that the effects of bombing British cities had confirmed his view that such raids could bomb Germany into submission. Other scientists disputed this, but Cherwell won the day. His policy caused the wasted deaths of tens of thousands of RAF men and German civilians, the misuse of vast resources and the senseless destruction of much of Germany's cultural heritage. After the war, analyses of the effects of Allied bombing on German war production confirmed Bernal and Zuckerman's findings: it had continued to rise despite the bombing.

In the same year, Lord Mountbatten, then Chief of Combined Operations, took Bernal and Zuckerman onto his staff to help with the planning of D-Day landings in Normandy. Bernal developed methods of determining the gradients of the beaches from aerial photographs of the breakers, and their consistency from geological data and his own observations made during holidays in Normandy. He worked closely with the military planners and was invited to join the troops landing there on D-Day in June 1944. His vivid diaries of the landings are a highpoint of the book. The day before, Bernal went to the military HQ:

> We lay on the grass and talked, estimating prospects of success or failure on the different beaches...I would not really influence things that were happening there...but I felt that I had to be there, and that I should never forgive myself if it did happen and I had not been...
>
> Already the whole expedition was on the sea. The wireless man came in. He was most astonished. 'They are well within range and they have not spotted us yet.' Everything very quiet and peaceful and on edge.

The next morning, D-Day, 'it seemed to be going so much better than we had hoped,' and Bernal crossed the Channel. As the boat approached Normandy, 'bombs began to fall...I found to my surprise that my teeth were chattering.' The next day he went ashore, and 'felt very irritated at the stupid slipshod construction' of German fortifications. The day after that they sailed along the coast: 'The battle of Bernières seemed to be over but we were told the town was full of snipers. I badly wanted to see the church but...had to admit that I would feel a fool if I were shot for the love of Norman architecture.' On the wall of a house in Courseulles he saw 'an eagle grasping a swastika-covered world in the middle, on one side "Unser Glaube ist der Sieg"' — 'Our faith is victory.' That evening he 'had to go back. I did not want to: I had got attached to the place and its strange inconsequent life, its near-peacefulness and distant dangers. I was taken back by the same torpedo boat that had taken me over...I fell asleep standing.'

After Bernal's death, Mountbatten wrote an appreciation of his war work, concluding:

> Desmond Bernal was one of the most engaging personalities I have ever known. I became really fond of him, and enjoyed my discussions and arguments immensely. He had a very clear analytical brain; he was tireless and outspoken. But perhaps his most pleasant quality was his generosity. He never minded slaving away on other people's ideas, helping to decide what could or could not be done, without himself being the originator of any of the major ideas on which he actually worked. This may be the reason why his great contribution to the war effort has not been properly appreciated, but those of us who really knew what he did have an unbounded admiration for his contribution to our winning the war.

After the war, Bernal had a hard time getting research going again at the heavily damaged Birkbeck College, but in 1953 things looked up with the coming of Rosalind Franklin, a skilful and patient experimenter, and Aaron Klug, an outstanding theoretician; they continued Bernal's pre-war research on the structure of viruses. As before, much of his own time went on political work and travels for the Cause.

In 1963, he suffered the first of several strokes which finally incapacitated him. When I visited him in London in 1967 to show him my three-dimensional map of haemoglobin, the climax of the work I had started under his supervision thirty years earlier, it gave him enormous pleasure without any sign of envy that I had solved the great puzzle of protein structure that he had set himself. He died in 1971.

People often ask me why Bernal did not win a Nobel Prize. The answer is that he started many things, but lacked the single-mindedness and patience to carry them through. He wasted his great talents on futile committees and travels. He was a prolific source of ideas and gave them away generously. During the long, lean years when most of my colleagues thought that I was wasting my time on an insoluble problem, Bernal would drop in like the advent of spring, imbuing me with enthusiasm and fresh hope. A Danish colleague once said to me: 'When Bernal comes to visit me, he makes me feel that my research is really worthwhile.' I loved him.

It Ain't Necessarily So

The Harvard University geneticist Richard Lewontin's book consists of review essays originally published in the *New York Review of Books* between 1981 and 1993, with epilogues to bring them up to date or to answer his critics. They range from developmental biology to intelligence tests, human sexual behaviour, cloning and the human genome. They are erudite, readable and they debunk a great deal of hype. They are also full of factual errors, prejudiced misjudgements and vital omissions.

As an early example of media hype, Lewontin quotes newspaper reports in 1899 that the biologist Jacques Loeb had created life and explained the Virgin Birth of Mary when he induced unfertilised sea urchin eggs to divide and develop. However, Lewontin fails to mention one important reason why the hype was misplaced: these eggs did not develop into adult sea urchins, but stopped short at the larval stage. Unfertilised frogs' eggs can also be induced to divide and develop, but development stops at an early tadpole stage, because cells grown from unfertilised eggs contain only one set of chromosomes and lack the essential complementary set contributed by the sperm. The same fundamental problem would have beset the Virgin Birth.

Lewontin condemns as biological determinism all attempts to link human intelligence and behaviour to race, class, physical characteristics or genes. He pokes fun at phrenology, the pseudoscience much in vogue in the 19th century that linked the size and shape of people's heads to their intelligence and character. Lewontin comments: "Since acquisitiveness is a product of a material organ, the brain, then highly developed acquisitiveness should be the manifestation of the enlargement of one region of the brain. On the not unreasonable (although factually incorrect) assumption that the skull will bulge a bit to accommodate a bulge in the cerebral hemisphere, we might well expect an enlarged 'bump of acquisitiveness' among the more successful members of the Exchange, not to mention Jews in general."

This tasteless joke ill matches his injunction against racial determinism. Recent studies have shown a correlation between IQ scores and brain size, but it is weak, and one of the most intelligent men I know also has the smallest head.

As a more recent form of biological determinism Lewontin attacks intelligence tests. He accuses the tests' practitioners of claiming that they "measure a single underlying innate thing which does not develop during the lifetime of the individual, but merely becomes crystallised by education." "It is the ability to learn, a fixed feature immanent to different degrees in every fertilised egg." "This biological determinism is the conjunction of political necessity with an ideologically formed view of nature, both of which arise out of the bourgeois revolution of the 17th and 18th centuries."

These strictures may have been justified in the past, but are they still true today? To find out, I turned to some of the recent literature on IQs (articles and books by U. Neisser, C. Jencks, N. J. Mackintosh, I. J. Deary, among others), and was surprised to find it open-minded, undogmatic and thoughtful. For example, studies of identical and non-identical twins brought up either together or separately have shown that no more than half the variation in IQ scores of different groups are genetically determined. Identical twins can actually differ in intelligence because of different blood supply in the womb. Again, the mean IQ scores of African-American children are typically 15 points below those of white American children, but investigators have found no evidence of their being genetically determined; they confess that they cannot explain them. Another example of the limitations of IQ tests comes from observations of Chinese- and Japanese-American children. Their average IQ scored slightly below the general average of 100, at about 97 or 98, yet the later occupational success of these children has turned out to be equivalent to those of white children with an average IQ of 120. The authors can offer no explanation for this anomaly.

Neither Lewontin nor his wordy critics ask the crucial practical question: what predictive value do IQ scores of children have for their future performance?

The correspondence between the IQs of the same groups of children at different ages is expressed statistically as a correlation coefficient. Coefficients of one imply complete correspondence, of zero none. Intermediate values signify different degrees of scatter of individual IQs from the mean; the lower the coefficient the greater the scatter. In Britain, the correlation coefficient between the IQs of groups of 11-year-olds and the results of school examination of the same groups at 16 is about 0.5, which means that there is a significant correlation, but with so large a scatter of individual values from the mean of the groups that they have little predictive value for the individual child: they should never have been used to assign 11-year-olds to either grammar or secondary modern schools. In America, a sample of children was tested repeatedly between the ages of three and 16. IQs at three and six showed only negligible correlation with later educational attainments. IQs between eight and 16 showed correlation coefficients of between 0.45 and 0.50, in agreement with British studies. About two-thirds of the differences in IQ scores were found to arise from causes other than family background. Children's IQ scores measured between the ages of 12 and 17 affected their future occupational status, but 60 to 80 per cent of that effect arose because most children with higher IQs also sought more education; those who failed to do this did no better as adults than children with lower IQs.

The authors of one of the studies conclude that "even if IQ scores were entirely explained by genes, which they almost certainly are not, the genes that do affect IQ scores have rather modest effects on occupational success, even though mental ability differences remain remarkably unchanged from childhood to old age." Ian Deary and others confirmed this when they compared the IQ scores at age 11 of Scottish children born in 1921 with scores of the identical tests administered to 101 of their survivors at the age of 77. I could find no biological determinism in any of this work.

Much that has been discovered about the brain's anatomical development is consistent with the psychologists' finding that nature and nurture are closely interwoven in the development of mental abilities. The main nerve fibres seen at age six have been laid down already before birth because they are genetically determined, but the branches which connect them to other nerves have multiplied. The growth of the branches and their different functions are also genetically controlled but the connections they make with other nerves, or the strength of these connections, seem to depend at least in part on external stimuli which include learning. Lewontin gives an interesting summary of the widely accepted theory that in the absence of stimuli, these branches make a multitude of random connections and that later only those survive that are strengthened by external stimuli. We do not yet know exactly what controls the connections, but the vital importance of external stimuli and maintenance of the right connections is well established.

Lewontin's black sheep include molecular biology, the human genome and gene therapy. He attacks Max Delbrück, the physicist who pioneered the immensely fruitful genetics of bacterial viruses, and whom he wrongly describes as a pupil of Schrödinger. He brands the phage group, the enthusiastic band of young people whom Delbrück assembled around himself, as "a political apparatus," and molecular biology as "a religion," which is absurd. He is right when he ridicules molecular biologist Walter Gilbert's vision of the human genome as its "holy grail" that will change our philosophic understanding of ourselves, but he then continues: "It is a sure sign of their alienation from revealed religion that a scientific community with a high concentration of Eastern European Jews and atheists has chosen for its central metaphor the most mystery-laden object of medieval Christianity." Remarks about people's race, religion and origin have no place in a book about science.

What does the human genome really tell us, now that it is almost completely known? Genes' only function is to code for the synthesis of proteins and nucleic acids that make up most of our bodies and perform nearly all their chemical functions. It takes about 31,000 genes to make a human; probably even more, because the same genes can be spliced in different ways to make more than one protein. The mouse genome contains about the same number of genes as the human, and 90 per cent of mouse genes are the same as human ones. I agree with Lewontin that this does not increase our philosophical understanding of what makes us different from mice. It takes as many as 19,000 genes to make a nematode worm that contains only a thousand cells and is only a millimetre long; 14,000 genes to make a fruitfly and 26,000 to make thale cress (*Arabidopsis thaliana*), a weed of the mustard family. Judging by the number of genes, plant organisms must be almost as complex on the molecular scale as those of mammals. A yeast cell needs about 6,000 genes and the humble *E. coli* bacterium 5,416, which makes one realise the staggering complexity of even the simplest single-cell organisms. The functions of sizeable proportions of the genes of all these organisms are still unknown. Two-fifths of the worm's, about half the fruitfly's, and 15 per cent of the weed's genes have human homologues, which testifies to the remarkable unity of life on the molecular scale.

David Baltimore writes in the issue of *Nature* (15 February 2001) announcing the completion of the human genome: "Understanding what does give us our complexity — our enormous behavioural repertoire, ability to produce conscious action, remarkable physical coordination (shared with other vertebrates), precisely tuned alterations in response to external variations of the environment, learning, memory...need I go on? — remains a challenge for the future.

"We wait with bated breath to see the chimpanzee genome. But knowing now how few genes humans have, I wonder if we will learn much about the origins of speech, the elaboration of the frontal lobes and the opposable thumb, the advent of upright posture or the sources of abstract reasoning ability, from a simple genomic comparison of human and chimp. It seems likely that these features and abilities have mainly come from subtle changes...that are not now easily visible to our computers and will require much more experimental study to tease out. Another half-century of work by armies of biologists may be needed before this key step is elucidated."

Only a small fraction of the diseases that affect us are inherited, or are due to inherited susceptibilities. Susceptibility genes for breast or colon cancer or Alzheimer's disease account for less than 3 per cent of all cases, and no preventive measures against them have so far proved safe and effective. The greatest medical advances from the human genome are expected in diagnosis of inherited conditions, in our knowledge of their prevalence in different populations, and of inherited predispositions in patients' response to drugs.

The genes for the most common inherited diseases that are due to mutations in single genes were already identified before scientists thought of sequencing the entire genome, but its analysis has already led to the identification of the genes for about 30 additional congenital diseases, among them one responsible for susceptibility to breast cancer, in addition to the ones already known, and another that causes a devastating lung disease. The causes of these diseases are still unknown, but knowledge of the gene for a disease can lead quickly to discovery of its molecular mechanism, which is the first essential step towards a therapy.

Unfortunately not a sure step. Sickle-cell anaemia is a hereditary disease that affects inhabitants of malarial regions of the world. It is a disease of haemoglobin, the protein of the red blood cell, a molecule on which I spent most of my life. As a result, we know in atomic detail the alteration of atomic structure caused by the sickle-cell mutation and understand exactly why that alteration causes anaemia, yet all our efforts to find a drug that would remedy this condition have failed. On the other hand, identification of the responsible gene has led to a successful therapy for haemophilia, which is caused by mutations in the genes for proteins required for blood clotting. Its most common form used to be treated by regular injections of the healthy protein isolated from donated blood, but there have been tragic cases of infection with HIV and hepatitis virus; knowledge of the gene has enabled scientists to manufacture the protein by recombinant DNA technology without any risk of infection.

In principle, many commonly inherited diseases should become curable by gene therapy either by introducing the healthy gene into the fertilised egg or by administering it to the patient; Lewontin condemns introduction as too risky and the recent creation of the first genetically modified monkey bears this out. In order to know whether the gene they introduced had taken, scientists chose one that codes for a fluorescent protein made in a jellyfish, believing that successful introduction would make the monkey fluoresce. They coupled the gene to a harmless virus which they injected into 222 monkey eggs; they then fertilised them with monkey sperm and incubated them. They implanted two each of 40 early embryos into the wombs of 20 surrogate mother monkeys. Only five of these resulted in pregnancies, one of them of twins. Only three monkey babies were born and only one carried the jellyfish gene, but the monkey does not fluoresce, because the gene, though spliced into the monkey's chromosomes, fails to express the protein for which it codes. This kind of gene transfer has now been practised in mice for several years with no greater success rate. It would be criminal to try it in humans and I see no good reason why it needed to be tried on our nearest relative, an intelligent monkey. Human cloning carries the same risks.

After many failures, gene therapy in patients succeeded last year for the first time. A French team has managed to cure a fatal inherited immune-deficiency disease. They took bone marrow from two baby boys who suffered from the disease and injected it with a harmless virus that carried the healthy gene. They then re-injected the infected marrow into the boys' bones. It restored complete immune function that was still active nine months later.

Another hopeful development has sprung from the identification, before the completion of the genome, of the gene for muscular dystrophy. Injection of fragments of the gene into a muscle of dystrophic mice has restored the muscle's normal function. This might, in due course, lead to successful treatment of human patients; both these successes are really great news. On the other hand, it has so far not been possible to cure one of the most common inherited disorders among Western Europeans, cystic fibrosis, which affects the lining of the lungs and airways. Patients who inhaled the healthy gene, carried either by a virus or contained in tiny fat droplets, derived little benefit; the treatment temporarily cured the lining of their noses, but it had no effect on their lungs. It is often proving extremely difficult to incorporate a gene in the correct place of the patients' chromosomes and express the required protein in sufficient quantity in the right tissues.

Before the completion of the Human Genome Project, identification of some of these genes required truly heroic efforts. The search for the Huntington's disease gene occupied up to a hundred people for about ten years. The same work could now be accomplished by few people in a fraction of the time. This is one of the Human Genome Project's important medical benefits. Another may be the rapid identification of promising new drug targets against diseases ranging from high blood pressure to a variety of cancers.

The title of Lewontin's book is misleading. The human genome is proving no illusion either medically or biologically. It has not been, as Lewontin alleges, initiated by financially interested scientists in order to extract money from the public purse for their own pockets. Some commercial companies have stepped in, but as John Sulston, the head of the Sanger Centre near Cambridge where a large part of the genome was sequenced, has written, the hundreds of devoted people who have contributed to it and made it speedy, efficient and effective have done this not for wealth, nor for unusual recognition, but to benefit mankind. I would add, also out of curiosity for the working of nature. It is, by any measure, a magnificent achievement. The man who invented the chemical method used for sequencing the genome was Fred Sanger, not Allan Maxam and

Walter Gilbert whom Lewontin wrongly credits with it. Sanger never patented his method, but it won him his second Nobel prize.

The greatest threats to health come not from our genes, but from infectious diseases. AIDS remains incurable and threatens to wipe out entire populations. In many countries of Africa and Asia, abuse of antibiotics has bred tubercle bacilli that are resistant to all known drugs. Travellers import them. They may make us once more helpless against the "white plague" of tuberculosis that killed Chekhov, the Brontës, Keats, Chopin, D.H. Lawrence, George Orwell and many thousands of others in the prime of life.

Photo Gallery

Fritz Haber as a young man (1891). Courtesy Archiv zur Geschichte der Max-Planck-Gesellschaft.

Fritz Haber (with outstretched arm) during World War I. Courtesy Archiv zur Geschichte der Max-Planck-Gesellschaft, Berlin-Dahlem.

V. Keilin using his microspectroscope in 1938.
Photograph by Dr. Emil Smith.

Fritz Strassmann, age 36.
Courtesy Irmgard Strassmann.

Enrico Fermi.
Reproduced from Biographical Memoirs of the Royal Society.

Otto Hahn and Lise Meitner at the Kaiser Wilhelm Institute in Berlin.
Courtesy Archiv zur Geschichte der Max-Planck-Gesellschaft, Berlin-Dahlem.

Glenn Seaborg and Lise Meitner. Presentation of the Fermi Prize (1965).
Press photograph, source unknown.

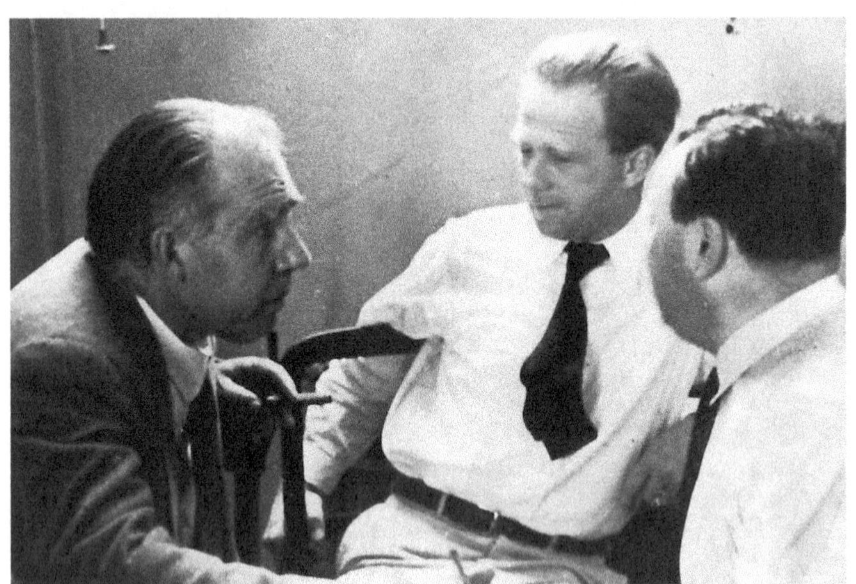

Niels Bohr, Werner Heisenberg, and Wolfgang Pauli (1934).
Niels Bohr Institute, courtesy AIP Emilio Segre Visual Archives.

The Sakharovs with Jewish refusenik leader Vladimir Slepak, circa 1978
from Sakharov's *Memoirs,* Knopf (1990). Courtesy Elena Bonner.

François Jacob.

Louis Pasteur (1857).

Peter Medawar.
Photograph by Sydney Weaver, London.

Linus Pauling, Max Delbrück, and Max Perutz in Pasadena (1976).
Press photograph. Source unknown.

John Desmond Bernal, circa 1935.
Photograph by Ramsey and Muspratt.

Oswald Avery at the Rockefeller Institute for Medical Research. Reproduced from *The Professor, Institute and DNA* by René J. Dubos, 1976, by copyright permission of The Rockefeller University Press.

Oswald Avery, aged 9, with his family, The Reverend and Mrs. Joseph Avery and brothers Ernest (standing) and baby Ray. Reproduced from *The Professor, Institute and DNA* by René J. Dubos, 1976, by copyright permission of the The Rockefeller University Press.

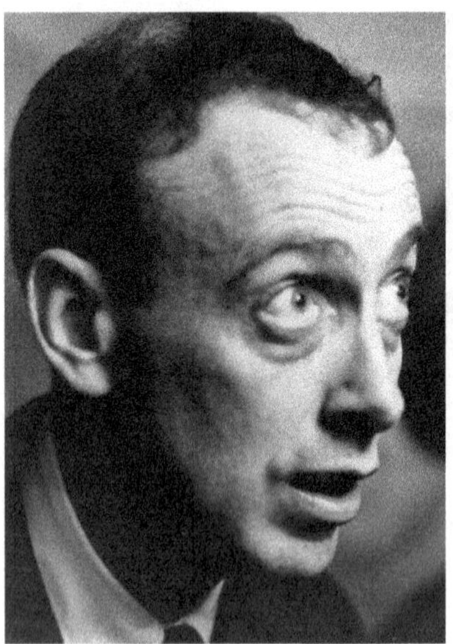

James Watson (1962).
Courtesy J.D. Watson.

Rosalind Franklin, circa 1950.
Photograph by Elliott & Fry. By courtesy of the National Portrait Gallery, London.

Francis Crick with his daughters Gabrielle and Jacqueline, circa 1957.
Courtesy Francis Crick.

William Lawrence Bragg, circa 1913. Source unknown.

Dorothy Crowfoot, later Hodgkin. Source unknown.

Jacques Monod (1976).

Carl Djerassi.
Reproduced from *The Pill, Pigmy Chimps and Degas' Horse*, by Carl Djerassi (Basic Books, 1992). Courtesy Carl Djerassi.

Rights and Wrongs

By What Right Do We Invoke Human Rights?[1]

Scientists the world over are united by a common purpose, ideally to discover Nature's secrets and put them to use for human benefit. Albert Szent-Györgyi, the discoverer of vitamin C, has said: "I feel closer to a Chinese colleague than to my own postman."

When a scientist who has committed no crime is imprisoned, we feel like the minister freeing the prisoners in *Fidelio* when he sings: "Es sucht der Bruder seine Brüder (Each brother searches for his brothers);" he or she is one of our brothers or sisters, and we feel a duty to appeal for his or her release. In doing so, we are now on strong legal grounds established by the United Nations Universal Declaration of Human Rights of 1948 and the conventions and covenants that followed it. They have the force of international law and are backed by courts and commissions to which individuals can appeal.

What do they say?

The International Covenant on Civil and Political Rights of 1966 "recognizes that these rights derive from the inherent dignity of the human person."

Its articles include the following:

- Everyone has the right to life, liberty, and security of the person.
- No one shall be subjected to torture or to cruel, inhuman, or degrading treatment or punishment.

- All are equal before the law.
- No one shall be subjected to arbitrary arrest, detention, or exile.
- Everyone has the right to freedom of thought, conscience, and religion.
- Everyone has a right to freedom of opinion and expression.[2]

Despite their many infringements, the enshrinement of these and other essential rights in international law is one of the great achievements of our civilisation.

In large part, we owe their formulation to the great jurist Hersch Lauterpacht, professor of international law in the University of Cambridge from 1937 to 1954. In 1945 he published a seminal book, *An International Bill of the Rights of Man*, which became the basis of much that is in the United Nations Declaration and the Conventions that followed it.[3]

According to him,

> The idea of the inherent rights of man, ultimately superior to the state itself, is the continuous thread in the historical pattern of legal and political thought. In antiquity, their substance has been a denial of the absoluteness of the State and its unconditional claim to obedience; the assertion of the value and freedom of the individual as against the State; the view that the power of the State and of its ruler is derived ultimately from the assent of those who compose the political community; the insistence that there are limits to the power of the State to interfere with man; the right to do what he considers his duty.[4]

Freedom's first conditions, the rule of law and equality before the law, stem from antiquity.

Aeschylus' *Oresteia*, written in 458 B.C., makes Pallas Athene, the goddess of Wisdom, admonish Athenians:

> Hold fast such upright fear of the law's sanctity,
> And you will have a bulwark of your city's strength.

They were her own laws and they provided for trial by jury.

Thucydides reports that Pericles, the Athenian statesman, said in his funeral speech commemorating the fallen in the first year of the Peloponnesian War against the Spartans: "Our constitution is called a democracy because power is in the hands… of the whole people. …Everyone is equal before the law."[5] In fact this was true only of male citizens, and it excluded women, barbarians, and slaves, but the principle survived and inspired future generations.

Plato and Aristotle were elitists. Plato advocated rule by the few wise over the stupid many, and Aristotle apparently approved of some men being born free and other as slaves, each to his station in life.

The Stoics took the next step by distinguishing natural law from men's law, where natural law meant the universal moral conscience common to all, an intuitive notion of what is meant by justice and goodness, by which the laws of states can be judged. This law applied to all human beings because they all bore within them a spark of the creative fire.

The Roman emperor Marcus Aurelius was a Stoic philosopher. In his *Meditations*, written in about 170 A.D., he favoured a polity in which there is the same law for all, a polity with equal rights and freedom of speech, and a kingly government that respects most of all the freedom of the governed.[6] These lofty thoughts did not hinder him from persecuting the Christians, but again, his thoughts inspired future generations. Epictetus, another Stoic philosopher of the first and second century A.D. and himself a slave in Rome, taught, not surprisingly, that slaves are the equals of other men, because all alike are the sons of God, thus anticipating St. Paul who wrote in his Epistle to the Galatians:"There is neither Jew nor Greek, there is neither slave nor free, there is neither male nor female; for you are all one in Christ Jesus." According to St. Thomas Aquinas, man's natural laws are God-given, therefore infallible and eternal.

The first step towards enshrining human rights in the laws of a state was Magna Carta, which King John of England granted in 1215 under duress from his barons. It assured freedom from arbitrary imprisonment, for which we are still fighting in many countries today and, less well known, also freedom to travel:

> Article 39. No freeman shall be captured or imprisoned or dispossessed or outlawed or exiled or in any way destroyed, nor will we go against him or send against him, except by the lawful judgment of his peers or by the law of the land.
>
> Article 42. Everyone shall henceforth be permitted… to leave our kingdom and to return in safety and security, by land or by water.. .
>
> Article 63. Wherefore we wish… that the men in our kingdom shall have and hold all the aforesaid liberties rights and grants well and in peace freely and quietly, fully and completely, for themselves and their heirs from us and our heirs… .[7]

This did not prevent many of King John's successors from trampling on the provisions of Magna Carta, but it provided laws to which their subjects were able to appeal.

After Magna Carta and up to the present day most concepts of human rights and their incorporation into law originated in the English-speaking world.

In 1628, Parliament presented King Charles I, who had disregarded Magna Carta at his peril, with a Petition of Rights that reasserted the freedom from arbitrary imprisonment and added freedom from arbitrary taxation.

> The lords spiritual and temporal, and commons in their present parliament assembled, concerning divers rights and liberties of the subject... do therefore humbly pray your most excellent majesty that no man hereafter be compelled to make or yield any gift, loan, benevolence, tax or such like charge without common consent by and of parliament; nor be called to make answer, take such oath, or to give attendance, or be confined, or otherwise molested or disquieted concerning the same, or for refusal thereof; and that no freeman, in any such manner as is before mentioned, be imprisoned or detained... .[7]

On the other hand, there was to be no nonsense about freedom of religion. In 1629, the Commons resolved that "whosoever shall bring in innovation of religion, or by favour or countenance seek to extend or introduce popery or Arminianism, or other opinion disagreeing from the true and orthodox church, shall be reputed a capital enemy to this kingdom and commonwealth."[7]

The seventeenth-century Dutch jurist Hugo Grotius first attempted to make natural law, inherent in the human conscience, independent of divine law as written in the Bible, the foundation for an international law. He postulated that "natural law would be valid, even if we were to concede, which we cannot concede without the utmost wickedness, that there is no God, or that the affairs of man are of no concern to Him."

In the 1640s, at the time of the Great Rebellion against King Charles I, the Levellers were a radical party within the Parliamentarians. One of them, Richard Overton, may have been the first to translate natural law into natural rights. In his pamphlet *An Arrow Against All Tyrants*, directed against Charles I, he wrote: "For by natural birth, all men are equally... borne to like propriety,[8] liberty and freedom... everyone equally and alike to enjoy his birthright and privileges."[9,10] These were revolutionary ideas for which he suffered imprisonment.

The greatest seventeenth-century protagonist of human rights was the English philosopher John Locke, whose *Essay Concerning the True, Origi-*

nal, Extent and End of Civil Government, published in London in 1689, overthrew the belief in the divine right of kings and put in its place a notion of the natural rights of man, rights that are universal and essential to all men, rights without which life is intolerable. They are the rights to life, liberty, and property.

He wrote:

> It having been shown:
>
> 1. That Adam had not either by natural Rights of Fatherhood or by positive Donation from God, any such Authority over his Children, or Dominion over the World as is pretended.
> 2. That if he had, his Heirs (i.e. Kings), yet, had no Right to it. To understand Political power right... we must consider what State Men are naturally in, and that is, a State of perfect Freedom (i.e. not subject to authority). That State of Nature has a Law of Nature to govern it, which obliges everyone: And Reason, which is that Law, teaches all Mankind, that being all equal and independent, no one ought to harm another in his Life, Health, Liberty and Possessions.

These principles did not hinder him from justifying the American settlers' seizures of the native Indians' lands, on the grounds that the Indians were hunters, not farmers, whence their land was not their property.

In the eighteenth century, Tom Paine extended Locke's natural rights to man's "intellectual rights, and also the rights of acting as an individual for his own comfort and happiness which are not injurious to the rights of others. Every civil right has for its foundation some natural right pre-existing in the individual."[11] Tom Paine's revolutionary talk and writing inspired the American Declaration of Independence of 1776, the American Bill of Rights, and the French Declaration of the Rights of Man and of the Citizen of 1791.

Other inspiration came from Montesquieu's *Spirit of the Laws*, first published in 1745. Montesquieu advocated a separation of the powers of the judiciary, the legislature, and the executive, trial by jury, and a two-party system, so that one party has the power to keep the other in check. He defined liberty as "the right to do anything which the law permits" and political liberty of the citizen as "that security of mind which derives from each person's view of his own security; and to enjoy that liberty, government must be such that no citizen fears another citizen." Montesquieu favoured freedom of speech not as a natural right—I found no mention of such rights—but as a safety valve. Writings that satirize the government should be allowed because "they satisfy the general malice, console discontents, diminish envy of those in high places, give people patience to bear their own sufferings and make people laugh at them."[12]

In Congress in Philadelphia on 4 July 1776 the thirteen United States of America declared unanimously:

> We hold that these truths are self-evident, that all men are created equal, that they are endowed by their Creator with certain inalienable rights, that among them are life, liberty and the pursuit of Happiness. That to secure these rights, governments are instituted among men, deriving their just powers from the consent of the governed.[2]

The Declaration failed to insert the word "white" between "all" and "men" because in fact none of these noble rights applied to their black slaves, for reasons Montesquieu had stated in this sarcastic passage thirty-one years earlier:

> The peoples of Europe, having exterminated the peoples of America, were bound to enslave those of Africa in order to exploit other countries. Sugar would be far too dear if the plant producing it was not cultivated by slaves. Besides… it is virtually impossible to feel compassion for people who are black from hand to foot and who have such flattened noses. And… how could it have come into the mind of God, who is a very wise being, to put a soul, still less a good soul, into an all black body? It is impossible to suppose that these people are human because, if we took them to be human, one would begin to believe that we ourselves are not Christians. Small-minded spirits exaggerate the injustice done to the Africans because, if this were really as bad as is said, would it not have entered the minds of European princes, who conclude so many useless treaties, to conclude one in favour of compassion and pity?[12]

In 1795, Condorcet echoed these views when he wrote: "There should not be different races, one destined to govern, the other to obey, one to lie, the other to be deceived; one must recognize that all have the same right to declare their interests, and that none of the powers established by them and for them is to have the right to hide any of these powers from them."

Condorcet anticipated Lord Acton's dictum that democracy consists in preventing revolution by timely reform. Decisions should be made by the majority of the people, but they must not infringe the rights of the individual, which he defines, following Locke, as the freedom to develop his faculties, dispose of his possessions, and look after his needs.[13]

Article I of The Bill of Rights of 15 December 1791, formulated chiefly by James Madison, stated that: "Congress shall make no law requesting an establishment of religion, or prohibiting the free exercise thereof, or abridging the freedom of speech, or of the press, or the right of people peaceably to assemble, and to petition the Government for a redress of grievances."[2] Madison failed to anticipate that the free exercise of religion could make people try to bring about the prophesied end of the world by spreading poison gas, or that free speech would be abused to incite to racial hatred.

The Declaration of the Rights of Man and of the Citizen was adopted by the French National Assembly in 1791; it was inspired by the American Bill of Rights, but it was more cautious and did not make these rights absolute.

> The representative of the peoples of France... considering that ignorance, neglect or contempt of human rights are the sole causes of public misfortune and corruption of Government, have resolved to set forth... these natural... inalienable rights
>
> 1. Men are born, and always continue free and equal in their rights.
> 7. No man should be accused, arrested or imprisoned except as determined by law.
> 10. No man ought to be molested on account of his opinions, provided that his avowal of them does not disturb the public order established by law.
> 11. The unhindered communication of thoughts and opinions being one of the most precious rights of man, every citizen may speak, write and publish freely, provided he can be held responsible for the abuse of this liberty, as determined by law.[2]

Sadly the Declaration failed to prevent the terror that followed.

Nineteenth-century liberal thought derived most inspiration from John Stuart Mill's essay *On Liberty*, first published in 1858.[14] In accordance with the intellectual climate of the day, he based his plea neither on the dignity of man nor on his inherent natural rights, but on utility and material progress:

> I regard utility as the ultimate appeal on all ethical questions, but it must be utility in the largest sense, grounded on the interests of man as a progressive being.... This, then, is the appropriate region of human liberty. It comprises, first, the inward domain of consciousness; demanding liberty of conscience, in the most comprehensive sense; liberty of thought and feeling; absolute freedom of opinion and sentiment on all subjects, practical or speculative, scientific, moral, or theological. The liberty of expressing and publishing opinions may seem to fall under a different principle, since it belongs to that part of the conduct of an individual

> which concerns other people; but, being almost of as much importance as the liberty of thought itself, and resting in great part on the same reasons, is practically inseparable from it. Secondly, the principle requires liberty of tastes and pursuits; of framing the plan of our life to suit our own character; of doing as we like, subject to such consequences as may follow: without impediment from our fellow-creatures, so long as what we do does not harm them, even though they should think our conduct foolish, perverse, or wrong.

About freedom of expression Mill has this to say:

> We have now recognised the necessity for the mental well-being of mankind (on which all their other well-being depends) of freedom of opinion, and freedom of the expression of opinion, on four distinct grounds; which we will now briefly recapitulate.
>
> First, if any opinion is compelled to silence, that opinion may, for aught we can certainly know, be true. To deny this is to assume our own infallibility.
>
> Secondly, though the silenced opinion be an error, it may, and very commonly does, contain a portion of truth; and since the general or prevailing opinion on any subject is rarely or never the whole truth, it is only by the collision of adverse opinions that the remainder of the truth has any chance of being supplied.
>
> Thirdly, even if the received opinion be not only true, but the whole truth; unless it is suffered to be, and actually is, vigorously and earnestly contested, it will, by most of those who receive it, be held in the manner of a prejudice, with little comprehension or feeling of its rational grounds. And not only this, but, fourthly, the meaning of the doctrine itself will be in danger of being lost, or enfeebled, and deprived of its vital effect on the character and conduct: the dogma becoming a mere formal profession, inefficious for good, but cumbering the ground, and preventing the growth of any real heart-felt conviction, from reason or personal experience.
>
> The only purpose for which power can be rightfully exercised over any member of a civilized community, against his will, is to prevent harm to others.

Mill lashed out against Calvinism, which today has lost much of its power, but if we substitute *the state* for God, his words apply with equal force to life in this century and the next. According to Calvin, the one great offence of man is self-will, but Mill objects strongly:

> All the good of which humanity is capable, is comprised in obedience. You have no choice; thus you must do, and no otherwise: "whatever is not a duty, is a sin." Human nature being radically corrupt, there is no

> redemption for any one until human nature is killed within him. To one holding this theory of life, crushing out any of the human faculties, capacities, and susceptibilities, is no evil: man needs no capacity, but that of surrendering himself to the will of God: and if he uses any of his faculties for any other purpose but to do that supposed will more effectually, he is better without them.
>
> It is not by wearing down into uniformity all that is individual in themselves, but by cultivating it and calling it forth, within the limits imposed by the rights and interests of others, that human beings become a noble and beautiful object of contemplation; and as the works partake the character of those who do them, by the same process human life also becomes rich, diversified, and animating... In proportion to the development of his individuality, each person becomes more valuable to himself, and is therefore capable of being more valuable to others.

Mill was one of the most prominent champions of the rights of women in the nineteenth century:

> A person should be free to do as he likes in his own concerns; but he ought not to be free to do as he likes in acting for another, under the pretext that the affairs of the other are his own affairs. The obligation is almost entirely disregarded in the case of the family relations, a case, in its direct influence on human happiness, more important than all others taken together. The almost despotic power of husbands over wives needs not be enlarged upon here, because nothing more is needed for the complete removal of the evil, than that wives should have the same rights, and should receive the protection of law in the same manner, as all other persons; and because, on this subject, the defenders of established injustice do not avail themselves of the plea of liberty, but stand forth openly as the champions of power.

Mill's plea was not generally heeded nor acted on in Britain until well into this century and is still being ignored in most countries of the world today. His powerful essay on *The Subjection of Women* is much less well known than that *On Liberty*, which deals with men.

Mill's justification of human rights on the grounds of utility for all, rather than for the individual, has been criticised because utility for all has often been invoked to justify repression of individual liberty, but since his essay is devoted entirely to the rights of the individual, this criticism hardly detracts from its merits.

The contemporary philosopher A. Gewirth has added the important rider that human rights must be justified demands *in relation to other people*,[15] i.e., they must not harm the legitimate claims of others.

Human rights have now been given the force of law in covenants adopted by the Council of Europe in 1950 and by the General Assembly of the United Nations in 1966, by the American Convention of 1969, and by the African Charter of 1981. In 1984, the United Nations also adopted a Convention against Torture and other Cruel, Inhuman or Degrading Treatment or Punishment.

As a schoolboy I believed that torture had gone out in Europe, at any rate, with the Inquisition. Later I learnt that Lenin revived it and Hitler followed suit. To my horror I have since come to realise that it is still being practised by many so-called civilized states. In London, the Medical Foundation for the Victims of Torture helps their rehabilitation and the Redress Trust tries to obtain redress for them. The Trust has supplied me with a list of states where there is widespread, constant use of torture, or of states occasionally practising Torture, or of states practising cruel, inhuman or degrading treatment or punishment.

Countries with Widespread, Constant Use of Torture (E.G., Electric Shock, Semi-Suffocation, Foot Beating, and Sexual Assault)

Algeria[16]
Angola
Bangladesh
Bosnia-Herzegovina
China
Colombia[16]
Ecuador[16]
Egypt[16]
Equatorial Guinea
Guatemala[16]
Haiti (perhaps no longer)
India
Indonesia (and East Timor)[16]
Iran
Iraq
Israel[16]
Liberia
Mexico[16]
Myanmar (Burma)
Pakistan
Papua New Guinea
Peru[16]
Saudi Arabia
Somalia[16]
Sri Lanka
Sudan[16]
Tunisia[16]
Turkey[16]
Venezuela[16]
Yemen[16]
Zaire

The abuse of psychiatry for the mental degradation of political prisoners has now ceased in Eastern Europe, but it is still being practiced in Cuba.

States From Which There Are No Reports of Inhuman Treatment of Suspects, or Cruel, Inhuman or Degrading Punishment Forms Only a Minority

Belgium	Netherlands
Denmark	New Zealand
Finland	Norway
Ireland	Sweden
Liechtenstein	Switzerland
Luxembourg	

In the face of these glaring violations of human rights I feel heartened that most of the world's scientific academies are resolved to defend the rights of our scientific colleagues wherever they have been infringed, but I find it tragic that just when the gradually evolving concept of human rights has at last been given the force of international law, the concept of human duties should have fallen into unprecedented disrepute, threatening the disintegration of our society and with it the collapse of our most precious heritage, European civilisation, which gave birth to the very concept of human rights. It is time to fight the fashionable notion that self-fulfilment, the development of one's personality and fulfilment of one's wishes at no matter what cost to one's family, friends, colleagues, and community, should be man's or woman's ultimate aim. Immanuel Kant's fundamental insight that the essence of morality consists in treating other people as ends in themselves rather than means, and the old fashioned virtues, love, loyalty, honesty, sense of duty, and compassion, which hypocrisy has brought into disrepute, are in bad need of revival, and deserve to be upheld along with human rights.

There have been many critics of human rights based on natural law. One of them was Justice of the American Supreme Court Oliver Wendell Holmes who regarded natural law not as law in any true sense, but rather as theology or morals, something existing in the mind of the Creator and in the mind of man when he exercised his reason in order to find it, something with a separate being of its own. Holmes argued that he never experienced it, and not believing in revelation, denied its eternal validity. The English Judge Frederick Pollock argued that natural law, being based on divine law, may by infallible, but there was no infallible way of finding out what it is.

It has also been argued that rights may conflict. For example, the right of an individual to speak, write, and publish freely may jeopardise everyone else's rights by compromising the security of the state. Rights are hard to maintain in the face of terrorism or widespread violent crime.

Other critics have attacked Mill for founding human rights on utility, because this can be abused, but then Mill may have invoked utility to make these rights acceptable to 19th century England. Nowadays he might have justified them instead as furthering the creation of wealth. In fact there is some truth in this. The British economist Partha Das Gupta has found that developing countries which respect human rights are more prosperous, on average, than those which don't.

I plead for human rights because many innocent men and women owe them their freedom and because they have had, and are having, a strong civilising influence. They are something to strive for to make a better world.

I thank Professors Kurt Lipstein, Peter Laslett, and Elihu Lauterpacht, and Mr. David Weigall for helping to introduce me to this subject, and Sir Isaiah Berlin, Sir Michael Atiyah, Sir Henry Chadwick, Sir Ernst Gombrich, Professor Edward Kenney, and Dr. Richard Tuck for advice.

*The Right to Choose**

The world population now increases by 1.7 percent (90 million) per year, while production of cereals is increasing by only 0.9 percent per year. During the past twenty years there have been about 200 million hunger-related deaths; the growing food deficit may raise that number fivefold in the next twenty years. The population of some of the poorest countries is growing fastest. Bangladesh, with a land area smaller than that of Wisconsin, now has a population of 114 million, which is expected to outstrip the present population of the United States, 240 million, in about thirty years' time. What will happen to these poor people? Even if by some miracle of science enough food could be produced to feed them, how could they find the gainful employment needed to buy it? These prospects are so grim to contemplate that both the Pope and the White House are reported to have forced the conference on the environment held at Rio in 1992 to ignore them.

Tragically, the population explosion is the result of the North's least controversial contribution to the South, the prolongation of life by modern medicine and hygiene, or, as Viktor Weisskopf put it, the introduction of death control without birth control.

Carl Djerassi, the inventor of the contraceptive pill, and Etienne-Emile Baulieu, the inventor of the abortion pill, have provided the means to avert

*A review of the books *The Pill, Pygmy Chimps, and Degas' Horse: The Autobiography of Carl Djerassi* by Carl Djerassi (Basic Books) and *The 'Abortion Pill'* by Etienne-Emile Baulieu (with Mort Rosenblum; Simon and Schuster).

or at least mitigate the catastrophe. Both their autobiographies express bitterness that religious and political pressures are preventing the introduction of these pills into some of the countries that need them most.

Djerassi, who was brought up in Vienna, came to New York just before World War II with twenty dollars in his pocket, and thirty years later he had become a world-famous scientist and millionaire. Fortunately when he arrived at the age of sixteen he had a first-class high-school education and a good knowledge of English. He had the cheek to ask Eleanor Roosevelt to find him a college scholarship and, even more remarkably, she responded by forwarding his letter to the Institute for International Education, which found him a scholarship at Tarkio College, Missouri, a Presbyterian school of twenty teachers and 140 students. After a year he was offered a room, board, and tuition scholarship at the Episcopalian Kenyon College in Gambier, Ohio, where, during the next year, he obtained the bachelor's degree in chemistry, needed in wartime, he tells us, in order to be drafted into the army as an officer rather than an enlisted man. However, a lame knee kept him out of the army, and he found a job with a pharmaceutical company in New Jersey instead. While working there he attended night classes in chemistry at New York University and the Brooklyn Polytechnic, helped to synthesize one of the first antihistamines, and got his name on a patent, all in one year. The next step was an Alumni Research Foundation Scholarship at the University of Wisconsin at Madison, where Djerassi obtained his Ph.D. in chemistry in two years at the age of only twenty-two.

In view of complaints of anti-Semitism at American colleges by the physicist Richard Feynman and others, I found it heartening that a penniless Jewish immigrant from Vienna was launched on a brilliant career by scholarships at two Protestant colleges and one university in the heart of the allegedly xenophobic Midwest. To judge by the time it took him to finish his studies, he must have possessed a phenomenal combination of talent and drive.

Djerassi returned to the pharmaceutical firm to work on antihistamines, but became restless for a greater challenge. The anti-arthritic properties of cortisone had just been discovered, but it had to be extracted from the adrenal glands of animals at a cost of $200 a gram. Djerassi wanted to attempt to make cortisone synthetically. When his firm showed no interest in such a forbiddingly difficult project, he left and joined Syntex, a newly formed pharmaceutical company run in Mexico City by young immigrant chemists from Europe.

Everyone had warned him that serious chemical research could never be done in such a remote place, but by May 1951 Djerassi's team had synthesized cortisone from a compound extracted from a Mexican yam. Their paper announcing that great feat arrived at the office of the editor of the American Chemical Society a few days ahead of papers by two famous chemists, R.B. Woodward and L.F. Fieser, who had achieved the same synthesis by different methods.

Nature makes its own contraceptive, progesterone, a steroid hormone that inhibits ovulation once pregnancy has begun. It can be administered to prevent pregnancy, but it is too weak when given by mouth: to work it must be injected. Djerassi set out to synthesize a contraceptive analogous to progesterone that would be active when administered orally. On October 15, 1951, Luis Miramontes, a Mexican chemistry student working under the direction of Djerassi, and the young head of the laboratory, George Rosenkranz, synthesized a compound, called norethindrone, that had the required chemical structure. Djerassi writes: "Not in our wildest dreams did we imagine that this substance would eventually become the active progestational ingredient of nearly half the oral contraceptives used worldwide." The team had accomplished this synthesis in less than six months, when Djerassi was twenty-eight years old. Eleven years later, after conducting a great many tests, the U.S. Food and Drug Administration approved norethindrone for contraceptive use.

Having been responsible for two spectacular inventions, Djerrassi resigned from Syntex to become associate professor of chemistry at Wayne State University in Detroit. His real interest lay in fundamental rather than applied research, and he wanted to explore the chemistry of the giant cacti that grow in Mexico.

His next original contribution was the development of an optical method for determining the chemical structure of asymmetric compounds. This won him a prize from the American Chemical Society and a chair as a professor of chemistry at Stanford University, where he has remained ever since, and where, while keeping up his connection with Syntex, he has brought new ideas to research and teaching. Having given a course of lectures on steroid chemistry, he asked each student to formulate examination questions, promising to distribute the questions randomly among the class. When he returned the questions, one student after another protested that he had been given his own question back by mistake. Gradually they realized this had been Djerassi's intention all along. Having tried to demonstrate their virtuosity by the complexity of their questions, each student was now faced with the grim task of answering them himself.

Despite Djerassi's fame and wealth his autobiography ends on a note of bitterness. In the United States, the development of contraceptives has become so expensive that most pharmaceutical firms, including Syntex, have abandoned it just when the world needs them most. A cheap, safe pill that can be taken once a month would ease contraception throughout the world, because many women, especially if they are uneducated, apparently find it hard to take the pill regularly every day at the correct interval after each period. Besides, many women cannot afford it. One factor that makes the extensive research needed for development of such a pill uneconomic in the United States is a widespread public misunderstanding of the effects of the prolonged use of any drug on a large population. Genetic diversity insures that there will always be some people in whom even the safest drug produces adverse effects. If the percentage of such people is substantial, then the adverse effects show up in clinical trials carried out on several hundred people before the drug is released on the market; but if the adverse effects manifest themselves in only, say, one person out of 10,000, then the probability of their showing up in clinical trials is negligible. When the drug is later used by millions, and affected persons sue the manufacturers for negligence, then litigation and damages may cost that firm many millions of dollars. These costs and the accompanying adverse publicity are some of the factors that have discouraged further development of contraceptives.

Other factors derive from the safety regulations imposed by the Food and Drug Administration, which, in reports of damage suits in the press, has sometimes been blamed for being careless in approving drugs. The FDA has responded by demanding extensive and prolonged animal trials before any contraceptive could be approved for human use, including two-year toxicity studies of rats, dogs, and monkeys, followed by seven-year toxicity studies at two, ten, and twenty-five times the human equivalent in monkeys. Djerassi points out that these rules make the development of new contraceptives excessively expensive; yet they could never ensure complete safety, because reproductive cycles and responses to steroids differ too widely among different animals and even among different primates to produce results that are meaningful in human beings. There is no way of introducing better contraceptives or indeed any new drugs without risk to some people, and if the press, Congress, and the courts refuse to accept this, efforts to develop new contraceptives will move elsewhere.

The most spectacular new development has come from France, where Etienne-Emile Baulieu invented an abortion pill call mifepristone, or RU 486. Like Djerassi's contraceptive pill, it is an analogue of the hormone progesterone that prevents ovulation once a woman has become pregnant. While Djerassi's pill mimics progesterone in its action, RU 486 blocks progesterone's uterine receptors so that this hormone cannot act on them. The uterus responds by menstruating and ejecting the embryo.

In France, RU 486 is sold in authorized family planning center pharmacies where every pill must be accounted for on a register, and doctors are supplied only with the exact amounts needed. Its use is allowed up to seven weeks after the last period, and only after a delay of seven days which is provided for the mother to reflect on whether she really wants to have the abortion. She then takes three pills of RU 486 in the doctor's presence, goes home and returns 36 to 48 hours later for the administration of a prostaglandin that promotes the expulsion of the embryo. After four hours of rest she is allowed to go home. By 1991 over 80,000 Frenchwomen had used RU 486. Its use has also been approved by the British National Health Service and by the government of China.

In Britain, about 160,000 pregnancies are terminated legally and free under the National Health Service each year; yet the abortion rate is only two fifths that of the United States—a fact that American opponents of abortion might reflect on. The National Health Service welcomes RU 486 because, unlike surgical termination, it needs neither operating theaters nor anesthesia and therefore saves money and doctors' time.

According to the traditional doctrine of the Catholic Church, making love is sinful except for the procreation of children; an unwanted child has therefore been conceived in sin. Abortion is held to be wrong not only because it kills a budding human life but also because it relieves the mother of the consequences of her sin and encourages sin in others. In Catholic Austria, for example, the law formerly included the notorious paragraph 144 that made abortion punishable by imprisonment for both the mother and whoever helped her. But, as Baulieu writes, such laws did not stop abortion, they only made it more dangerous to the mother, especially if she was poor.

As a medical man, Baulieu has no moral scruples about the abortion pill. He writes that more than 50 million abortions are performed in the world each year, about half of them illegally. In the former Soviet Union abortion was the main method of birth control. According to official figures 7 to 8 million were performed a year. In Poland abortion has been legal if approved by two doctors and a psychiatrist. Since these are hard to find in a poor country, about half a million of the 600,000 abortions a year are performed illegally and at great danger to the mother; recent legislation, moreover, threatens to make virtually all abortions illegal. In Japan, doctors have such a strong vested interest in abortion that they have prevented the introduction of contraceptives. In the United States 1.6 million abortions are performed annually, half of them on girls under twenty; the U.S. has the highest rate of teen-age pregnancy in the developed world, four times that of the Netherlands. In Romania, the Communist dictator Ceauçescu made abortion punishable by death, but this Draconian law made the birth rate rise only briefly. Instead of producing more babies, the law resulted in the highest rate of dead mothers in Europe. Opponents of abortion might well reflect on that experience and on the appalling mortality and morbidity of women caused by badly done abortions in all countries where the operation is illegal or medical help is unavailable.

Baulieu writes that 200,000 women throughout the world die each year from botched abortions. For every one who dies, twenty to thirty others suffer infections, perforations of the uterus, or lasting injuries that can lead to infertility. "Medicine's mission is to help people," he writes. "Why should 200,000 women die each year for lack of a better way? If science can make it otherwise, why must a woman's decision to terminate an unintended pregnancy be accompanied by pain and punishment?" Baulieu tells of encountering poor women who injured themselves with a stick, and of brutal surgeons who tell their assistants not to bother with anesthesia when repairing the damage, in order "to teach them a lesson." He set out early in his career to find a better and more humane way.

Anti-abortionists have branded RU 486 a death pill or a chemical time bomb, and Baulieu has been compared to "an amalgam of Joseph Stalin and Adolf Hitler," and accused of plotting the death of several billion human beings. Like Djerassi, he is a doctor's son, but unlike Djerassi he was never employed by industry and has spent his entire career in university clinics. He began research on sex hormones and their receptors with support from the Ford Foundation and the Population Council in New

York in the 1960s, and it took him fifteen years to develop RU 486. He receives, as he emphasizes, no royalties from its sales. (Baulieu quotes an older colleague's description of himself as "optimistic, enthusiastic, high-spirited, open-minded, cultured, serious, but always aware of the comedy, philosophic but never ponderous." Yet he sounds rather solemn when he writes: "I have always been motivated by my scientific curiosity and, at the same time, I want to take part in advancing science to benefit society.")

The German pharmaceutical firm Hoechst owns a controlling interest in Roussel-Uclaf, the French manufacturers of RU 486. According to Baulieu, Wolfgang Hilber, the Catholic president of Hoechst, opposed the introduction of the pill, partly from conviction and also because he feared reprisal against Hoechst by opponents of abortion in the U.S. When RU 486 was first introduced in France, abortion opponents bombarded the French embassy in Washington with intimidating letters and threatened to boycott all Roussel-Uclaf's goods. Faced with those threats and vociferous Catholic protests at home, the directors of Roussel-Uclaf decided in 1988 to suspend distribution of RU 486 just as Baulieu was departing for the World Congress of Gynaecology and Obstetrics in Rio to report on his new drug. He was shattered by the decision, but a few days after his arrival there he heard that Claude Evin, the French Minister of Health, had told Roussel-Uclaf that if they refused to produce RU 486, the rights would go to another company that would. Evin declared: "From the moment governmental approval for the drug was granted, RU 486 became the moral property of women, not just the property of a drug company." His courageous decision reopened the way to the marketing of the drug in France.

A social worker in California, Ms. Leona Benten, in collaboration with prochoice organizations, tried to bring the issue of the introduction of RU 486 into the United States to a head. She became pregnant, flew to London to purchase twelve pills of RU 486, and notified the Food and Drug Administration that she would import them at Kennedy Airport in order to use them on herself under her doctor's supervision.

The Food and Drug Administration had ruled some years ago that a person is allowed to import untested drugs not approved by them if they are for personal use. Under political pressure the FDA officials later excepted RU 486 from that rule. When Ms. Benten arrived at Kennedy Airport, FDA officials were there to meet her and confiscate the pills of RU 486. Ms. Benten, her doctor, and lawyer then applied to the federal court for the Eastern District of New York for return of the drugs and also asked the court to enjoin the enforcement of a ban on the drug. Judge Charles Sifton ruled that the pills should be returned to Ms. Benten, commenting that she could "hardly be faulted for relying on her own physician, in her own state," to perform the supervision necessary to ensure a safe and successful outcome. But he denied the broader application sought by the plaintiffs. A few days later, the 2nd Circuit Court of Appeals blocked his ruling. An emergency request to the Supreme Court to annul the Appeals Court's decision was turned down on the grounds, *The Washington Post* reported, "that the petitioners had failed adequately to make their case that the FDA's confiscation was illegal."

The opposition of the FDA and the Supreme Court to introducing RU 486 into the U.S. derives in part from the fear that it would further increase the number of abortions, but such an increase has not taken place in France where the abortion pill was introduced in 1988. By 1989 about one third of the total number of French abortions were performed with it. According to the statistics of the Institut National de la Santé et de la Recherche Médicale, the number of abortions since 1980 has varied from a peak of 182,862 in 1983 to a low point of 162,352 in 1987. The total was 163,090 in 1989 and 169,303 in 1990, well within the usual fluctuations. The number of abortions per hundred live births has remained steady since 1986 at just above twenty-one. The number of abortions per head of population performed annually in France is slightly larger than in Britain, despite the fact that France is predominantly Catholic and Britain predominantly Protestant.

In 1965, the great Scottish gynecologist Sir Dugald Baird gave a lecture with the title "A Fifth Freedom?". He began it by reminding his audience of a speech by Franklin D. Roosevelt on January 6, 1941, about the four freedoms for which World War II was being fought: freedom of speech and expression; of worship; and freedom from want and fear. Baird suggested that it was time to consider a fifth freedom, from the tyranny of civilized man's excessive fertility.[1] Unwanted pregnancies are a sad fact of life, and the decision whether an abortion is to be performed at all and whether by surgery of by hormone treatment should be a personal and medical decision rather than a judicial or political one.

Carl Djerassi pleads for continued intensive research on better and cheaper contraceptives as the most effective and humane way not only of reducing the number of abortions but of helping to limit population growth, particularly in poor countries; such research and the contraceptives that result from it, however, will not have a wide impact if intelligent programs of sex education are not also provided for young people. The low rate of teen-age pregnancies in the Netherlands is said to be a direct outcome of the excellent education about sexual behavior that children receive at school. Indian medical authorities have made it clear that educating women about sex, pregnancy, and contraception is essential for lowering the birth rate. Such education could be the most effective way of reducing the number of unwanted teen-age pregnancies everywhere.

Swords into Ploughshares: Does Nuclear Energy Endanger Us?[*]

Here in Britain we are all criminals: guilty of conniving at a crime against humanity committed by a government that is polluting the Irish Sea, the British Isles, and the entire globe with the radioactive discharges from its nuclear plants at Sellafield, a village in northwest England on the Irish Sea. According to Marilynne Robinson, the author of the book *Mother Country: Britain, the Welfare State and Nuclear Proliferation,* "The earth has been under nuclear attack [from Sellafield] for almost half a century." This book is aflame with indignation at the diabolical practices of the British Atomic Energy Authority, at the irresponsibility of our National Radiological Protection Board, at the careless indifference of our venal members of Parliament and of the British public, at the American press for failing to warn unsuspecting tourists of the deadly dangers threatening their health if they set foot on these poisoned isles, and the American government for wasting its armed forces on their protection.

Since reports of scandalous happenings that at first seemed beyond belief have often turned out to be true, I approached these accusations with an open mind. I had read of an accidental release of radioactive

*A review of the book *Mother Country: Britain, the Welfare State and Nuclear Proliferation* by Marilynne Robinson (Farrar, Straus and Giroux).

smoke from Sellafield and of radioactive wastes being discharged into the Irish Sea, but without knowing how much these discharges had added to the natural radioactivity that surrounds us, I had not been able to judge how dangerous they were.

The nuclear plants at Sellafield were constructed shortly after the end of World War II by the Labour government of Clement Attlee, in the first instance to produce plutonium for atomic bombs. Attlee and a few of his close associates reached that decision because the war had left Britain without allies. The United States had entered the war against Germany only after being attacked by Japan, and the war had ended without any treaty pledging the United States and Britain to come to each other's aid in case of another attack. Attlee feared that Britain might again find itself alone, as it did in 1940 and decided that having the ultimate weapon was essential for its security.

Under an agreement between Franklin Roosevelt and Winston Churchill signed at Quebec in August 1943, the first atomic bomb was developed at Los Alamos by a joint Anglo-American-Canadian team. According to this agreement,

> any post-war advantages of an industrial or commercial character should be dealt with as between the United States and Britain on terms to be specified by the President of the United States to the Prime Minister of Great Britain.

Doubts about postwar collaboration left by this agreement were allayed by an *aide-mémoire* signed by Roosevelt and Churchill at Hyde Park in September 1944, promising that full atomic collaboration between the two countries for military and commercial purposes should continue after the war, unless and until terminated by joint agreement. Seven months later Roosevelt died, and it seems that no other American officials knew of that agreement until they were told of it by the British. After the victory over Japan, Attlee and President Truman signed another document stating: "We desire that there should be full and effective cooperation in the field of atomic energy between the United States, the United Kingdom and Canada," but the following year Congress made most forms of atomic collaboration with other countries, including Britain and Canada, illegal.[1]

Nuclear reactors use the fission of uranium atoms to produce heat and plutonium. Natural uranium consists of two kinds of atoms, one having 235 and the other 238 times the weight of a hydrogen atom. For each atom of the former there are 140 atoms of the latter. Every so often an atom of uranium 235 splits up spontaneously into two lighter atoms with the emission of neutrons. If one of these neutrons collides with and is absorbed by another atom of uranium 235, that atom in turn splits, with the emission of more neutrons. In a large lump of pure uranium 235, this sets up an uncontrolled chain reaction leading to an atomic explosion.

In natural uranium, chain reactions do not occur, at least not on Earth, because the atoms of uranium 235 are too thinly spread and most of the neutrons emitted by them travel so fast that they escape without being absorbed. In nuclear reactors, that escape is prevented by a "moderator," a substance made of light atoms that bounce the neutrons back and forth until they have lost most of their speed and therefore have a better chance of being absorbed. The first American reactor for plutonium production at Hanford in the state of Washington consisted of a pile of uranium rods immersed in water that acted both as a moderator and as a coolant, and thus allowed a controlled chain reaction to take place. In that reaction, neutrons captured by uranium 235 generated more neutrons, together with radioactive fission products and energy, while neutrons captured by uranium 238 generated plutonium that was later extracted from the uranium rods in a chemical processing plant. The reactor required a large supply of very pure water, a safe way of discharging it, and a safe distance from large centers of population. No suitable site of this kind could be found in Britain.

The British team that returned from Los Alamos had to design their first atomic piles and the chemical separation plant for the extraction of plutonium with knowledge of only part of the American experience. They decided to use an as yet untried system: a pile of uranium rods interspersed with rods of graphite (pure carbon) as a moderator was cooled by a stream of air drawn in from below the reactor; the air was discharged, after being filtered, from 400-foot-high chimney stacks. The atomic piles were built at Windscale, the site of a wartime ordnance factory near the village of Sellafield, on the Cumberland coast. The first pile went into operation in November 1950, and the first British atomic bomb was exploded in Australia in November 1952, the same month as the first American hydrogen bomb.

Under the neutron bombardment the graphite rods in the Sellafield plant gradually became brittle. That brittleness could be cured by allowing the pile to warm up above its normal working temperature for several hours. In 1957, during one such operation, some of the fuel rods overheated and caught fire. While the operators tried to cool the rods by blowing more air through the pile, highly radioactive vapor escaped through the chimney stacks; finally the fire was extinguished by flooding the pile with water. Most of the dangerous radioactivity that resulted came in the form of radioiodine that contaminated the nearby countryside and made the milk from the cows grazing there unfit to drink for several weeks. More came in the form of polonium (the radioactive element Marie Curie named after her native land). At the request of the prime minister, Harold Macmillan, the Medical Research Council (an autonomous body equivalent to the National Institutes of Health) set up an independent committee to consider the consequences of the accident on the workers at Windscale and on the public, but the committee was not told about the release of polonium.

Radioiodine can give rise to cancer of the thyroid, but monitoring of the radioactivity of the thyroids of workers at Windscale and of people living nearby showed that none of them had received dangerous doses. The committee concluded "that it is in the highest degree unlikely that any harm has been done to anyone in the course of this incident."[2]

Before 1957 exposure to radioactivity below a certain threshold was generally believed to be harmless, but in the years that followed scientists became increasingly concerned about the biological effects of the radioactive fallout from atomic weapons tests. They found that the probability of a mouse developing cancer, or of a fruit fly's offspring being affected by a genetic mutation, increased if it received a dose of radiation, however small. It may increase only from one in 50,000 to one in 49,999, but this means that absorption of the same small dose by each of 50 million people may give rise to a hundred additional cases of cancer.[3]

In the light of these findings the National Radiological Protection Board, an autonomous body set up by the British government in 1970, later reevaluated the likely aftereffects of the Windscale fire. A plume of radioactive iodine and polonium spreading out from Windscale over parts of Britain and Northern Europe would have caused in many people traces of radioiodine to be taken up by the thyroid glands and traces of polonium by the lungs. Even though most of them would have received only minute doses of each, the probability that some of them would later develop cancer was thereby increased. Calculations showed that, in the forty years following the fire, there might be about 260 cases of thyroid cancer over and above the 27,000 or so naturally occurring ones in the affected populations. Of these additional cases about thirteen might prove fatal. Nine cases of other fatal cancers might be caused by the fallout of polonium.[4]

However, according to Rosalyn Yalow, the American physicist who received the Nobel Prize for Medicine for her invention of radioimmunoassays, an important and widely used tool in diagnostic medicine, there is no trustworthy experimental evidence to support these views. On the contrary, a great variety of observations indicate that our bodies are well equipped to withstand moderate doses of radiation. For example, no increased incidence of cancer or genetic abnormalities has been found in populations living in regions where the natural background radiation is abnormally high.

People in the Rocky Mountain states in the U.S. receive twice as much natural radiation as the rest of the American population, but cancer rates there are lower than average. In certain districts of India and Brazil, people's exposure to natural background radiation over a period of twenty-five years equals the acute exposure of Hiroshima and Nagasaki survivors, yet no deleterious health effects could be found there.[5] If Rosalyn Yalow is right, there would have been no additional cases of cancer, nor any other deleterious effects as a result of the Windscale fire.

Marilynne Robinson writes that the Windscale fire bore "an uncanny, not to say unnerving, similarity" to the nuclear accident at Chernobyl. In fact, the two reactors were quite different and so were the accidents. The atomic piles at Windscale were air-cooled, while those at Chernobyl were cooled by water under high pressure. At Chernobyl, the cooling water turned into steam that reacted with hot metals and graphite rods, producing hydrogen and carbon monoxide, while the nuclear reaction was still continuing. The hydrogen and carbon monoxide ignited, causing a tremendous explosion that lifted the roof off the building. There followed a meltdown of the reactor that could have contaminated the ground water of the region had it not been contained by the heroic efforts of the workers who excavated a tunnel underneath the reactor and filled it with concrete.[6] At Windscale, the uranium and graphite rods caught fire after the nuclear reaction had already been shut down, and the smoke from the fire escaped through the chimney stacks. There was no explosion and no meltdown. Extinguishing the Windscale fire with water could have initiated the same dangerous reaction between the steam and the graphite rods as at Chernobyl, but fortunately it did not, and the fire was put out.

Doses of ionizing radiation are measured in units called sieverts (after the Swedish radiation physicist Rolf Sievert). The dose received by an entire population is obtained by multiplying the dose received by one typical individual by the number of people in the population; the product of these two numbers is called a man-sievert.[7–9] On that basis the dose released by the accident at Three Mile Island amounted to 20 man-sieverts, the one at Windscale to 1,300 man-sieverts, and the one at Chernobyl to 150,000 man-sieverts, or over a thousand times greater than at Windscale.

So much for the unnerving similarity between the two accidents. For comparison, in 1963, the year of the atmospheric test ban, the radioactive fallout from nuclear weapons tests had caused the atmosphere to release 500,000 man-sieverts to people at ground level, and in 1986 it was still releasing 50,000 man-sieverts. The global exposure to natural background radiation amounts to twelve million man-sieverts per year. Distributed evenly over the world's 6,000 million people, this gives each of them an annual dose of two millisieverts, but in fact the distribution is very uneven. The International Commission on Radiological Protection recommends that the average annual exposure of members of the public to man-made radiation should not exceed one millisievert per person and that the annual dose to the most exposed workers should, on average, not exceed fifteen.

Robinson is indignant that no one was evacuated from Windscale. She writes:

> Comparison in this regard is to the advantage of the Russians who only delayed evacuation and who only temporized for a few days about the severity of the accident.

In fact there was no case for evacuation, and there is none today, not even with hindsight, because even those most exposed did not receive more than ten millisieverts. Robinson alleges that the staff at the reactor had undertaken an experiment whose nature has never been revealed. As we have seen, the cause of the fire was a routine maintenance procedure whose danger was not appreciated. It was described in detail in the White Paper published in 1957. Robinson writes that the Magnox reactors at Calder Hall next to Sellafield are similar to the one at Chernobyl. This is untrue; the piles of the Magnox reactors are cooled with carbon dioxide, a gas used to extinguish fires, thus avoiding the danger of fire as well as of explosion, while the Russian reactors are cooled by water under high pressure. According to Robinson the type of reactor that caught fire is still being used. In fact that reactor was never repaired and its twin was closed down immediately after the fire.

Originally the reprocessing plant at Sellafield was constructed to separate plutonium for military purposes only, but it was used later also to reprocess the spent fuel of civil reactors in Britain and other countries. The ensuing radioactive waste is separated into three categories of different radioactivity: high, intermediate, and low. The first two categories are stored. After treatment and further reduction of radioactivity, low-level liquid waste is discharged through a two-mile-long pipe into the Irish Sea. One of its components is plutonium, whose compounds are practically insoluble in water, are as dense as gold, and were expected to sink to the bottom and get covered with sediment. Another component is caesium 137, which resembles sodium in its chemical properties. Its salts are soluble and were expected to be diluted and dispersed without causing any perceptible rise in radioactivity of the oceans.

Between 1957 and 1982 British Nuclear Fuels discharged into the Irish Sea about a quarter of a ton of plutonium dioxide, which corresponds to 17,000 curies (not 50,000, as Robinson writes) and 650,000 curies of caesium 137, in addition to other smaller quantities of long-lived fission products and other radioactive elements. In 1982, the Atomic Energy Establishment at Harwell and the National Radiological Protection Board discovered that measurable quantities of plutonium and americium were getting washed ashore and were carried inland by the wind, even though their concentration in sea water is very low. In one mile along the coast from Sellafield the concentration of plutonium was seventy times greater than that deposited there by the atmospheric nuclear weapons tests of the 1950s and the early 1960s; three miles away it was ten times greater, seven miles away five times greater, and twenty miles away and beyond it was undetectable. The excess of caesium 137 was only five times above background level at its highest and fell to that level three to nine miles inland and beyond. In 1982, the total radioactivity deposited on land amounted to twice that deposited on the same small area by the nuclear weapons tests.[10–12]

In response to concern that radioactive elements might be taken up and concentrated by marine life, the Minister of Agriculture, Fisheries and Food commissioned regular annual studies of the fish, crabs, mussels, and seaweed near Sellafield. This showed the concentrations of plutonium in shellfish and seaweed to be up to a thousand times greater than in the seawater. Even so, heavy consumers of seafood caught near Sellafield would have been exposed to only about a third of the annual dose of one millisievert recommended as a safe limit by the International Commission on Radiological Protection.[13,14]

What about the caesium 137 that was poured into the Irish Sea? In 1987 contamination of the Irish Sea with caesium 137 produced a radioactivity of one-tenth of a becquerel per kilogram of seawater, except near the northwest coast of England where the activity rose to half a becquerel per kilogram.[14] For comparison, the natural radioactivity of seawater amounts to twelve becquerels per kilogram, nearly all of it from potassium 40. Hence the discharge from Sellafield has increased the radioactivity of the Irish Sea by just under 1 percent, which can hardly be called a danger.[15]

All the same, the buildup of radioactivity could not be allowed to continue. The Sellafield plants have been modernized and the outflux of radioactive waste reduced to near zero at a projected cost of over three billion dollars.

Robinson reports that in the village of Seascale, a few miles from Sellafield, one child in sixty died of lymphoid leukemia. Between 1955 and 1984 there have been seven deaths from leukemia among 1,068 children born there, or one in 152. This is ten times the national average. On the other hand, mortality among 1,546 children living there, but born elsewhere, was normal. No case of leukemia or lymphoma was reported among them; nine of the ten deaths that did occur were caused by accidents. The additional frequency of lymphoid leukemia expected from the levels of radioactivity determined at Sellafield in a long series of painstaking measurements should be not one in 152, but one in 50,000.[16–21]

This disturbing discrepancy has stimulated statistical analyses of the incidence of leukemia near and far from nuclear installations. They showed a significantly raised incidence of lymphoid leukemia near Sellafield and Dounreay, two nuclear installations built before 1955, and a significantly lowered incidence near other nuclear installations. The study was headed by Sir Richard Doll, the distinguished epidemiologist and codiscoverer of the association between smoking and lung cancer. The increased incidences of lymphoid leukemia were too large to be owing to chance, but they could not be explained by the observed levels of radiation.[22] Nevertheless, in response to a lawyer's advertisement which offered his services, several of the families at Sellafield whose children contracted cancer sued British Nuclear Fuels for damages.

Robinson heaps scorn on an enquiry headed by Sir Douglas Black,[23] a former chief medical officer of health whose "lines of reasoning was ingenuous rather than persuasive," because he argued that the very low level of additional radiation from the nuclear plant could not account for the high incidence of leukemia. She chooses to ignore the results of two other independent inquiries that confirmed his conclusions.[18,19]

Despite these finds, public suspicion that some hitherto undiscovered effect of radiation is responsible may persist until an alternative explanation has been found. Leo Kinlen of the Cancer Research Campaign Epidemiology Unit in Edinburgh has advanced such an explanation and tested it by making a bold prediction.[24] Both Sellafield and Dounreay were small isolated villages until the building of the nuclear plants brought large influxes of people. Such movements of people into isolated rural areas are liable to bring infections with diseases to which larger populations have become immune. If we suppose childhood leukemia was caused by some unidentified virus, as has often been suspected, then an influx of population into an isolated rural community, distant from any nuclear installation, should have given rise to an increased incidence of leukemia similar to that seen at Sellafield and Dounreay.

Before 1948, Glenrothes in Scotland, with a population of 1,100 people, was a relatively isolated rural community, far from any nuclear installation. By 1961, the founding of the new town of Glenrothes had raised the population to 12,750. Kinlen predicted that, if his hypothesis were right, this rise should have led to an increase in childhood leukemia. Examination of the medical records did indeed show a significant excess of leukemia deaths below the age of twenty-five; ten observed deaths compared to 3.6 expected, seven of them below the age of five; and six occurred between 1954 and 1959. After 1968 there was no excess, and indeed a significant deficit. Kinlen found no such excess in other regions where the population had not increased. The cluster of cases of childhood leukemia in Glenrothes is the first to be predicted by a hypothesis that was formulated before such data were collected. There is as yet no direct evidence for viral origin of most human leukemias, but the similarity between them and animal leukemias known to be caused by viruses has long been suggestive. Kinlen's important result will intensify the search for a possible virus, especially since unexplained clusters of lymphoid leukemia have also been found in other places far from nuclear installations.[24]

In 1985, a Committee at the Department of Health discussed another conceivable explanation. Perhaps children had some special pathway, not found in adults, that would cause traces of plutonium to be selectively absorbed and concentrated in their bone marrow where it would give rise to leukemia. The minutes of that meeting were leaked to Greenpeace who informed the House of Commons Environment Committee: "Unbelievably, it was suggested that Cumbrian children should be fed contaminated food and monitored to see what effect it had on them in terms of concentration within their bodies." The Environment Committee reports: "Not surprisingly, we were very shocked by this." Journalists were equally shocked. *The Times* carried a front page article headlined "'PLUTONIUM FOOD' SOUGHT FOR CHILDREN," the *Daily Mail* headlined "SHOCK OVER 'NUCLEAR TEST' CHILDREN." Other articles followed. The committee's report states:

> We questioned Greenpeace witnesses closely on their statement and found that under examination they began to shift their ground. The experiments became "voluntary"—as if parents would submit their children to these risks. However we were assured by Greenpeace that their claim could be fully substantiated. At our insistence they sent us a confidential copy of minutes taken at the meeting in the Department of Health at which the proposal for the experiment was allegedly made. We examined these carefully and could find no reference which could be construed as supporting the claim. The nearest we could come to it was a discussion that the only way in which incontrovertible evidence could be obtained of the effects of ingestion of contaminated shellfish on the human system was by finding a group which had never eaten shellfish, such as children. But, it was added, such an experiment would be wholly unacceptable. Thus on the most generous of interpretations, Greenpeace stretched a passing reference to the point of extreme distortion, just for the sake of sensation or, more seriously, in order to mislead the Committee.

The report of the House of Commons Committee has been published,[25] but this does not deter Robinson from gleefully citing the reported intention to feed plutonium to children as a prime example of "the moral aphasia" of British society and alleging Britain to be so contaminated with plutonium dust that many children would have eaten it already.

Is Britain really "befouled" by radioactivity? Table 1 shows that the exposure to radioactivity of the average American is half as much again as that of the average Briton, because in the U.S. exposure is greater both to medical X-rays and to radon. Radon is a natural radioactive gas given off

Table 1. Contributions to Average Radiation Exposure (millisieverts)

UK 1988	Source	US 1987
1.2	*Radon isotopes from earth materials (98% from indoor exposure)	2.0
0.35	Terrestrial gamma-rays	0.28
0.30	Medical exposures	0.53
0.30	*Ingested natural radionuclides	0.39
0.25	*Cosmic rays	0.27
0.01	Fallout (including Chernobyl)	0.0006
0.01	Miscellaneous sources	0.05–0.13
0.005	Occupational exposure	0.009
0.001	Radioactive effluent discharges	0.0005
Total: 2.5		**Total: 3.6**

*Natural sources

by certain rocks. Recent research has shown that it can accumulate dangerously in people's houses. Radioactive effluents and fumes from nuclear installations account for only 0.02 percent of the exposure in either country. Radon in houses, not nuclear effluent, presents the greatest single radiation risk in both countries.[26]

Some of Robinson's most venomous diatribes are directed at the British National Radiological Protection Board. She calls it "the incredibly feckless agency responsible for monitoring public exposure to radiation," "small and besieged," "struggling with a shrinking budget,"

> a creature of the state, funded, shielded and patronized by the government and flourishing in the balmy atmosphere of Crown Immunity, where no acts of parliament apply, and under the protection of laws affecting national defense and commercial confidentiality as well as the Official Secrets Act.

In fact, the board is an independent advisory body set up by an act of Parliament and answerable for its own actions. It does not have crown immunity and is only partly funded by the government. The board regularly publishes detailed reports of radioactivity throughout the British Isles and all its publications are freely available in the Cambridge University Library, where I went to study them.[27,28] Robinson alleges that British doctors "are legally prohibited from giving out information that is not officially authorized." In fact, British doctors' contracts with the Department of Health contain no such restrictions; doctors do not have to sign the Official Secrets Act. Robinson's book abounds with scientific errors and unfounded allegations, reminding me of the lines in Heinrich Kleist's play *The Broken Jug*, where the judge says to the defendant: "In your head science and error are kneaded together intimately as in a dough; with every slice you give me some of each."

I wonder why Robinson turned a blind eye to American nuclear plants that have polluted the countryside with radiochemicals. I read that at Hanford an estimated 15 million gallons of high-level liquid waste containing plutonium has been pumped into rocks saturated with water beneath the Hanford reservations. In 1988 water from a local spring that flows into the Columbia river was found to contain 350 becquerels of plutonium per liter. Compare this with one-tenth of the becquerels of caesium and one-thousandth of a becquerel per liter of plutonium in the Irish Sea.[29] I was alerted to the American discharges by a British newspaper report alleging the discharge of 4 million kilos of plutonium into the rocks below Hanford, which seemed absurd. When I remonstrated with the editor, he checked with his Washington correspondent, who told him that it was four kilos. Nearly all of Robinson's more than three hundred cited sources are newspaper reports, but she apparently never checked what she read there.

There is much to be criticized about the operation of the plutonium factories at Sellafield and the misleading information they issued repeatedly about their radioactive discharges. Both the original piles and the reprocessing plant were built hurriedly and suffered from technical defects. The government's original decision to pour low-level radioactive waste into the Irish Sea was taken in 1950, at a time when less was known than today about the harmful biological effects of radiation and the possible buildup of radionuclides in living creatures, but the discharge should not have been continued for over thirty years. It was inexcusable that the government concealed the escape of polonium during the Windscale fire from the Medical Research Council Committee set up the 1957 to study the health effects of the accident and from the National Radiological Protection Board's reassessment of its impact in 1982. That escape became known only after publication of the board's report, and was the subject of an addition to the report published in 1983, twenty-six years after the event.[30]

In 1983, Greenpeace found radioactive debris washed up on a beach near Sellafield. This turned out to be caused by faulty separation of the effluent from the reprocessing plant about which the firm had kept quiet. It was the National Radiological Protection Board that Robinson so maligns that told the government to warn the public about the contamination of the beaches and urged it to clean them up. These and other scandalous malpractices shook public confidence in the management of Sellafield, but none of them had the severe ecological consequences of global significance that Robinson attributes to them.

Robinson's account of Sellafield forms the second part of her book. In the first part she presents a social history of England from the fourteenth century to the present day intended as background for her discussion of contamination at Sellafield. She writes:

> For decades the British Government has presided over the release of deadly toxins into its own environment... Such behaviour... has... a history in which the inhibitions which expedite it and the relations it expresses evolve together... The core of British culture is the Poor Law... A very important article of faith was that wages of workers should not exceed subsistence.

This last claim is broadly true. In the seventeenth century, Sir William Temple said that high wages would make the poor "loose, debauched, insolent, idle and licentious," and in the next century Adam Smith declared that "everyone but an idiot knows that the lower classes must be kept poor or they will never be industrious." He might have added "ignorant," for England did not introduce universal schooling until 1880, while Prussia had introduced it already in 1763.

However, these snobbish attitudes were not the real cause of widespread poverty. In preindustrial times most people were poor everywhere because insufficient wealth was produced to keep everyone housed, fed, and clothed. The Englishman Gregory King showed this vividly in the *Scheme of the Income and Expense of Social Families of England* that he published in 1688.[31] At the top of the scale, he lists 186 families of spiritual and temporal Lords with annual incomes of $6,000 per head (calculated on the basis of one pound sterling in 1688 being equivalent to $75 at present prices). At the bottom there are 850,000 families of laborers, servants, cottagers, paupers, soldiers, and seamen with yearly incomes of between $150 and $500 per head, and finally 30,000 vagrants, gypsies, thieves, and beggars without any income. Only 2.7 million people earned more than they needed for their bare subsistence, while 2.8 million earned less and had to be helped by the others.

The distribution of as much as four-fifths of the income of the 60,000 wealthiest people would have raised the annual income of the nearly three million poorest by only $200 per head. This might have doubled the income of some of them, but it would still have left them desperately poor. In 1622, a preacher declared that laborers "are scarce able to put bread into their mouths at the week's end, and clothes on their backs at the year's end." The Cambridge historian Peter Laslett writes:

> It is probably safe to assume that at all times before the beginning of industrialization, a good half of all those living were judged by their contemporaries to be poor, and their standards must have been extremely harsh, even in comparison with those laid down by the Victorian poor law authorities.

This was true of much of the rest of Europe, where the annual GNP per person was about the same as in India today ($235). The Poor Law, contrary to Ms. Robinson's view, was a safety net, designed to keep people within the village community from starvation; it was a Christian institution that compelled the more fortunate half of the population to help the other half to survive.[32]

Robinson alleges that high infant mortality rates and short life expectancy have been particularly characteristic of England, but we know them to have been universal before the present century. In 1693, the English astronomer Edmund Halley published a study of life expectancy in the German city of Breslau, where good records of births and deaths were kept. Of every hundred children born, only fifty-one were alive at the age of ten and only thirty-six survived to the age of forty.[33] Most cities used to be death traps, where life expectancy was no more than twenty years, because people lived crowded together and perished from infections. In England, child mortality was lower and life expectancy longer than in Breslau: between 1550 and 1800 about three quarters of English children born survived to the age of ten, probably because well over four-fifths of the population lived in villages, and people were therefore less exposed to infections. Robinson's scathing picture of nineteenth-century England would make one expect that Americans lived longer than Englishmen. To my surprise I found the contrary. In 1850, the average life expectancy in the state of Massachusetts was 38.3 years for males and 40.5 for females;[34] in England it was 40 years for males and 42 for females.[35]

Robinson alleges that the present Welfare State defrauds the poor. According to her,

> The British government turns a profit on the National Insurance System which goes into the treasury. So those who pay National Insurance [which includes all those employed] are taxed at a rate that subsidizes other activities of government.

The truth is that National Insurance Contributions almost exactly balance benefits that include pensions, unemployment and sickness pay, and others. The National Health Service from which everyone benefits is financed almost entirely out of taxes which the poorest do not pay.

Robinson's social history lacks historical perspective because she fails to compare social conditions and attitude in England to those prevalent throughout Christian Europe at the time. Through much of her history, Robinson confuses literary impressions and distortions with historical fact and ignores modern research based on numerical analyses of historical records. Sneers about every aspect of English character and institutions fill page after dreary page, with tedium enough to turn an IRA man into a Loyalist. Her account would make one believe that social deprivation has never existed in the United States. Has she never read John Steinbeck's *Grapes of Wrath*?

Robinson's account of Sellafield is based on press reports and antinuclear pamphlets. Knowing no science, she has spurned study of the abundant technical literature that would have saved her from monstrous exaggerations of the danger presented by Sellafield. In the middle Seventies, when discharges from Sellafield were at their height, the collective annual dose from its atmospheric discharges amounted to six man-sieverts,[36] compared to about 100,000 from the atmospheric atomic weapon tests carried out until 1963. We have seen that the discharges into the Irish Sea have raised its radioactivity by less than 1 percent. That can hardly be called "a nuclear attack" on our planet. Nor is Britain "the largest source, by far, of radioactive contamination of the World's environment." That source is natural radioactivity; next comes fallout from nuclear weapons tests and from Chernobyl. By contrast, radioactive contamination from nuclear plants accounts for less than one-thousandth of the average Briton's radioactive exposure, and away from the British Isles it is barely detectable.

*What If?**

At dawn on May 10, 1940, Hitler's armies broke into Belgium and Holland. That same afternoon Winston Churchill took office as prime minister of Great Britain. At 7:30 AM on May 15, Paul Reynaud, the French premier, woke Churchill with the news that German tanks were pouring into France across the Ardennes at Sedan. France, he said, was beaten.[1]

Churchill realized at once the deadly threat to Britain that this posed. On that same afternoon he wrote to President Roosevelt:

> As you are no doubt aware, the scene has darkened swiftly. If necessary, we shall continue the war alone and we are not afraid of that. But I trust you realise, Mr. President, that the voice and the force of a United States may count for nothing if they are withheld too long. You may have a completely subjugated, Nazified Europe established with astonishing swiftness, and the weight may be more than we can bear.

Roosevelt sent a friendly but noncommittal reply to which Churchill answered two days later: "We are determined to persevere to the very end whatever the result of the great battle raging in France may be.... But if American assistance is to play any part it must be available soon." Churchill should have written "I am determined to persevere...," because he had yet to persuade his colleagues in the War Cabinet—the inner group set up to decide war policy—that this was the right course. In his *War Memoirs*

*A review of the book, *Five Days in London, May 1940* by John Lukacs (Yale University Press).

Churchill generously concealed that battle so as not to embarrass his former colleagues. John Lukacs has extracted it from the dry official records and transformed it into a gripping historical drama, *Five Days in London.* He shows that during those crucial five days in May 1940, the fate of Europe and indeed of much of the world depended on the outcome of an argument between just three men. Lukacs makes the drama unfold against a background of many British people's slow and often placid reactions to the disasters in France and the imminent threat of an enemy invasion.

Churchill's predecessor, Neville Chamberlain, had been forced out of office by a revolt in the House of Commons, but he still enjoyed strong support in the Conservative Party. The party leaders first offered to the foreign secretary, Lord Halifax, that they would recommend him to the King as Chamberlain's successor, but he declined. The next choice fell on Churchill, really because there was no one else for the job.

Churchill shared his responsibilities with four colleagues in the War Cabinet. They included Neville Chamberlain, who had agreed to the dismemberment of Czechoslovakia at Munich in 1938 in return for Hitler's promise that a slice of that country was his last territorial demand, and the foreign secretary, Lord Halifax, a pillar of the Church and ex-viceroy of India, who had naively judged Hitler to be just another nationalist leader like Gandhi. Lukacs writes that in July 1938, Hitler's adjutant, Captain Fritz Wiedemann, visited Halifax in his office. According to Wiedemann, Halifax bade him goodbye, saying that he "would like to see as the culmination of his work the Führer entering London at the side of the King amid the acclamations of the English people." Conscious of his exalted rank on inheriting his viscountcy, he had instructed his daughters to address him as "Lord Halifax."[2] The other two members of the War Cabinet were the rather silent Clement Attlee and Arthur Greenwood, both leaders of the Labour Party who had just been invited to join the government in the interest of national unity.

The King and most members of the Conservative Party trusted Chamberlain and regarded Churchill as an untrustworthy adventurer. Churchill was aware of this and did not yet feel secure in his job. Hitler was confident that Churchill would not last; perhaps the many British peers who pandered to Hitler in the 1930s, like Lord Darlington and his friends described in Kazuo Ishiguro's *Remains of the Day*, had given him that illusion. In the novel Lord Darlington invites them to his country seat for the weekend to meet Ribbentrop, Hitler's ambassador; Lord Darlington swallows the Nazi creed so far as to dismiss his two hard-working and innocent Jewish maids. In the real world, Lord Astor organized house parties for Ribbentrop and influential British Nazi sympathizers at Cliveden, his country seat on the Thames; Chamberlain rented his house in elegant Eaton Square to Ribbentrop while he lived at his official residence; and Lord Rothermere, the owner of a newspaper empire that included the conservative *Daily Mail*, telegraphed Hitler in 1938: "Mein Führer, your star is rising higher and higher and I wish you every success." Lukacs does not mention that highly placed British and American officials dismissed reports of concentration camps and Nazi atrocities as Jewish propaganda right up until 1945, when the advancing Allied armies confirmed them.

According to Reynaud's memoirs, the French disaster need not have happened. The German General Gunther Blumentritt, who was in charge of the breakthrough at Sedan, recalled meeting no serious resistance there. Reynaud wrote in his memoirs that, despite several warnings, this sector had been manned by a poorly officered army corps equipped with neither antitank nor antiaircraft guns. King Leopold III of the Belgians had warned General Maurice Gustave Gamelin, the French commander-in-chief, that the Germans' main thrust would take place around Sedan. French intelligence had informed Gamelin of intense air reconnaissance over the area and a buildup of military supplies on the German border with Luxembourg, and the French military attaché in Bern informed Reynaud that the attack was planned for between May 8 and 10, but, according to Reynaud's memoirs, "Gamelin did not change his plans by one iota." Few people take notice of what they are told.

The Allies could muster nearly twice as much heavy artillery, nearly half as many and better tanks, and nearly a third more warplanes than the Germans, and they were confident of their superior strength. Hitler expected a protracted war, and the German High Command doubted that their attack would succeed.[3] Its success was due to the weakness of the French defenses near Sedan and to General Heinz Guderian's new strategy of concerted attacks by tanks and aircraft, to which the ponderous French war machine was too slow to respond. Field Marshal Gerd von Rundstedt, who was in overall command, called its success miraculous. Ernest May writes in his recent book, *Strange Victory*: "The essential thread in the story of Germany's victory over France hangs on the imaginativeness of German war planning and the corresponding lack of imaginativeness on the Allied side.... They neglected to prepare for the possibility of surprise."[4]

On the day after receiving Reynaud's grim news, Churchill flew to Paris, where he saw the ominous spectacle of the Foreign Ministry's archives on fire in the garden behind the Quai d'Orsay. Gamelin told him that German armored divisions had broken through on a fifty-mile-wide front, and had been followed immediately by truckloads of infantry. When Churchill asked where his strategic reserves and his maneuverable troops were deployed, Gamelin replied that he had none. In response to Reynaud's desperate plea for help Churchill telegraphed London (using Hindustani as a code), asking for another ten fighter squadrons of the Royal Air Force to be sent to France. He did so reluctantly because he foresaw that they would soon be needed for the defense of Britain, and he later refused to send any more.

One of Churchill's first acts was to dismiss from the larger twenty-five-member Cabinet the chief appeasers, Sir Samuel Hoare and Sir John Simon, and he also dismissed Sir Horace Wilson, the *éminence grise* in the Cabinet Office, and he replaced them with men resolved to fight the Germans. As the news from France got worse, Churchill flew to Paris again on May 22, accompanied by chief of the General Staff, Sir John Dill. They heard that German Panzer divisions had reached the English Channel. This meant that the British army of some 200,000 men, the Belgian army, and a large French army were surrounded on three sides and faced with being either annihilated or taken prisoner unless they could escape to England from Dunkirk, the only port still open. Reynaud had by now replaced Gamelin with the seventy-three-year-old World War I veteran General Maxime Weygand, who presented Churchill and Dill with a joint plan of action. French, British, and Belgian armies were to attack the German bulge from the north and another French army was to attack it from the south.

But was it not too late? Alone with Reynaud afterward, Churchill complained that Lord Gort, the commander of the British Expeditionary Force, had been left without orders from the French High Command for an entire week, so that the chance for a successful counterattack had been missed. Faced with the alternative of mounting a counterattack that was doomed to fail, or saving the only army Britain had, Gort ordered it to retreat to Dunkirk, and the War Cabinet supported his decision a few hours later.

In London, on May 25, Halifax took the initiative in what he conceived as the only way to save Britain. He invited Signor Giuseppe Bastianini, the Italian ambassador, to the Foreign Office to see him. Using diplomatic circumlocution, Halifax sounded out Bastianini to explore with what concessions—perhaps over Gibraltar and Malta—Mussolini could be bribed to keep Italy out of the war, and to intervene with Hitler in order to call a conference for a "general European settlement." The same idea had led Halifax to visit Hitler, Goering, Goebbels, and Schacht in 1937 when they all pulled the wool over his eyes. Hitler had assured Halifax of his peaceful intentions, although a few days before, he had assembled his top generals and admirals and told them to prepare for a major war within five years. Neither in his diaries nor in his memoirs does Halifax mention his approach to Bastianini.

At 10:00 PM on May 25, Churchill called a meeting of the Defence Committee where he said that he would not be at all surprised if the Germans made a peace offer to the French. Lukacs writes that "this was extraordinary. Churchill knew nothing about the sorry deliberations of the French high council, which had adjourned in Paris only an hour or so before." That council had debated whether France was bound by its treaty of alliance with Britain not to enter into unilateral negotiations with Germany. Weygand and the eighty-four-year-old Marshal Pétain, hero of Verdun in World War I, whom Reynaud had invited to join his government, favored negotiations before the army was completely destroyed. Reynaud proposed flying to London to tell the British that France would continue the struggle if only to save its honor, but Weygand insisted that the army must be preserved as the last instrument of order (not to save the lives of his soldiers).

On Sunday, May 26, Reynaud and a French delegation came again to confer with the War Cabinet. Reynaud told the Cabinet that Mussolini was about to declare war, which would force France to divert parts of its army to defend its frontier with Italy and expose its ships in the Mediterranean to Italian attacks. He asked whether they could not try to persuade Mussolini to change his mind by offering him a formula that would satisfy his *amour propre* in the event of an Allied victory, because Mussolini would find himself in difficulties if the Germans lost the battle for France. Reynaud reported that both Weygand and Pétain favored an armistice, but Churchill, "with the courage of a lion," rejected any concession to Mussolini. At lunch alone with Churchill, Reynaud told Churchill of the near hopelessness of the French military position and hinted that if he (Reynaud) refused to sign the peace terms imposed by the Germans, he might be forced out of office.

In Paris later that evening, Paul Henri Spaak, the Belgian foreign minister, awaited Reynaud at the airport to tell him that King Leopold of the Belgians was going to capitulate, which would exacerbate the perils to the British and French armies that had come to Belgium's rescue. When the War Cabinet met again that evening Churchill said:

> If France could not defend herself, it was better that she should get out of the war rather than that she should drag us into a settlement which involved intolerable terms. There was no limit to the terms which Germany would impose upon us if she had her way.

Halifax, on the other hand, and to some extent Chamberlain still believed that, with French mediation, Italy could be bought off by territorial concessions and prevented from entering the war on Germany's side, and that it would be in Mussolini's interest to arrange a conference for a "general European settlement." Toward the end of the May 26 meeting Halifax presented a draft which he attributed to Reynaud, but which Lukacs suspects was his own:

> If Signor Mussolini will co-operate with us in securing a settlement of all European questions which safeguard[s] the independence and the security of the Allies, and could be the basis of a just and durable peace for Europe, we will undertake at once to discuss, with the desire to find solutions, the matters in which Signor Mussolini is primarily interested.

Lukacs concludes that Reynaud had hoped to buy off Mussolini, while Halifax had wanted him to mediate with Hitler. Alexander Cadogan, the permanent secretary of the Foreign Office, summed the meeting up in his diary:

> He [Churchill] is against final appeal, which Reynaud wanted, to Muss. He may be right there. Settled nothing much. W.S.C. too rambling and romantic and sentimental and temperamental. Old Neville still best of the lot.[5]

Churchill opposed the approach to Mussolini because he believed that Hitler would have regarded it with contempt, that the Dominions and the rest of the world would have interpreted it as Britain suing for peace, and that at home it would have broken the will to fight. Churchill was convinced that Hitler would have imposed humiliating conditions and would not have agreed to any settlement that did not leave him in complete control of Europe. Besides, as Chamberlain found out to his cost after Munich, and as Stalin would discover in the year that followed, agreements with Hitler were worthless. An approach would have been a fatal step.

Where did "Old Neville" stand? This was crucial, because Churchill's position would have become untenable if both Halifax and Chamberlain had opposed him. Chamberlain was still leader of the Conservative Party with a large majority in the House of Commons. After he had been discredited for having believed Hitler's promise that a takeover of the German-speaking part of Czechoslovakia constituted "his last territorial demand," Chamberlain realized that Hitler could not be trusted, and in that he agreed with Churchill. On the other hand, Halifax believed, and historians such as John Charmley still maintain today,[6] that to save the British Empire and prevent a long-drawn-out war, a peace with Hitler should have been attempted, especially in the spring of 1941, after the Battle of Britain and the Blitz on British cities had shown Hitler that Britain could not be defeated. Goebbels' record of a telephone conversation with Hitler on June 25, 1940, suggests what this peace would have been like:

> Call from the Führer.... Does not yet know for certain whether he will proceed against England. Believes that the Empire must be preserved if at all possible. For if it collapses, then we shall not inherit it, but foreign and even hostile powers will take it over. But if England will have it no other way, then she must be beaten to her knees. The Führer, however, would be agreeable to peace on the following basis: England out of Europe, colonies and mandates returned. Reparations for what was stolen from us after the World War.... England must not be allowed to get off easily this time.[7]

Hitler's idea of ruling the Empire was made clear by his advice to Halifax at Berchtesgaden in 1937 concerning the British troubles in India: "Shoot Gandhi, and if that does not suffice to reduce them to submission, shoot a dozen leading members of Congress; and if that does not suffice, shoot 200 and so on until order is established." Hitler thought that this was the way a superior race must behave.[8]

Hitler may actually have planned to fulfill Halifax's dream in 1938 by setting up a puppet king and government in Britain after a successful invasion. There is no direct evidence for this, but in his memoirs, Walter Schellenberg, the head of Hitler's foreign intelligence service, writes that in July 1940, Ribbentrop summoned him to convey Hitler's order either to kidnap the Duke of Windsor (the deposed King Edward VIII) and his American-born wife or to lure them into Hitler's orbit. At the time the duke and duchess were on a visit to Portugal, and Spanish friends had invited them to a hunt near the Spanish border. Schellenberg was to contact the duke there and offer him fifty million Swiss francs if he agreed to dissociate himself from the British royal family and to move to Spain or Switzerland; if he refused, Schellenberg was to remove him and the duchess by force, but making sure not to injure them. Hitler's farcical plot failed. The duke canceled his participation at the hunt, and the Portuguese assigned an extra twenty police to guard him. An emissary of Churchill's arrived to escort the duke and duchess on board a vessel bound for the Bahamas, where Churchill had appointed the duke governor to have him safely out of the way.[9]

Sunday, May 26, was made a National Day of Prayer; a service in Westminster Abbey was attended by the King and Queen and all the top brass. John Betjeman later satirized the uplifting spirit of such occasions in a poem, quoted by Lukacs, that ends with the stanza:

Now I feel a little better,
  What a treat to hear Thy Word,
Where the bones of leading statesmen
  Have so often been interr'd.
And now, dear Lord, I cannot wait
  Because I have a luncheon date.

On Monday, May 27, evacuation of the British army from Dunkirk had begun, but, as Lukacs writes, so far only 7,700 men had been shipped home to England. Churchill sent a message to King Leopold of the Belgians imploring him not to surrender, because it would divide his nation and deliver it into Hitler's hands, quite apart from its disastrous consequences for his allies. Ignoring this, the King asked the Germans for a cease-fire starting at midnight.

Churchill issued a stern message to his ministers to use confident language in their public pronouncements, because most of the people would refuse to accept defeat, but *Mass Observation*, an opinion poll of the period, reported that some of the younger housewives would have welcomed Hitler, because "it couldn't be worse, they'd at least have their husbands back." Little did they know.

In the morning of May 27, Chamberlain reminded the War Cabinet of the chief of staff's advice that Britain's ability to hold out depended on full financial and economic support from the United States, since otherwise Britain would no longer be able to pay for the arms it needed for its defense, but there was as yet no sign of such support, partly because Roosevelt did not yet trust Churchill. Apparently, Roosevelt expected that if

England were defeated on land, the British fleet would come to North America, but Churchill warned him not to count on that. Did Churchill mean that a Nazi-backed puppet government might hand the fleet over to Germany?

At the afternoon meeting, the conflict between Halifax and Churchill became acute. Lukacs writes that Halifax confronted Churchill with his own draft memorandum, "A Suggested Approach to Signor Mussolini," asking him to mediate with Hitler. Chamberlain supported Halifax, believing it to be important for the sake of the French, who should at least be given the chance to negotiate with Italy, but Churchill disagreed and was backed up by Sir Archibald Sinclair, the leader of the Liberal Party, who had joined the meeting, because "any weaknesses on our part would encourage the Germans and the Italians, and it would tend to undermine morale both in this country and in the Dominions." Greenwood said, "If it got out that we had sued for terms at the cost of ceding British territory, the consequences would be terrible...."

> After more discussions Churchill said that he was increasingly oppressed with the futility of the suggested approach to Mussolini, which the latter would certainly regard with contempt.... The best help we could give to M. Reynaud was to let him feel that, whatever happened to France, we were going to fight it out to the end.... The approach proposed was not only futile, but involved us in a deadly danger.

The argument continued until Halifax finally asked: "Suppose Herr Hitler, being anxious to end the war..., offered terms to France and England, would the Prime Minister be prepared to discuss them?" Churchill gave the conciliatory reply that he "would not join France in asking for terms; but if he were told what the terms offered were, Churchill would be prepared to consider them." At one point, apparently in the heat of argument with Halifax, Churchill said he would be willing to accept an offer of peace on terms of restoration of German colonies and overlordship of Central Europe; but this may have been no more than a tactical move in his argument with Halifax, suggesting that the Cabinet should wait for such an offer rather than initiating negotiation with Hitler by asking Mussolini to mediate. In any case, Churchill thought such an offer unlikely, and it is a firmly documented historical fact that he refused to approach Mussolini.

At the end of the meeting Halifax told Sir Alexander Cadogan: "I can't work with Winston any longer." But Cadogan said: "Nonsense; his rhodomontades probably bore you as much as they do me, but don't do anything silly under the stress of that."[10] Like Churchill, a descendant of the great duke of Marlborough, Cadogan, son of the 5th Earl Cadogan, and also Halifax were aristocrats whose forebears had made history, and they were unimpressed by Churchill. Halifax complained that he talked "the most frightful rot" and Cadogan was irritated by his being "theatrically bulldogish." He does seem to have addressed the War Cabinet as if it had been a public meeting. Halifax now asked Churchill to come out into the garden with him. What went on between them is unrecorded, except that Halifax told Cadogan afterward that Churchill had been very affectionate, but Lukacs believes that Churchill impressed upon Halifax that Halifax's resignation would open up the gravest national crisis. On the other hand, Churchill could not convince him that asking Mussolini to mediate with Hitler would be futile.

On May 26, Cadogan wrote in his diary: "It is a strain—daily and hourly looking the ugliest facts in the eye.... A non-stop nightmare.... God grant that I can go on without losing faith or nerve. V. tired, but how these others— Chiefs-of-Staff etc. stand up to it, I can't think." But of Churchill himself his private secretary, John Colville, wrote in his diary: "Winston's ceaseless industry is impressive." He is said to have thrived on crises, but his manner suggests that the strain told on him too. On June 27 his wife, Clementine, wrote to him:

> One of the men in your entourage (a devoted friend) has been to me and told me that there is a danger of your being generally disliked by your colleagues and subordinates because of your rough sarcastic and overbearing manner—It seems your Private Secretaries have agreed to behave like schoolboys and "take what's coming to them" and then escape out of your presence shrugging their shoulders—Higher up, if an idea is suggested (say at a conference) you are supposed to be so contemptuous that presently no ideas, good or bad, will be forthcoming.... Except for the King, the Archbishop of Canterbury and the Speaker you can sack anyone and everyone—Therefore with this terrific power you must combine urbanity, kindness and if possible Olympic calm.[11]

His daughter Mary Soames, who edited the letters of her mother and father, believes that Churchill took this advice to heart, writing that "although during the years of his greatest power he could be formidable and unreasonable, many of the people who served directly under him during those dire years have put on record not only their admiration of him as a chief, but also their love for a warm and endearing human being."

On May 28 the argument between Churchill and Halifax, reported in detail by Lukacs, reached a climax. Halifax said that the Italian embassy wanted the government to give a clear indication that it would welcome a mediation by Italy. Churchill said it was clear that the French purpose was to see Signor Mussolini acting as an intermediary between ourselves and Herr Hitler, but he was determined not to get into this position. He said that Hitler's terms, if accepted, would put us completely at his mercy; nations which went down fighting rose again, but those which surrendered were finished. Halifax argued that nothing in his suggestion could even remotely be described as ultimate capitulation, but Churchill thought that the odds were a thousand to one against decent terms being offered to Britain. Realizing that he could make no more headway, Churchill then asked all twenty-five members of his Cabinet, other than those in the War Cabinet, to meet him in his room in the House of Commons.

After informing them of the difficulties of extricating the army from Dunkirk and stressing the futility of negotiations with Hitler, Churchill said, "Of course, whatever happens at Dunkirk, we shall fight on." In his *War Memoirs* he described the scene that followed:

> There occurred a demonstration which, considering the character of the gathering—twenty-five experienced politicians and Parliament men, who represented all the different points of view, whether right or wrong, before the war— surprised me. Quite a number seemed to jump up from the table and come running to my chair, shouting and patting me on the back. There is no doubt that had I at this juncture faltered at all in the leading of the nation I should have been hurled out of office. I was sure that every Minister was ready to be killed quite soon, and have all his family and possessions destroyed, rather than give in.[12]

Hugh Dalton's *A Labour Minister's Memoirs* confirm Churchill's description of the scene.[13]

By contrast, Dalton reports that on one of his first visits to Halifax at the Foreign Office during those critical days, Halifax asked him with a placid air: "Have you heard any of the stories about the possibility of a German invasion? That would be a great bore."

At seven o'clock that evening the War Cabinet met again. When Churchill told them of his meeting with the other ministers, Halifax brought up a plan of Reynaud's to appeal to Roosevelt for help. Churchill

> thought that an appeal to the United States at the present time would be altogether premature. If we made a bold stand against Germany, that

> would command their admiration and respect; but a grovelling appeal, if made now, would have the worst possible effect. He therefore did not favour making any approach on the subject at the present time.

Chamberlain did not object. That settled it.

Lukacs concludes that Hitler was never closer to ultimate victory than during those five days in May 1940, and that the one man in Hitler's way was Churchill. He and Britain could not have won the war without the Soviet Union and the United States, but in May 1940 Churchill, supported by the Cabinet, was the one who did not lose it. Churchill embodied Britain's undaunted, defiant spirit; thanks to him, Britain became the symbol of hope to millions in Nazi-occupied Europe.

To younger readers, those events may now seem almost as remote as the French Revolution, but to those of us who lived through them, the nightmare that Hitler was going to obtain unchallenged domination of Europe is still as fresh as if it had happened yesterday. Lukacs's book has made me aware that our debt to Churchill for preventing that nightmare from coming true is even greater than I realized.

On June 4 Dunkirk fell to the Germans, but miraculously nearly 220,000 British and 123,000 French soldiers as well as 34,000 vehicles had by then been evacuated to England. All other equipment was lost, but at least the core of trained men was intact. On June 10 Mussolini declared war on France and Britain. On June 17 Marshal Pétain ousted Reynaud, who had been as adamant as Churchill for fighting on; he was to spend four and a half years in German prison camps. In September the Royal Air Force won the Battle of Britain and lifted the threat of an imminent Nazi invasion. Goebbels' diaries are full of his own and Hitler's frustration at England's continued resistance, and their violent hatred of Churchill. On June 16, 1941, Goebbels wrote in his diary: "Bolshevism must be destroyed, and with it England will lose its last possible ally on the continent of Europe."

To cause England that loss seems to have been one of Hitler's motives for his attack on the Soviet Union. On June 22, 1941, the German armies invaded Russia, and Britain was no longer fighting alone. On December 11 Hitler assured Churchill's ultimate triumph by declaring war on the United States.

Lukacs's story is not new. P.M.H. Bell's book *A Certain Eventuality*, published in 1974, contains a brief account of the arguments between Churchill and Halifax,[14] but Lukacs has transformed it into a memorable drama. I missed a map showing Sedan, and the positions of the French, British, Belgian, and German armies at the beginning of the battle, because without one I could not quite grasp the deadly threat posed to these armies by the German breakthrough. Lukacs confines himself largely to British accounts of events. He mentions the exchange of visits between Churchill and Reynaud, but to read what went on between them I had to turn to Reynaud's well-documented memoirs. Quotations from French and German sources would have provided a fuller picture of events, but Lukacs's exciting account of a decisive event in history has the virtue of brevity.

More about Discoveries

The Second Secret of Life

Why grasse is greene, or why our
blood is red,
Are mysteries which none have reach'd
unto
In this low forme, poore soule, what
wilt thou doe?

JOHN DONNE, *Of the Progresse of the Soule*

In 1937, I chose hemoglobin as the protein whose structure I wanted to solve, but it proved so much more complex than any solved before that it eluded me for more than 20 years. First success came in 1959, when Ann F. Cullis, Hilary Muirhead, Michael G. Rossmann, Tony C.T. North, and I first unraveled the architecture of the hemoglobin molecule in outline. We felt like explorers who have discovered a new continent, but it was not the end of the voyage, because our much-admired model did not reveal its inner workings—it provided no hint about the molecular mechanism of respiratory transport. Why not? Well-intentioned colleagues were quick to suggest that our hard-won structure was merely an artifact of crystallization and might be quite different from the structure of hemoglobin in its living environment, which is the red blood cell.

Hemoglobin is the vital protein that conveys oxygen from the lungs to the tissues and facilitates the return of carbon dioxide from the tissues back to the lungs. These functions and their subtle interplay also make hemoglobin one of the most interesting proteins to study. Like all proteins, it is made of amino acids strung together in a polypeptide chain. The amino acids are of 20 different kinds and their sequence in the chain is genetically determined. A hemoglobin molecule is made up of four polypeptide chains, two alpha chains of 141 amino acid residues each and two beta chains of 146 residues each. The alpha and beta chains have different sequences of amino acids but fold up to form similar three-dimensional structures. Each chain harbors one heme, which gives blood its red color. The heme consists of a ring of carbon, nitrogen, and hydrogen atoms called porphyrin, with an atom of iron, like a jewel, at its center. A single polypeptide chain combined with a single heme is called a subunit of hemoglobin or a monomer of the molecule. In the complete molecule

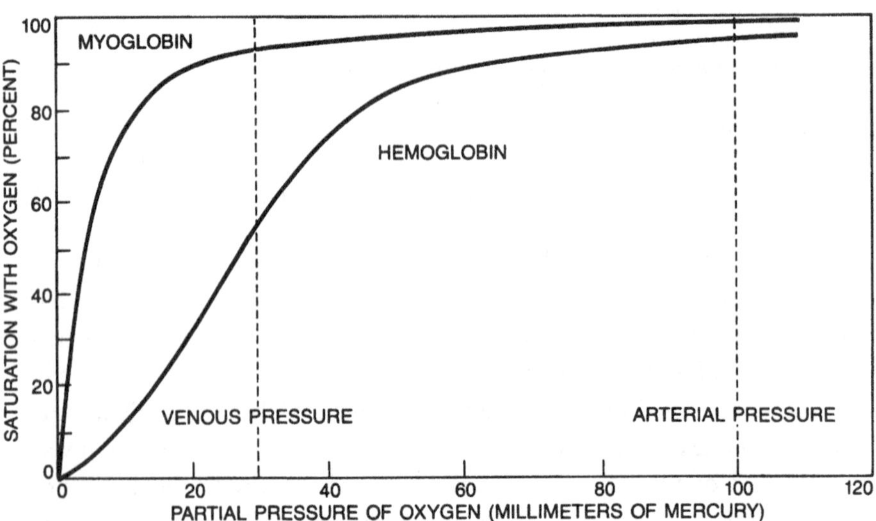

Figure 1. Equilibrium curves measure the affinity for oxygen of hemoglobin and of the simpler myoglobin molecule. Myoglobin, a protein of muscle, has just one heme group and one polypeptide chain and resembles a single subunit of hemoglobin. The vertical axis gives the amount of oxygen bound to one of these proteins, expressed as a percentage of the total amount that can be bound. The horizontal axis measures the partial pressure of oxygen in a mixture of gases with which the solution is allowed to reach equilibrium. For myoglobin (*top curve*) the equilibrium curve is hyperbolic. Myoglobin absorbs oxygen readily but becomes saturated at a low pressure. The hemoglobin curve (*bottom curve*) is sigmoid: initially hemoglobin is reluctant to take up oxygen, but its affinity increases with oxygen uptake. At arterial oxygen pressure both molecules are nearly saturated, but at venous pressure myoglobin would give up only about 10 percent of its oxygen, whereas hemoglobin releases roughly half. At any partial pressure myoglobin has a higher affinity than hemoglobin, which allows oxygen to be transferred from blood to muscle.

four subunits are closely joined, as in a three-dimensional jigsaw puzzle, to form a tetramer.

Hemoglobin Function

In red muscle there is another protein, called myoglobin, similar in constitution and structure to a beta subunit of hemoglobin but made up

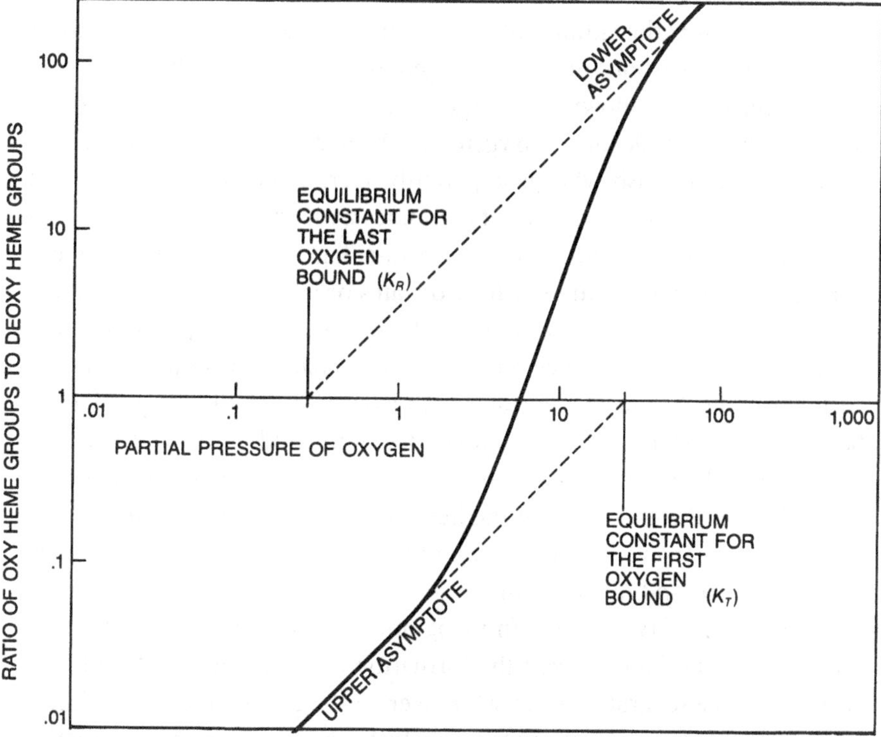

Figure 2. Sigmoid shape of the oxygen equilibrium curve appears more pronounced when the fractional saturation and partial pressure of oxygen are plotted on logarithmic scales. On such a graph the equilibrium curve for myoglobin becomes a straight line at 45 degrees to the axes. The hemoglobin curve begins and ends with straight lines, called asymptotes, at the same angle. Their intercepts with the horizontal line drawn where the concentrations of deoxyhemoglobin and oxyhemoglobin are equal give the equilibrium constants for the first and last oxygen molecules to combine with hemoglobin. In the allosteric interpretation of the curve these are respectively the equilibrium constants of the *R* structure (K_R) and the *T* structure (K_T). For the curve shown, the two constants are respectively 30 and 0.3, indicating that the affinity for the last oxygen bound is 100 times the affinity for the first. This ratio determines the free energy of the heme-heme interaction, which is a measure of the influence exerted by the combination of any one of the four hemes with oxygen on the oxygen affinity of the remaining hemes. If the beginning and end of the curve cannot be measured accurately, the maximum slope of the curve, known as Hill's coefficient, indicates the degree of the heme-heme interaction.

of only one polypeptide chain and one heme. Myoglobin combines with the oxygen released by red cells, stores it, and transports it to the subcellular organelles called mitochondria, where the oxygen generates chemical energy by the combustion of glucose to carbon dioxide and water. Myoglobin was the first protein whose three-dimensional structure was determined—the structure was solved by my colleague John C. Kendrew and his collaborators.

Myoglobin is the simpler of the two molecules. This protein, with its 2,500 atoms of carbon, nitrogen, oxygen, hydrogen, and sulfur, exists for the sole purpose of allowing its single atom of iron to form a loose chemical bond with a molecule of oxygen (O_2). Why does Nature go to so much trouble to accomplish what is apparently such a simple task? Like most compounds of iron, heme by itself combines with oxygen so firmly that its bond with an oxygen atom, once formed, is hard to break. This happens because an iron atom can exist in two states of valency: ferrous iron, carrying two positive charges, as in iron sulfate, which anemic people are told to eat, and ferric iron, carrying three positive charges, as in iron oxide, or rust. Normally, ferrous heme reacts with oxygen irreversibly to yield ferric heme, but when ferrous heme is embedded in the folds of the globin chain, it is protected so that its reaction with oxygen is reversible. The effect of the globin on the chemistry of the heme has been explained with the discovery that the irreversible oxidation of heme proceeds by way of an intermediate compound in which an oxygen molecule forms a bridge between the iron atoms of two hemes. In myoglobin and hemoglobin, the folds of the polypeptide chain prevent the formation of such a bridge by isolating each heme in a separate pocket. Moreover, in the protein the iron is linked to a nitrogen atom of the amino acid histidine, which donates negative charge that enables the iron to form a loose bond with oxygen.

An oxygen-free solution of myoglobin or hemoglobin is purple like venous blood, and when oxygen is bubbled through such a solution, it turns scarlet like arterial blood. If these proteins are to act as oxygen carriers, then hemoglobin must be capable of taking up oxygen in the lungs, where it is plentiful, and giving it up to myoglobin in the capillaries of muscle, where it is less plentiful: myoglobin in turn must pass the oxygen on to the mitochondria, where it is still scarcer.

A simple experiment shows that myoglobin and hemoglobin can accomplish this exchange because there is an equilibrium between free oxygen and oxygen bound to heme iron. Suppose a solution of myoglobin is placed in a vessel constructed so that a large volume of gas can be mixed with it and so that its color can also be measured through a spectroscope. Without oxygen only the purple color of deoxymyoglobin is observed. If a little oxygen is injected, some of the oxygen combines with some of the deoxymyoglobin to form oxymyoglobin, which is scarlet. The spectroscope measures the proportion of oxymyoglobin in the solution. The injection of oxygen and the spectroscopic measurements are repeated until all the myoglobin has turned scarlet. The results are plotted on a graph with the partial pressure of oxygen on the horizontal axis and the percentage of oxymyoglobin on the vertical axis. The graph has the shape of a rectangular hyperbola—it is steep at the start, when all the myoglobin molecules are free, and it flattens out at the end, when free myoglobin molecules have become so scarce that only a high pressure of oxygen can saturate them (Figure 1).

To understand this equilibrium one must visualize its dynamics. Under the influence of heat the molecules in the solution and in the gas are whizzing around erratically and are constantly colliding. Oxygen molecules are entering and leaving the solution, forming bonds with myoglobin molecules and breaking away from them. The number of iron-oxygen bonds that break in one second is proportional to the number of oxymyoglobin molecules. The number of bonds that form in one second is proportional to the frequency of collisions between myoglobin and oxygen, which is determined in turn by the product of their concentrations. When more oxygen is added to the gas, more oxygen molecules dissolve, collide with and bind to myoglobin. This raises the number of oxymyoglobin molecules present and therefore also the number of iron-oxygen bonds liable to break, until the number of myoglobin molecules combining with oxygen in one second becomes equal to the number that lose their oxygen in one second. When that happens, a chemical equilibrium has been established.

The equilibrium is best represented by a graph in which the logarithm of the ratio of oxymyoglobin molecules (Y) to deoxymyoglobin molecules ($1-Y$) is plotted against the logarithm of the partial pressure of oxygen. The hyperbola now becomes a straight line at 45 degrees to the axes. The intercept of the line with the horizontal axis drawn at $Y/(1-Y) = 1$ gives the equilibrium constant K. This is the partial pressure of oxygen at which exactly half of the myoglobin molecules have taken up oxygen. The greater the affinity of the protein for oxygen, the lower the pressure needed to achieve half-saturation and the smaller the equilibrium constant. The 45-degree slope remains unchanged, but lower oxygen affinity shifts the line to the right and higher affinity shifts it to the left (Figure 2).

If the same experiment is done with blood or with a solution of hemoglobin, an entirely different result is obtained. The curve rises gently at first, then steepens and finally flattens out as it approaches the myoglobin curve. This strange sigmoid shape signifies that oxygen-free molecules (deoxyhemoglobin) are reluctant to take up the first oxygen molecule but that their appetite for oxygen grows with the eating. Conversely, the loss of oxygen by some of the hemes lowers the oxygen affinity of the remainder. The distribution of oxygen among the hemoglobin molecules in a solution therefore follows the biblical parable of the rich and the poor: "For unto every one that hath shall be given, and he shall have abundance: but from him that hath not shall be taken away even that which he hath." This phenomenon suggests some kind of communication between the hemes in each molecule, and physiologists have therefore called it heme-heme interaction.

In a logarithmic graph the equilibrium curve begins with a straight line at 45 degrees to the axes because oxygen molecules are so scarce at first that only one heme in each hemoglobin molecule has a chance of catching one of them, and all the hemes therefore react independently, as in myoglobin. As more oxygen flows in, the four hemes in each molecule begin to interact and the curve steepens. The tangent to its maximum slope is known as Hill's coefficient (n), after the physiologist A. V. Hill who first attempted a mathematical analysis of the oxygen equilibrium. The normal value of Hill's coefficient is about 3; without heme-heme interaction it becomes unity. The curve ends with another line at 45 degrees to the axes because oxygen has now become so abundant that only the last heme in each molecule is likely to be free, and all the hemes in the solution react independently once more (Figure 2).

Cooperative Effects

Hill's coefficient and the oxygen affinity of hemoglobin depend on the concentration of several chemical factors in the red blood cell: protons (hydrogen atoms without electrons, whose concentration can be measured as pH); carbon dioxide (CO_2); chloride ions (Cl^-); and a compound of glyceric acid and phosphate called 2,3-diphosphoglycerate (DPG). Increasing the concentration of any of these factors shifts the oxygen equilibrium curve to the right, toward lower oxygen affinity, and makes it more sigmoid. Increased temperature also shifts the curve to the right, but it makes it less sigmoid. Strangely, none of these factors, with the exception of temperature, influences the oxygen equilibrium curve of myoglobin, even though the chemistry and structure of myoglobin are related closely to those of the individual chains of hemoglobin.

What is the purpose of these extraordinary effects? Why is it not good enough for the red cell to contain a simple oxygen carrier such as myoglobin? Such a carrier would not allow enough of the oxygen in the red cell to be unloaded to the tissues, nor would it allow enough carbon dioxide to be carried to the lungs by the blood plasma. The partial pressure of oxygen in the lungs is about 100 millimeters of mercury, which is sufficient to saturate hemoglobin with oxygen whether the equilibrium curve is sigmoid or hyperbolic. In venous blood, the pressure is about 35 millimeters of mercury; if the curve were hyperbolic, less than 10 percent of the oxygen carried would be released at that pressure, so a man would asphyxiate even if he breathed normally.

The more pronounced the sigmoid shape of the equilibrium curve is, the greater the fraction of oxygen that can be released. Several factors conspire to that purpose. Oxidation of nutrients by the tissues liberates lactic acid and carbonic acid; these acids in turn liberate protons, which shift the curve to the right, toward lower oxygen affinity, and make it more sigmoid. Another important regulator of the oxygen affinity is DPG. The number of DPG molecules in the red cell is about the same as the number of hemoglobin molecules, 280 million, and probably remains fairly constant during circulation. A shortage of oxygen, however, causes more DPG to be made, which helps to release more oxygen. With a typical sigmoid curve nearly half of the oxygen carried can be released to the tissues. The human fetus has a hemoglobin with the same alpha chains as the hemoglobin of the human adult but different beta chains, resulting in a lower affinity for DPG. This gives fetal hemoglobin a higher oxygen affinity and facilitates the transfer of oxygen from the maternal circulation to the fetal circulation.

Carbon monoxide (CO) combines with the heme iron at the same site as oxygen, but its affinity for that site is 150 times greater; carbon monoxide therefore displaces oxygen, which explains why it is so toxic. In heavy smokers, up to 20 percent of the oxygen combining sites can be blocked by carbon monoxide, so that less oxygen is carried by the blood. In addition carbon monoxide has an even more sinister effect. The combination of one of the four hemes in any hemoglobin molecule with carbon monoxide raises the oxygen affinity of the remaining three hemes by heme-heme interaction. The oxygen equilibrium curve is therefore shifted to the left, which diminishes the fraction of the oxygen carried that can be released to the tissues.

If protons lower the affinity of hemoglobin for oxygen, then the laws of action and reaction demand that oxygen lowers the affinity of hemoglobin for protons. Liberation of oxygen causes hemoglobin to combine with protons and vice versa; about two protons are taken up for every four molecules of oxygen released, and two protons are liberated again when four molecules of oxygen are taken up. This reciprocal action is known as the Bohr effect and is the key to the mechanism of carbon dioxide transport (Figure 3). The carbon dioxide released by respiring tissues is too insoluble to be transported as such, but it can be rendered more soluble by combining with water to form a bicarbonate ion and a proton. The chemical reaction is written:

$$CO_2 + H_2O \rightleftarrows HCO_3^- + H^+$$

In the absence of hemoglobin this reaction would soon be brought to a halt by the excess of protons produced, like a fire going out when the chimney is blocked. Deoxyhemoglobin acts as a buffer, mopping up the protons and tipping the balance toward the formation of soluble bicarbonate. In the lungs the process is reversed. There, as oxygen binds to hemoglobin, protons are cast off, driving carbon dioxide out of solution so that it can be exhaled. The reaction between carbon dioxide and water is

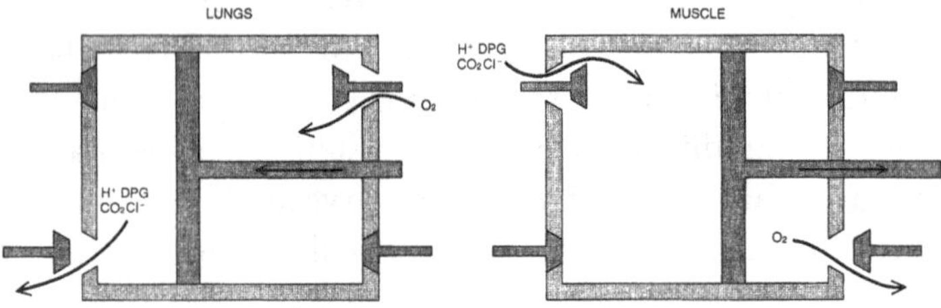

Figure 3. Reciprocating engine serves as a model of the cooperative effects of hemoglobin. The piston is driven to the right by the energy liberated in the reaction of hemoglobin with oxygen (O_2) and to the left by the protons (H^+) and carbon dioxide (CO_2) liberated by respiring tissues. Diphosphoglycerate (DPG) and chloride ions (Cl^-) are passengers riding in company with protons and carbon dioxide.

catalyzed by carbonic anhydrase, an enzyme in the red cells. The enzyme speeds up the reaction to a rate of about half a million molecules per second, one of the fastest of all known biological reactions.

There is a second, but less important, mechanism for transporting carbon dioxide. The gas binds more readily to deoxyhemoglobin than it does to oxyhemoglobin, so that it tends to be taken up when oxygen is liberated and cast off when oxygen is bound. The two mechanisms of carbon dioxide transport are antagonistic. For each molecule of carbon dioxide bound to deoxyhemoglobin either one or two protons are released, which oppose the conversion of other molecules of carbon dioxide to bicarbonate. Positively charged protons entering the red cell draw negatively charged chloride ions in with them, and these ions too are bound more readily by deoxyhemoglobin than by oxyhemoglobin. DPG is synthesized in the red cell itself and cannot leak out through the cell membrane. It is strongly bound by deoxyhemoglobin and only very weakly bound by oxyhemoglobin.

Heme-heme interaction and the interplay between oxygen and the other four ligands are known collectively as the cooperative effects of hemoglobin. Their discovery by a succession of physiologists and biochemists took more than half a century and aroused many controversies. In 1938, Felix Haurowitz of the Charles University in Prague made another vital observation. He discovered that deoxyhemoglobin and oxyhemoglobin form different crystals, as though they were different chemical substances, which implied that hemoglobin is not an oxygen tank but a molecular lung because it changes its structure every time it takes up oxygen or releases it.

Theory of Allostery

The discovery of an interaction among the four hemes made it obvious that they must be touching, but in science what is obvious is not necessarily true. When the structure of hemoglobin was finally solved, the hemes were found to lie in isolated pockets on the surface of the subunits (Figure 4). Without contact between them how could one of them sense whether the others had combined with oxygen? And how could as heterogeneous a collection of chemical agents as protons, chloride ions, carbon dioxide, and diphosphoglycerate influence the oxygen equilibrium curve in a similar way? It did not seem plausible that any of them could bind directly to the hemes or that all of them could bind at any other common

Figure 4. Hemoglobin molecule, as deduced from X-ray diffraction studies, is shown from above (*top*) and side (*bottom*). The drawings follow the representation scheme used in three-dimensional models built by the author and his co-workers. The irregular blocks represent electron-density patterns at various levels in the hemoglobin molecule. The molecule is built up from four subunits: two identical alpha chains (*light blocks*) and two identical beta chains (*dark blocks*). The letter *N* in the top view identifies the amino ends of the two alpha chains; the letter *C* identifies the carboxyl ends. Each chain enfolds a heme group (*disk*), the iron-containing structure that binds oxygen to the molecule.

site, although there again it turned out we were wrong. To add to the mystery, none of these agents affected the oxygen equilibrium of myoglobin or of isolated subunits of hemoglobin. We now know that all the cooperative effects disappear if the hemoglobin molecule is merely split in half, but this vital clue was missed. Like Agatha Christie, Nature kept it to the last to make the story more exciting.

There are two ways out of an impasse in science: to experiment or to think. By temperament, perhaps, I experimented, whereas Jacques Monod thought. In the end our paths converged.

Monod's scientific life had been devoted to finding out what regulates the growth of bacteria. The key to this problem appeared to be regulation of the synthesis and catalytic activity of enzymes. Monod and François Jacob had discovered that the activity of certain enzymes is controlled by switching their synthesis on and off at the gene. They and others then found a second mode of regulation that appeared to operate switches on the enzymes themselves.

In 1965, Monod and Jean-Pierre Changeux of the Pasteur Institute in Paris, together with Jeffries Wyman of the University of Rome, recognized that the enzymes in the latter class have certain features in common with hemoglobin. They are all made of several subunits, so that each molecule includes several sites with the same catalytic activity, just as hemoglobin includes several hemes that bind oxygen, and they all show similar cooperative effects. Monod and his colleagues knew that deoxyhemoglobin and oxyhemoglobin have different structures, which made them suspect that the enzymes too may exist in two (or at least two) structures. They postulated that these structures should be distinguished by the arrangement of the subunits and by the number and strength of the bonds between them.

If there are only two alternative structures, the one with fewer and weaker bonds between the subunits would be free to develop its full catalytic activity (or oxygen affinity); this structure has therefore been labeled *R*, for "relaxed." The activity would be damped in the structure with more and stronger bonds between the subunits; this form is called *T*, for "tense" (Figure 5). In either of these structures the catalytic activity (or oxygen affinity) of all the subunits in one molecule should always remain equal. This postulate of symmetry allowed the properties of allosteric enzymes to be described by a neat mathematical theory with only three independent variables: K_R and K_T, which in hemoglobin denote the oxygen equilibrium constants of the *R* and *T* structures, respectively; and *L*, which stands for

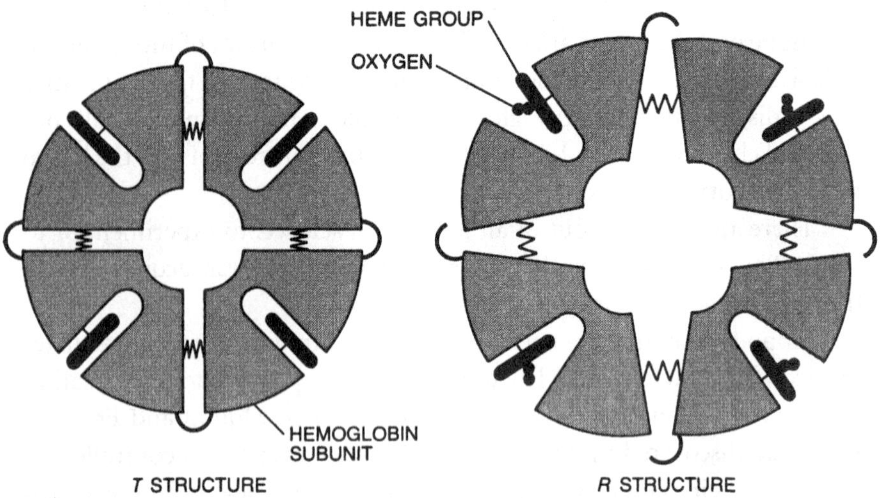

Figure 5. Allosteric theory explains heme-heme interaction without postulating any direct communication between the heme groups. The hemoglobin molecule is assumed to have two alternative structures, designated *T* for tense and *R* for relaxed. In the *T* structure, the subunits of the molecule are clamped against the pressure of springs and their narrow pockets impede the entry of oxygen. In the *R* structure, all the clamps have sprung open and the heme pockets are wide enough to admit oxygen easily. Uptake of oxygen by the *T* structure would strain the clamps until they all burst open in concert and allow the molecule to relax to the *R* structure. Loss of oxygen would narrow the heme pockets and allow the *T* structure to re-form.

the number of molecules in the *T* structure divided by the number in the *R* structure, the ratio being measured in the absence of oxygen. The term allostery (from the Greek roots *allos*, "other," and *stereos*, "solid") was coined because the regulator molecule that switches the activity of the enzyme on or off has a structure different from that of the molecule whose chemical transformation the enzyme catalyzes.

This ingenious theory simplified the interpretation of the cooperative effects enormously. The progressive increase in oxygen affinity illustrated by the parable of the rich and the poor now arises not from any direct interaction between the hemes, but from the switchover from the *T* structure with low affinity to the *R* structure with high affinity. This transformation should take place either when the second molecule of oxygen is bound or when the third is bound. Chemical agents that do not bind to the hemes might lower the oxygen affinity by biasing the equilibrium between the two structures toward the *T* form, which would make the transition to the *R* structure come after, say, three molecules of oxygen have been bound rather than after two molecules have been bound. In terms of allosteric theory, such agents would raise L, the fraction of molecules in the T structure, without altering the oxygen equilibrium constants K_T and K_R of the two structures (Figure 5).

Atomic Structures

My own approach to the problem was also influenced by Haurowitz's discovery that oxyhemoglobin and deoxyhemoglobin have different structures. Gradually I came to realize that we would never explain the intricate functions of hemoglobin without solving the structures of both crystal forms at a resolution high enough to reveal atomic detail.

In 1970, 33 years after I had taken my first X-ray-diffraction pictures of hemoglobin, that stage was finally reached. Hilary Muirhead, Joyce M. Baldwin, Gwynne Goaman, and I got a good map of the distribution of matter, not in oxyhemoglobin, but in the closely related methemoglobin, of horse, in which the iron is ferric and the place of oxygen is taken by a water molecule. William Bolton and I got a map of horse deoxyhemoglobin, and Muirhead and Jonathan Greer got one of human deoxyhemoglobin. These maps served as guides for the construction of three atomic models, each a jungle of brass spokes and steel connectors supported on brass scaffolding, edifices of labyrinthine complexity nearly four feet in diameter. At first it was hard to see the trees for the forest.

In allosteric terms, our methemoglobin model represented the R structure and our two deoxyhemoglobin models the T structure. We scanned them eagerly for clues to the allosteric mechanism, but could not see any at first because the general structure of the subunits was similar in all three models. The alpha chains included seven helical segments and the beta chains eight helical segments interrupted by corners and nonhelical segments. Each chain enveloped its heme in a deep pocket, which exposed only the edge where two propionic acid side chains of the porphyrin dip into the surrounding water (Figure 6).

The heme makes contact with 16 amino acid side chains from seven segments of the chain. Most of these side chains are hydrocarbons; the two exceptions are the heme-linked histidines which lie on each side of the heme and play an important part in the binding of oxygen. The side chain of histidine ends in an imidazole ring made of three carbon atoms, two nitrogen atoms, and either four or five hydrogen atoms. One of these histidines, called the proximal histidine, forms a chemical bond with the

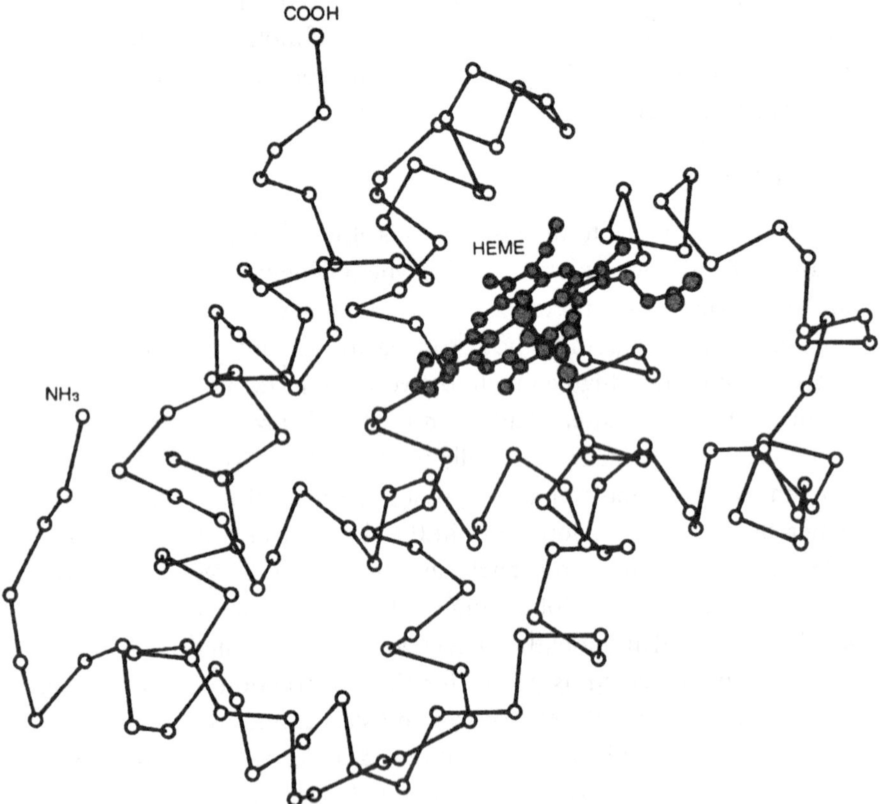

Figure 6. Subunit of hemoglobin consists of a heme group (gray) enfolded in a polypeptide chain. The polypeptide is a linear sequence of amino acid residues, each of which is represented here by a single dot, marking the position of the central (alpha) carbon atom. The chain begins with an amino group (NH_3) and ends with a carboxyl group (COOH). Most of the polypeptide is coiled up to form helical segments but there are also nonhelical regions. The computer-generated diagram of a horse-hemoglobin subunit was prepared by Feldmann and Porter.

heme iron (Figure 7). The other histidine, called the distal one, lies on the opposite side of the heme, in contact with it and with the bound oxygen, but without forming a covalent chemical bond with either. Apart from these histidines, most of the side chains in the interior of the subunits, like those near the hemes, are hydrocarbons. The exterior of the hemoglobin molecule is lined with side chains of all kinds, but electrically charged and dipolar ones predominate. Thus each subunit is waxy inside and soapy outside, which makes it soluble in water but impermeable to it.

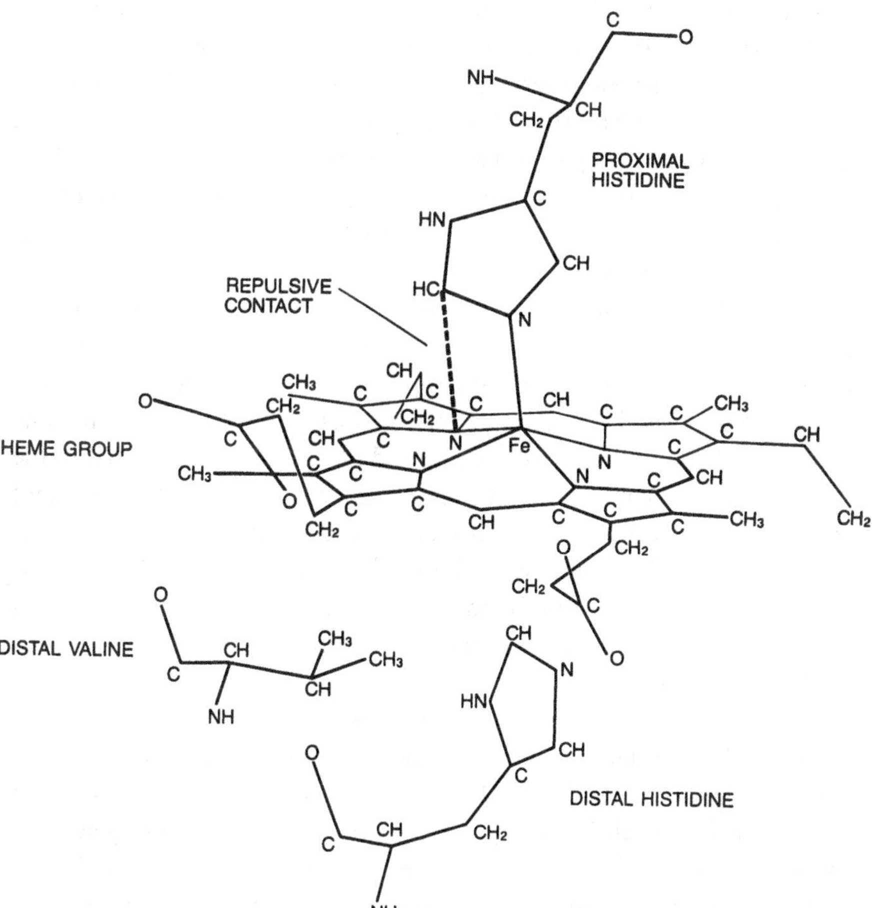

Figure 7. Chemical structure of the heme group and surrounding amino acids is shown by a skeleton of lines connecting the centers of atoms. The only chemical bond between the heme and the protein that engulfs it is the link between the iron atom and the amino acid at the top, called the proximal histidine. The two amino acids at the bottom (the distal histidine and the distal valine) touch the heme but are not bonded to it. The proximal histidine is the principal path for communication between the heme and the rest of the molecule. In the deoxy state shown the iron protrudes above the porphyrin and may be hindered from returning to a centered position by repulsion between one corner of the proximal histidine and one of the porphyrin nitrogen atoms.

The four subunits are arranged at the vertices of a tetrahedron around a two-fold symmetry axis. Since a tetrahedron has six edges, there are six areas of contact between the subunits. The twofold symmetry leaves four distinct contacts, which cover about a fifth of the surface area of the subunits. Sixty percent of that area is made up of the alpha$_1$-beta$_1$ and alpha$_2$-beta$_2$ contacts, each of which includes about 35 amino acid side chains tightly linked by 17 to 19 hydrogen bonds. Hydrogen bonds are made between atoms of nitrogen (N) and oxygen (O) through an intermediate hydrogen atom (H), for instance N-H...N, N-H...O, O-H...O or O-H...N. The hydrogen is bonded strongly to the atom on the left and weakly to the one on the right.

The numerous hydrogen bonds between the alpha$_1$-beta$_1$ and alpha$_2$-beta$_2$ subunits make them cohere so strongly that their contact is hardly altered by the reaction with oxygen, and they move as rigid bodies in the transition between the *T* and the *R* structures. On the other hand, the contact alpha$_1$-beta$_2$ in the *R* structure looked quite different from that in the *T* structure. This contact includes fewer side chains than alpha$_1$-beta$_1$ and is designed so that it acts as a snap-action switch, with two alternative stable positions, each braced by a different set of hydrogen bonds. We wondered at first whether these bonds were stronger and more numerous in the *T* structure than they are in the *R* structure, but that did not seem to be the case (Figure 8).

Where, then, were the extra bonds between the subunits in the *T* structure that allosteric theory demanded? We spotted them at the ends of the polypeptide chains. In the *T* structure the last amino acid residue of each chain forms salt bridges (which is a bond between a nitrogen atom carrying a positive charge, and an oxygen atom carrying a negative charge) with neighboring subunits. In our maps of the *R* structure the last two residues of each chain were blurred. At first I suspected this to be due to error, but improved maps made by my colleagues Elizabeth Heidner and Robert Ladner have convinced us that the final residues remain invisible because they are no longer tethered and wave about like reeds in the wind.

Geometrically, the transition between the two structures consists of a rock-and-roll movement of the dimer alpha$_1$-beta$_1$ with respect to the dimer alpha$_2$-beta$_2$. Joyce Baldwin has shown that if one dimer is held fixed, the movement of the other one can be represented by a rotation of some 15 degrees about a suitably placed axis together with a small shift along the same axis. The movement is brought about by subtle changes in the internal structure of the subunits that accompany the binding and dissociation of oxygen (Figure 9).

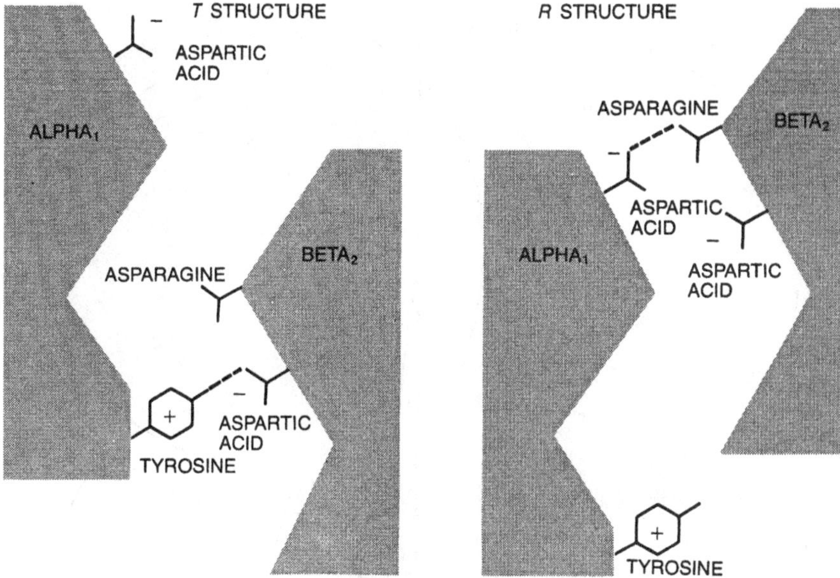

Figure 8. Contact between the two dimers has two stable conformations, one for the *T* structure and the other for the *R* structure. On transition between the structures the dimers snap from one position to the other. They are stabilized by alternative sets of hydrogen bonds formed between amino acid side chains attached to the opposing faces of the dimers. The two bonds shown here were first discovered by X-ray crystallography. In 1975 Leslie Fung and Chien Ho at the University of Pittsburgh demonstrated the presence of these bonds in solution. This provided evidence that the two structures found in crystals are the same as the structures in red blood cells.

Function of the Salt Bridges

The salt bridges at the ends of the polypeptide chains clearly provide the extra bonds between the subunits in the *T* structure predicted by Monod, Changeux, and Wyman. They also explain the influence on the oxygen equilibrium curve of all the chemical factors that had puzzled us so much. All agents that lower the oxygen affinity do so either by strengthening existing salt bridges in the *T* structure or by adding new ones. Not all these extra bonds, however, are between the subunits; some are within the subunits and oppose the subtle structural changes the subunits undergo on combination with oxygen (Figure 10).

The salt bridges explain both the lowering of the oxygen affinity by protons and the uptake of protons on release of oxygen. Protons increase the number of nitrogen atoms carrying a positive charge. For example, the

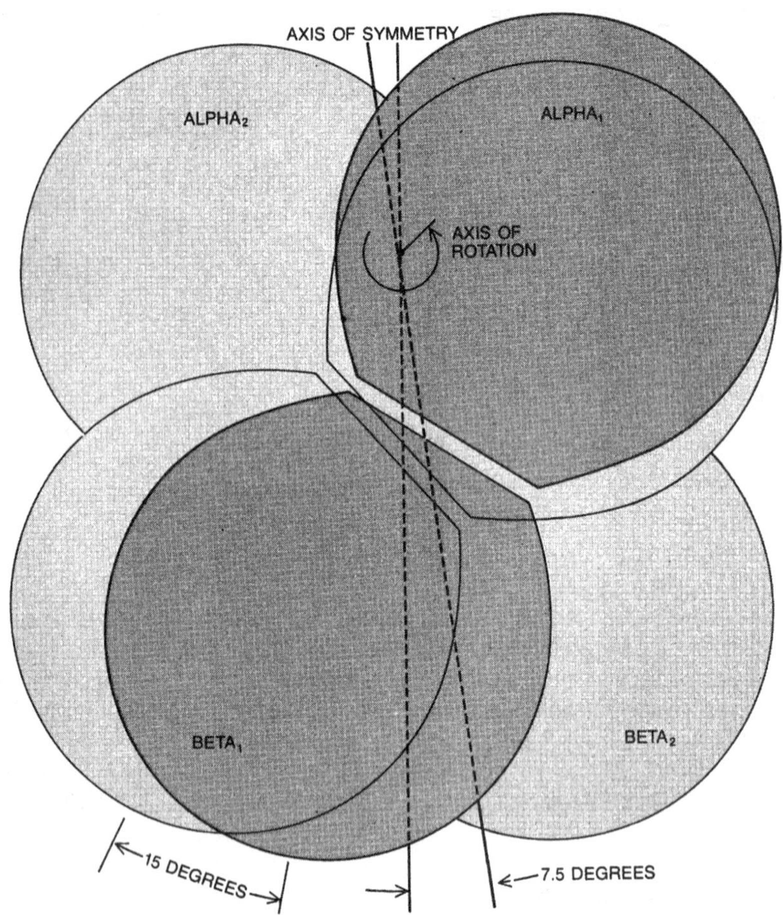

Figure 9. Rearrangement of subunits during the transition from the *T* structure to the *R* structure consists mainly in a rotation of one pair of subunits with respect to the other pair. Each alpha chain is bonded strongly to one beta chain, and the dimers formed in this way move as rigid bodies. If one dimer is held fixed, the other turns by 15 degrees about an off-center axis and shifts slightly along it. The twofold symmetry of the molecule is preserved, but the axis of symmetry is rotated by 7.5 degrees. The diagram is based on one prepared by Baldwin.

imidazole ring of the amino acid histidine can exist in two states, uncharged when only one of its nitrogen atoms carries a proton and positively charged when both do. In neutral solution each histidine has a 50 percent chance of being positively charged. The more acid the solution, or in other words the higher the concentration of protons, the greater the chance of a histidine becoming positively charged and forming a salt

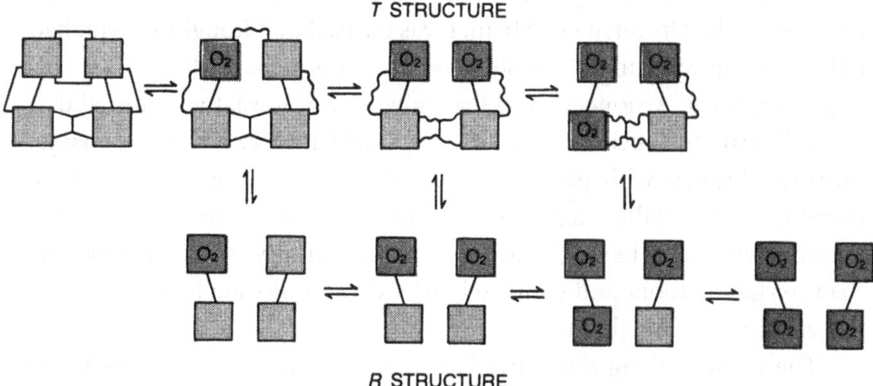

Figure 10. Transition from the *T* structure to the *R* structure increases in likelihood as each of the four heme groups is oxygenated. In this more realistic model, salt bridges linking the subunits in the *T* structure break progressively as oxygen is added, and even those salt bridges that have not yet ruptured are weakened, a process that is represented here by making the lines wavy. The transition from *T* to *R* does not take place after a fixed number of oxygen molecules have been bound, but it becomes more probable with each successive oxygen bound. The transition between the two structures is influenced by several factors, including protons, carbon dioxide, chloride, and DPG. The higher their concentration is, the more oxygen must be bound to trigger the transition. Fully saturated molecules in the *T* structure and fully deoxygenated molecules in the *R* structure are not shown because they are too unstable to exist in significant numbers.

bridge with an oxygen atom carrying a negative charge. Conversely, the transition from the *R* structure to the *T* structure brings negatively charged oxygen atoms into proximity with an uncharged nitrogen atom and thereby diminishes the work that has to be done to give the nitrogen atom a positive charge. As a result, a histidine that has no more than a 50 percent chance of being positively charged in the *R* structure has a 90 percent chance in the *T* structure, so that more protons are taken up from the solution by hemoglobin in the *T* structure.

Hemoglobin includes one other set of groups that behave in this way: they are the amino groups at the start of the polypeptide chains, but their nitrogen atoms take up protons only if the concentration of carbon dioxide is low. If it is high, these nitrogens are liable to lose protons and to combine instead with carbon dioxide to form a carbamino compound. The physiologists F. J. W. Roughton and J. K. W. Ferguson proposed in 1934 that this mechanism plays a part in the transport of carbon dioxide, but their proposal was treated with skepticism until it was confirmed 35 years later by my colleague John Kilmartin, working with Luigi Rossi-Bernardi at the University of Milan. I was pleased that Roughton, who had fathered their experiment, was still alive to see his ideas vindicated. My colleague Arthur R. Arnone, now at the University of Iowa, then showed that in the *T* structure such carbamino groups, which carry a negative charge, form salt bridges with positively charged groups of the globin and are therefore more stable than they are in the *R* structure. This finding explains why deoxyhemoglobin has a higher affinity for carbon dioxide than oxyhemoglobin and conversely why carbon dioxide lowers the oxygen affinity of hemoglobin.

The most striking difference between the *T* and *R* structures is the width of the gap between the two beta chains. In the *T* structure, the two chains are widely separated and the opening between them is lined by amino acid side chains carrying positive charges. This opening is tailor-made to fit the molecule of 2,3-diphosphoglycerate and to compensate its negative charges, so that the binding of DPG adds another set of salt bridges to the *T* structure. In the *R* structure the gap narrows, and DPG has to drop out.

The Trigger

How does combination of the heme irons with oxygen make the subunits click from the *T* structure to the *R* structure? Compared with the hemoglobin molecule, an oxygen molecule is like the flea that makes the elephant jump. Conversely, how does the *T* structure impede the uptake of oxygen? What difference between the two structures is there at the heme that could bring about a several-hundred-fold change in oxygen affinity?

In oxyhemoglobin the heme iron is bound to six atoms: four nitrogen atoms of the porphyrin, which neutralize the two positive charges of the ferrous iron; one nitrogen atom of the proximal histidine, which links the heme to one of the helical segments of the polypeptide chain (helix *F*); and one of the two atoms of the oxygen molecule. In deoxyhemoglobin the oxygen position remains empty, so that the iron is bound to only five atoms (Figure 7).

I wondered whether the heme pockets might be narrower in the *T* than in the *R* structure, so that they had to widen to let the oxygen in. This widening might be geared to break the salt bridges, rather like the childish mechanism shown in Figure 5. When the atomic model of horse deoxyhemoglobin emerged, Bolton and I saw some truth in this idea because in the beta subunits a side chain of the amino acid valine next to the distal histidine blocked the site that oxygen would have to occupy. The alpha subunits, however, showed no such obstruction. Then we noticed the odd positions of the iron atoms. In methemoglobin, which has the *R* structure, the iron atoms had been displaced very slightly from the porphyrin plane toward the proximal histidine, but in deoxyhemoglobin (with the *T* structure) the displacement stood out as one of the most striking features of our maps. In each subunit, the iron atom had carried the proximal histidine and helix *F* with it so that they too had moved away from the porphyrin plane. I quickly realized that this might be the long-sought trigger.

Each iron atom is displaced by 0.50 (± 0.1) angstrom unit from the mean plane of the porphyrin. (One angstrom unit is 10^{-10} meter.) The nitrogen atom of the proximal histidine to which the iron atom is bound, lies at a distance of 2.7 (± 0.1) angstroms from that plane. In oxyhemoglobin the iron lies within 0.1 and the nitrogen of the histidine within 2.1 angstroms of the porphyrin plane, which means that the nitrogen would be 0.6 angstrom closer to the porphyrin plane than it is in deoxyhemoglobin. This shift triggers the transition from the *T* structure to the *R* structure (Figure 11).

How is this movement transmitted to the contacts between the subunits and to the salt bridges? One might as well puzzle out how a cat jumps off a wall from looking at one picture of the cat on the wall and another of it on the ground because our static models of deoxyhemoglobin and methemoglobin do not show what happens in the transition between the *T* and the *R* structures.

If movement of the proximal histidine and the iron toward the porphyrin puts into motion a set of levers that loosens and breaks the salt bridges, then the making of the bridges must cause the same set of levers to go into reverse and move the histidine and the iron away from the porphyrin. The oxygen molecule on the other side cannot follow because it bumps against the four porphyrin nitrogen atoms, and so the iron-oxygen bond is stretched until it finally snaps.

To be guided by the atomic models toward the molecular mechanism of respiratory transport seemed like a dream. But was it true? Would the mechanism stand the cold scrutiny of experiment? It has been said that scientists do not pursue truth, it pursues them. It pursued me for 25 years until Massimo Paoli, a graduate student of Professor Guy Dodson at York University, found the answer. He immersed crystals of human deoxyhe-

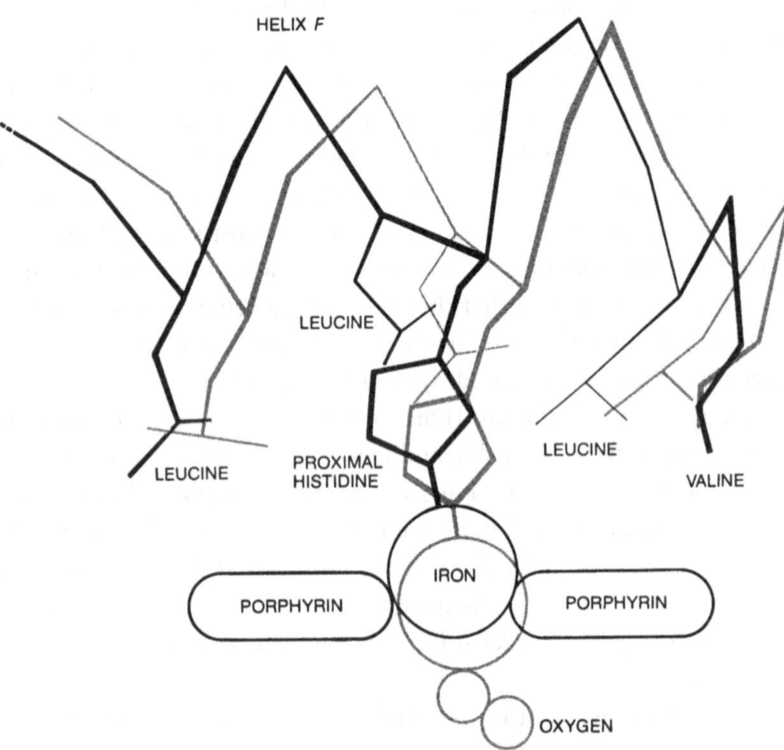

Figure 11. Triggering mechanism for the transition from the *T* to the *R* structure is a movement of the heme iron into the plane of the porphyrin ring. In the *T* structure (*black lines*), the center of the iron atom is about 0.5 angstrom unit above the plane. (One angstrom unit is 10^{-10} meter.) When the molecule switches to the *R* structure (*gray lines*), the iron moves into the plane, pulling with it the proximal histidine and helix *F*. Once the iron has descended into the plane it can readily bind an oxygen molecule. In the reverse transition (from *R* to *T*), the iron is pulled out of the plane and the oxygen cannot follow because it bumps against the porphyrin nitrogen atoms. The iron-oxygen bond is thereby weakened and easily breaks. These movements of the iron atoms are transmitted to the contacts between the subunits and promote the transitions between the *T* and *R* structures. Diagram is based on a drawing by John Cresswell of University College London.

moglobin in a medium which clamped them so firmly that the molecules maintained the T-structure even when all the iron atoms had combined with oxygen. To bind the oxygen, the iron atoms moved towards the porphyrin plane, but the tension of the T-structure kept the histidines back, so that many of the bonds between the histidines and the irons got broken. That tension must come from the salt bridges, because rupture of any of them raises the oxygen affinity of the T-structure.

In 1937, when I started my X-ray work on haemoglobin, I had no inkling that such an intricate, complex molecular mechanism underlies its deceptively simple physiological function, let alone that my work would lead me to unravel it. I am thankful that I have lived long enough to see it vindicated.

How W.L. Bragg Invented X-ray Analysis

Bragg's invention of X-ray analysis for finding the arrangement of atoms in crystals and his determinations of the atomic structures of the rocks that make up the bulk of the Earth's crust revolutionised the foundations of chemistry, mineralogy, and metallurgy.[1–4]

I first met Bragg in Cambridge in the autumn of 1938 when he had just been appointed Rutherford's successor as Cavendish Professor of Experimental Physics. One day I burst into his room announcing proudly, "I have received an honour that you can't match, I have had a glacier named after me." "I have one that you can't match" retorted Bragg, "I have had a cuttlefish named after me." He then told me that as a boy in Adelaide he had been a keen collector and found a new species which his seniors promptly named *Sepia Braggi*.

Bragg's father had studied mathematics at Cambridge and finished 3rd Wrangler. On the strength of that he was appointed professor of mathematics and physics at the recently founded University of Adelaide when he was only 23, never mind that he had learnt no physics. He read Deschanel's *Electricity and Magnetism* on the boat going out and remained in Adelaide until 1909, when he moved back to England to become professor of physics at Leeds.[5] Willie Bragg was born in 1890; aged 15 he entered Adelaide University to read mathematics, and graduated there at 18. The following year he entered Cambridge University to read mathematics and physics. One day he wrote to his father at Leeds: "Dear Dad, I'm so glad you liked the notes on Jeans. I got an awful lot from a Dane who had seen me asking Jeans questions. He was awfully sound, and most interesting, his name was Böhr or something that sounds like it." That was the start of his lifelong friendship with Niels Bohr. Bragg took his Cambridge degree in 1911. He records in his biography: "Then came a time of research in the Cavendish. It was a sad place. There were too many young researchers (about 40) attracted by its reputation, too few ideas for them to work on, too little money, and too little apparatus. We had to made practically everything for ourselves, and even at that the means were meagre. There were a few senior people who had built little kingdoms for themselves with good equipment, but most of us were breaking out hearts trying to make bricks without straw. J.J.Thomson did his best to think of ideas for us all and guide us, but there were too many of us, and he was the only leader of research. C.T.R. Wilson (the inventor of the cloud chamber) liked doing everything on his own, and no other member of the staff was interested in research" (unpublished memoirs).

After a frustrating year he joined his family on the Yorkshire coast for the summer holidays and found his father excited about a paper by Friedrich, Knipping, and Laue that had just appeared in Munich. Bragg's father had regarded X-rays as "minute bundles of energy, tiny entities which move like material particles, but with the speed of light." On the other hand, Max von Laue, a theoretical physicist at the University of Munich, believed that they are electromagnetic waves. It occurred to him that the wavelength of X-rays might be of the same order as the distance between atoms in crystals, in which case crystals would act as diffraction gratings for X-rays. Friedrich and Knipping verified this prediction by the discovery of X-ray diffraction patterns given off by crystals of copper sulphate, zincblende, and other simple compounds (Figure 1).[6] Bragg's father thought the Germans' X-ray patterns might have been due, not to diffraction, but to neutral particles running down different channels in their crystals. On returning to Cambridge the son continued to mull over Laue's results and soon convinced himself that they must be due to diffraction. To his father he wrote: "I have just got a lovely series of reflections of the rays in mica plates with only a few minutes' exposure! Huge joy" and he signed himself: "Your affectionate son, W.L. Bragg." Those were formal days. And the father wrote: "My dear Rutherford, my boy has been getting beautiful X-ray reflections from mica sheet just as simple as the reflections

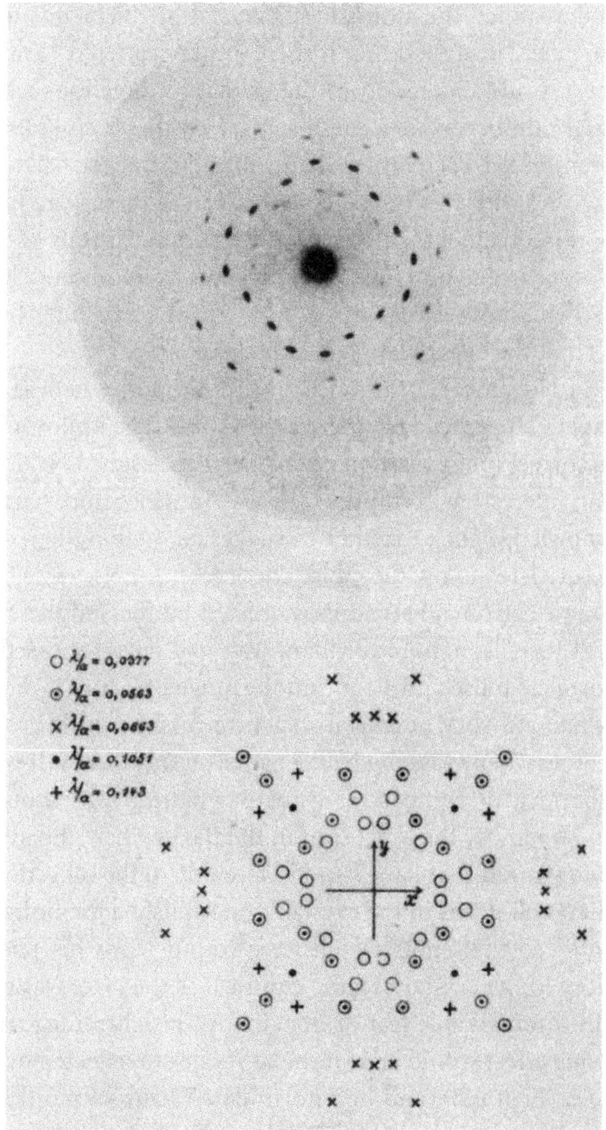

Figure 1. Friedrich, Knipping, and Laue's X-ray diffraction picture of zincblende, taken with the X-ray beam along one of the cube axes, together with their assignment of the spots to five distinct wavelengths.[6]

of light in a mirror," but in Cambridge the son was teased for having disproved his father's corpuscular theory when he focused X-rays by reflecting them from a bent sheet of mica.

Laue had assumed the atoms in his crystal of zincblende to lie at the corners of a cube. He argued that if these atoms scattered X-rays, the diffracted X-rays would emerge from the crystal in directions where atoms lay an integral number of wavelengths apart, so that their scattering contributions reinforced each other. Laue himself noticed that there was something wrong with this interpretation because there were many directions where reinforcements of X-rays diffracted by crystals of zincblende should have occurred, but the relevant spots were absent; he tried to explain this by assuming that the X-rays consisted of only five distinct wavelengths that the crystal lattice had picked out.[6]

On the 11 November, 1912, only four months after he had first heard of Laue's papers, Bragg read a paper to the Cambridge Philosophical Society with the correct interpretation of the German results.[7] He describes his success as "an interesting example of the way in which apparently unrelated bits of knowledge click together to suggest something new. J.J. Thomson had lectured to us on the pulse theory of X-rays, which explained them as being electromagnetic pulses created by the sudden stopping of electrons. C.T.R. Wilson, in his brilliant way, had talked about the equivalence of a formless pulse and a continuous range of 'white' radiation. Pope and Barlow had a theory of crystal structure, and our little group had an evening meeting when Gossling read a paper on this theory. It was the first time that the idea of a crystal as a regular pattern was brought to my notice. I can remember the exact spot in the Backs where the idea suddenly leapt into my mind that Laue's spots were due to the reflection of X-ray pulses by sheets of atoms in the crystal" (unpublished memoirs).

Bragg noticed that spots which were round when his photographic plate was close to the crystal became elliptical as the plate was moved further away. By a remarkable feat of imaginative insight Bragg realised that such a focusing effect would arise if the X-rays were reflected by successive atomic planes (Figure 2), and he reformulated Laue's conditions for diffraction into what became known as Bragg's Law, which gives a more direct relationship between the crystal structure and its diffraction pattern ($n\lambda = 2d \sin\theta$). He then noticed something else. The German group had tilted the crystal away from its symmetrical position by 3°. If the X-rays had consisted of five discreet wavelengths as Laue believed, then the spots should have disappeared as the conditions for diffraction for the planes from which these wavelengths were reflected no longer held true. In fact the same spots moved by 6° and changed in intensity. This led Bragg to

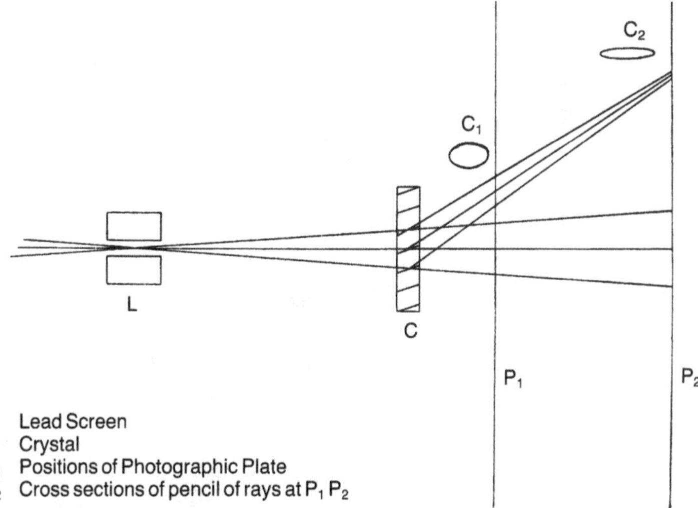

Figure 2. Change of shape of the X-ray reflexions as the photographic plate was moved away from the crystal. Reflexions that were found when the plate was near the crystal became drawn out in the horizontal direction further away. Bragg pointed out that reflexion by the lattice planes of an incident cone of X-rays of continuously varying wavelength would come to a focus in the vertical direction, but would spread out in the horizontal direction.[7]

recognise that sets of parallel lattice selected from a continuous spectrum (or pulse, as he called it) those wavelengths which corresponded to integral multiples of the path difference between reflexions from successive atomic planes, so that each Laue spot was made up of several harmonics of some selected wavelength. Finally, he demonstrated that the presence of spots with certain combinations of indices, and the absence of others in the X-ray diffraction pattern of zincblende could be accounted for by assuming a face-centred rather than a primitive cubic lattice. With that assumption the entire diffraction pattern fell into place (Figure 3).[7]

Why did this 22-year-old student succeed in correctly interpreting the diffraction pattern predicted and discovered by an accomplished theoretician 11 years his senior and two experimental physicists? Bragg himself modestly attributes it to a "concatenation of fortunate circumstances," but his formidable paper soon convinces you that its success owed more to Bragg's astute powers of penetrating through the apparent complexities of physical phenomena to their underlying simplicity.

Bragg's first paper was quickly followed by another, written in collaboration with his father, on their newly developed X-ray spectrometer, and a third, written by himself alone, solving the structure of common salt and

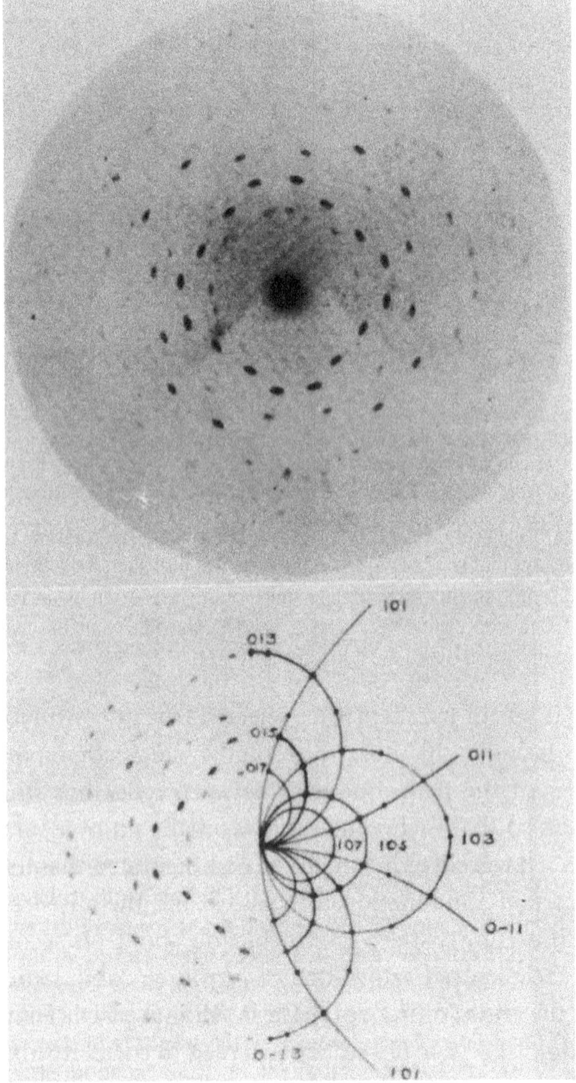

Figure 3. W.L. Bragg's re-interpretation of the Germans' X-ray diffraction photograph of zincblende. He indexed the reflexions by assigning a face-centred cubic lattice to zincblende and a continuous spectrum to the X-rays. He showed that the reflexions lie on the intersections of the photographic plate with a series of cones, each cone containing the reflexions from planes parallel to a zone axis.[7]

showing how the Laue pictures of several simple minerals could be indexed. There follows the structure of diamond, solved, as he relates, largely by his father, and the structures of fluorspar, zincblende, iron pyrites, calcite, and dolomite solved by himself alone. Finally, on 16 July, 1914, he communicated a paper on the structure of metallic copper.[8–11]

In view of this published record and the fact that for most of the relevant period the father was at Leeds and the son at Cambridge, it seems hardly believable that the scientific public tended to attribute most of the credit for these discoveries to the father, sometimes with the undertone that the son had cashed in on the father's success. The son must have suffered a great deal from these thoughtless and lazy judgments. Lazy, because people could not be bothered to read the literature.

Bragg wrote many years later: "Inevitably the results with the spectrometer, especially the solution of the diamond structure, were far more striking and far easier to follow than my elaborate analysis of Laue photographs, and it was my father who announced the new results at the British Association, the Solvay Conference, lectures up and down the country and in America, while I remained at home." Andrade wrote: "It was always a delight to his hearers to note the affection that came into Sir William Bragg's voice when, in lectures, he found occasion to deal with some one or other piece of work which had been carried out by 'my boy'."[12] But the "boy's" reaction to this patronising was: "My father more than gave me full credit for my part, but I had some heartaches."

So their great discoveries, which brought them the Nobel Prize for Physics in 1915, are said to have strained relationships between them for the rest of their lives. In his many lectures on the development of X-ray analysis, W.L. Bragg was fond of defining the exact roles played by himself and his father, but he never hinted at those strains until a few days before his death when he wrote to me: "I hope that there are many things your son is tremendously good at which you can't do at all, because that is the best foundation for a father-son relationship."

In most of the earliest structures of elements or simple compounds solved by the Braggs, crystal symmetry had so restricted the choice of atomic arrangements that only very few atomic parameters were left open. For example, the structure of diamond, published in the *Proceedings of the Royal Society* by "Professor W.H. and Mr W.L. Bragg" in 1913, was determined like this. The crystals were cubic. The presence on the Laue photographs of certain spots and the absence of others showed that the atoms of carbon must lie on a face-centred cube. The length of the cube edges could be measured from the angles at which the diffracted rays emerged. This gave the volume of the cube. The volume multiplied by the density of the crystals showed that it contained eight carbon atoms rather than four. Therefore there must be two sets of four carbon atoms, each occupying the corners and face centres of a cube. How far were they shifted relative to each other? This was the only unknown parameter. The Braggs showed that it can be deduced simply from the orders of reflexion that are reinforced and those that are extinguished by interference.[11]

The unravelling of the structures of minerals containing several different kinds of atoms presented challenging new problems that could not be solved simply by looking for present and absent reflexions. Bragg described his ingenious new methods for solving such structures in a seminal paper on "A Technique for the X-ray Examination of Crystal Structures with Many Parameters," published with J. West in the *Zeitschrift für Kristallographie* in 1928, and in the following paper on their application to diopside.[13,14]

In the 1920s and 1930s most crystallographers recorded the X-ray diffraction patterns photographically, which told them the *relative* intensities of the X-ray reflexions. They were content with qualitative data, but Bragg, together with R.W. James and C.H. Bosanquet, began his post-war research at Manchester with the introduction of quantitative ones. He used an X-ray spectrometer, the forerunner of today's diffractometer, with which he recorded the *absolute* intensities of the X-ray reflections, *i.e.* the fraction of the incident intensity diffracted by the crystal. This provided him with far more meaningful data for solving structures and testing whether they were correct than those used by most other workers in the field.

Diopside is a silicate mineral that was believed to contain molecules of $CaSiO_3$ and $MgSiO_3$. In 1928, solution of its structure seemed a more formidable undertaking than anything done before because it involved the determination of 14 independent parameters. Compare this with the 36,000 atomic parameters of the structure of the photochemical reaction centre for which H. Michel, J. Deisenhofer, and R. Huber shared the Nobel Prize for Chemistry in 1988.

Diopside crystals are monoclinic with a face-centred unit cell. This is the name given by crystallographers to the smallest volume containing the atomic pattern that repeats itself in all directions. In diopside, that unit of pattern consists of four molecules of $CaMg(Si_3)_2$. Crystal symmetry restricts the calcium and magnesium atoms to four alternative positions but does not tell which is the right one; the silicon and oxygen atoms can lie anywhere.

The way to find the silicon atoms may be illustrated by considering the 804 reflexion, which is too weak to observe (Figure 4). There is a contribution of +47 from the four Ca and Mg atoms, another of between +44 and –44 from eight silicon atoms, and another of between +41 and –41 from 24 oxygen atoms. The oxygen contribution is unknown within these limits. The silicon atoms cannot be making a positive contribution to F(804), for in that case, even if all the oxygen atoms were making negative contributions, there would be a positive resultant that would be observed. On the other hand, a negative contribution of any amount by silicon is possible. The planes (804) are now drawn and parallel strips are shaded in which atoms making positive contributions might be situated. These areas are forbidden to the silicon atoms. Repetition of this procedure for many reflexions led Bragg to the exclusion of all but four possible positions for the silicons (Figure 5). He then found that in three of them, neighbouring, symmetry-related atoms would be so close together that they would overlap. This left 4 and 4′ as the only possible silicon positions. Knowing where

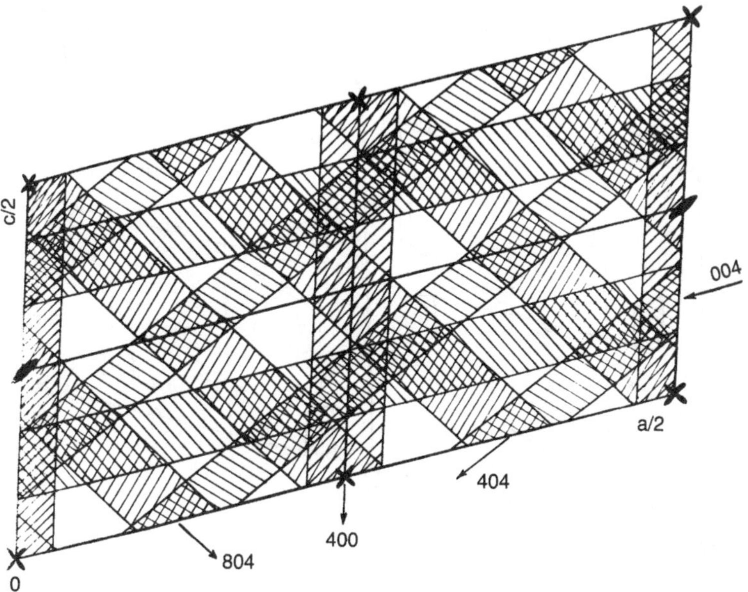

Figure 4. Unit cell of diopside projected along the b-axis with shaded areas forbidden to the silicon atoms.[13] The crosses mark centres of symmetry, the black elipse signs two-fold rotation axes.

the silicons were, Bragg was now able to decide which of the four possible positions for Ca and Mg is the right one. The 14 0 0 reflexion was so strong that all the atoms had to scatter in phase (Figure 6). If the Ca and Mg atoms were at either B or D they would scatter out of phase with the Si atoms; hence these two positions could be excluded. 406 is equally strong. If the Ca and Mg atoms were at either A or B they would also scatter out of phase with the Si atoms. Hence the only positions not excluded were C which lie on the axes of two-fold symmetry.

That was easy! The difficulty began with finding the positions of the six oxygen atoms, so that the sum of their scattering contribution, together with those from the Ca, Mg, and Si atoms, equalled the observed amplitude of each of a hundred reflexions. It was an intricate game of chess where every move made to satisfy agreement with the observed amplitude of one reflexion could spoil the agreement with 10 others. If the calculated amplitude of only a single one of the 100 reflexions came out radically different from the observed amplitude, then your structure was wrong and you had to start all over again. But Bragg got it right (Figure 6).

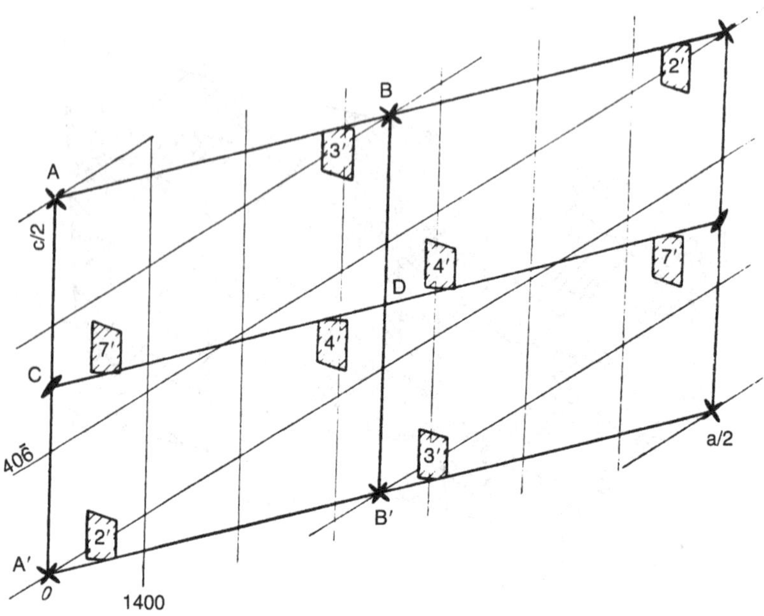

Figure 5. Unit cell of diopside showing areas allowed to the silicon atoms. A, B, C, and D mark positions of the calcium and magnesium ions allowed by the crystal symmetry. The observed intensities of the 406 and 1,400 reflexions exclude A, B, and D.[13]

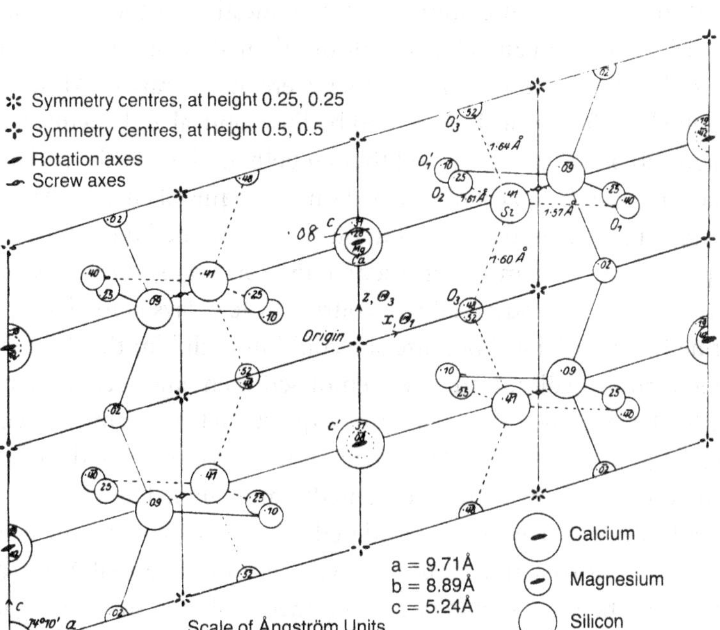

Figure 6. Atomic positions of diopside projected along the b-axis. The numbers give the y-coordinates of the atoms.[13]

Was the answer worth such exertion? Or did Bragg just play a sophisticated intellectual game, like some of the people working on artificial intelligence or topology today? In the notes that Bragg left me with his collection of reprints, he wrote:

> The analysis of diopside was a turning point in our ideas about silicate structures. I showed that the "SiO_3" which appears in the chemical formula does not represent SiO_3 acid groups but a string of SiO_4 groups joined by shared oxygen atoms. It was a crucial step in showing that silicon always occurs in a tetrahedral group of oxygen atoms.

To appreciate the novelty of Bragg's results you have to put your mind back 70 or 80 years and ask what was the body of knowledge in inorganic chemistry and mineralogy in those days. I looked at textbooks published early in this century and tried to recall the lectures in inorganic chemistry that I attended as an undergraduate in Vienna. In J.R. Partington's textbook of inorganic chemistry published in 1925, 12 years after Bragg had solved the structure of common salt, the question of the atomic arrangement of the sodium and chlorine atoms, their state of ionisation or of the forces that hold the crystal together were never raised. Minerals were described by their morphology, and by their optical and chemical properties, but no one asked what held them together. Many of the chemical formulae given in the textbooks turned out to be mistaken, like that of diopside. Partington's outdated section on silica reminded me of my viva with my professor of Chemistry in Vienna, the formidable Ernst Späth, at the conclusion of my undergraduate courses. A few days beforehand, I heard that he had failed a girl because she could not tell him the different crystalline forms of silica. I quickly memorised them and recited them at the viva to the professor's satisfaction: α-quartz, left or right handed below 575°, β-quartz from 575–800°, tridymite above 800°, and crystoballite above 1,470°. Späth purred contentedly and invited me to become his research student. He never wondered what atomic structures underlay these several forms and he must have been unaware that X-ray analysis had shown them to be made up of tetrahedra of SiO_4 sharing corners, but stacked in different ways.

Few chemists took much notice of X-ray crystallography's new insights until 1939 when Pauling published *The Nature of the Chemical Bond*, and some ignored them until 1945 when A.F. Wells published his *Structural Inorganic Chemistry*. T.M. Lowry's textbook of inorganic chemistry, published in 1922, was a notable exception. It includes a section on the crystal structures that had been solved and points out that crystals of common salt do not contain molecules of sodium chloride, but ions of sodium and chlorine.

Bragg and his followers showed that most crystals of inorganic compounds do not contain discrete molecules, but a continuum of alternate positive and negative ions. The positive ones are small and surround themselves with the larger negative ones arranged at the corners of polyhedra so that they are tightly packed and all electric charges are locally compensated. The silicates that form the bulk of the Earth's crust are made of SiO_4 tetrahedra that are either separate or share corners or edges, and their structures explain each mineral's strength or weakness. Thus Bragg's ingenious and immensely laborious puzzle-solving made people understand for the first time the atomic structure of the ground we stand on, and that surely was worthwhile.

Was there an easier way? When Bragg's father delivered the Bakerian

Lecture to the Royal Society in 1915, he suggested that the periodic repeat of atomic patterns in crystals could be represented by Fourier series.[16]

> If we know the nature of the periodic variation of the density of the medium we can analyse it by Fourier's method into a series of harmonic terms. The medium may be looked on as compounded of a series of harmonic media, each of which will give the medium the power of reflecting at one angle. The series of spectra which we obtain for any given set of crystal planes may be considered as indicating the existence of separate harmonic terms. We may even conceive the possibility of discovering from their relative intensities the actual distribution of the scattering centres, electrons and nucleus, in the atom; but it would be premature to expect too much until all other causes of the variations of intensity have been allowed for, such as the effects of temperature, and the like.

The American physicist R.J. Havighurst later used a triple Fourier series to deduce the electron density distribution in a crystal sodium chloride along the cube edges and the cube diagonals, using the absolute intensities measured by Bragg, James, and Bosanquet.[17]

Bragg extended the Fourier series to two dimensions. Each of the shaded stripes of his trial and error work on diopside now became a sinusoidal wave. Symmetry dictated that each wave must have either a crest or a trough at the positions of the magnesium and calcium ions, and it was easy to decide which was right. The sum of all the waves took the form of a map that revealed the positions of the oxygen atoms even though these had not been used in deciding whether a crest or trough was to be assigned to any particular wave (Figure 7).[15]

The amount of labour Bragg had to perform to calculate the Fourier projections of diopside on three principal planes was gigantic. For the projection on the *b*-axis alone he had to calculate the value of each of 26 different waves at 288 separate points and then had to sum 7,488 numbers. For the other two projections he had to sum 3,360 and 6,912 numbers, respectively, or 17,760 numbers in all. This was the birth of the Fourier projections which were used to solve hundreds of crystal structures for the next 30 years, until the advent of digital computers made it possible to calculate Fourier series in three dimensions.

Bragg's paper on *The Determination of Parameters in Crystal Structures by Means of Fourier Series* is by himself alone. He mentions no collaborators. How did he do all these tedious summations when adding machines had not yet been invented? We shall never know.

Incidentally, reports of continual tension between himself and his father are belied by the final paragraph of that paper.[15]

> It is with great pleasure that I acknowledge my indebtedness to my father, Sir William Bragg, for suggestions which materially contributed to the work described in this paper. At the time when I was following up the connection between our usual methods of analysis and the analysis by Fourier series, a connection briefly treated in the paper by Mr. West and myself, my father showed me some results which he had obtained by using relative values of the first few terms of two- and three-dimensional Fourier series to indicate the general distribution of scattering matter in certain organic compounds. It was largely as a result of his suggestions that I was encouraged to make all the computations for this two-dimensional series, using the extensive absolute measurements which we had made on certain crystals.

In the notes he left me, Bragg wrote: "This paper should really have been written with my father. He produced a crucial idea about two-dimensional Fourier series; I happened to have all the experimental data which showed how such a series could be used. It was the first paper in which Fourier series were used for parameter determination."

Bragg's application of the Fourier method to diopside required knowledge of the calcium, magnesium, and silicon positions to determine whether any particular set of waves had a peak or a trough at the calcium plus magnesium positions. In 1927, Bragg's American postdoctoral student J.M. Cork showed that this ambiguity of sign can be solved by the method of isomorphous replacement with heavy atoms.[18] The alums form an isomorphous series of the general formula $AB(SO_4)_2 \cdot 12H_2O$ where "A" can be any alkali metal and "B" can be a trivalent metal such as aluminium. Cork determined the signs of the Fourier terms in a one-dimensional series by analysing the changes in intensity that substitution of one metal ion by another brought about in the reflexions from one set of planes. With the solution of the alums Bragg's school laid the foundation for the method of isomorphous replacement that I used 25 years later to solve the structure of haemoglobin.

Peter Medawar wrote that "Every discovery, every enlargement of the understanding, begins as an imaginative understanding of what the truth might be."[19] Bragg's success in solving structures was based on a remarkable imaginative insight into the workings of natural phenomena, especially those concerned with optics and the properties of matter. According to Karl Popper and Peter Medawar, research consists of the formulation of

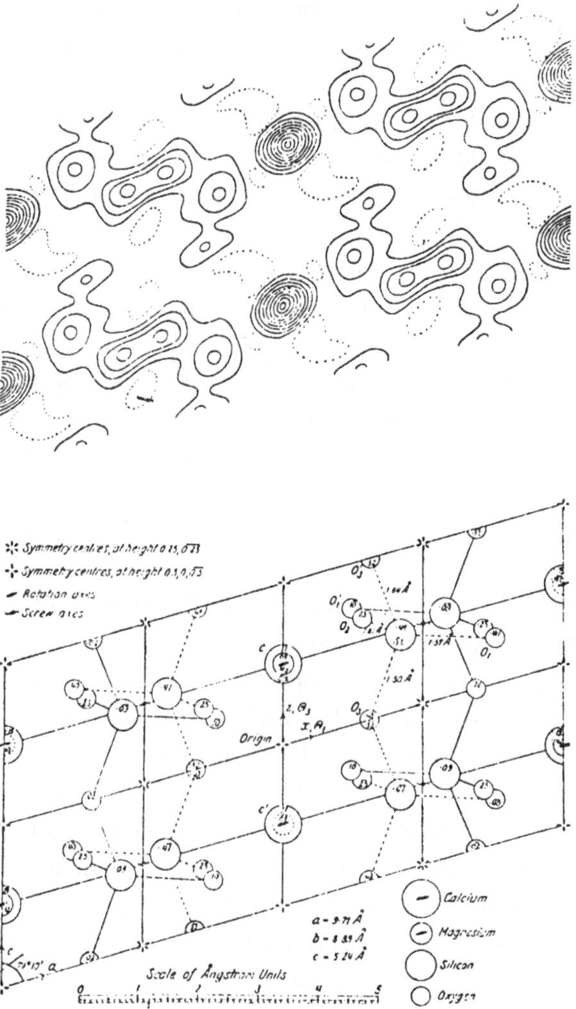

Figure 7. Comparison of Fourier map calculated from the known positions of the calcium, magnesium, and silicon ions alone, and the complete structure solved by trial. The lower peaks in the Fourier map coincide with the oxygens found by trial.[15]

imaginative hypotheses that are open to falsification by experiment. This is exactly how Bragg went about finding where the atoms lay, but he combined imagination with a phenomenal amount of hard work. Popper and Medawar argue further that no hypothesis can ever be completely proved, but that it can only be disproved experimentally so that it gradually corresponds more and more closely to the truth. However, Bragg's structures are not preliminary approximations subject to revision; any student setting out to redetermine the structures of calcite, quartz, or beryl will be disappointed.

T.S. Kuhn argued that science advances by a succession of paradigms,[20] but the perusal of old textbooks of chemistry and mineralogy have convinced me that there was no paradigm for the atomic structure of solid matter before 1912. The results of X-ray analysis opened a new world that had not even been imagined before.

When reviewing scientific work I sometimes paraphrase people's papers, but when I tried to paraphrase Bragg's I always found that he had said it much better. Bragg's superb powers of combining simplicity with rigour, his enthusiasm, liveliness and charm, and his beautiful demonstrations conspired to make him one of the best lecturers on science that ever lived.

Bragg united C.P. Snow's two cultures because his approach to science was an artistic, imaginative one. He thought visually rather than mathematically, generally in terms of concrete models that could be either static, like his crystal structures, or dynamic, like the interaction between crystals and electromagnetic waves or the order-disorder transitions and mobile dislocations in metals. His artistic gifts surfaced in his delicate sketches and water colours, and in his limpid prose.[21]

His scientific output in the 1920s and 1930s was prodigious, yet I am told that he was never rushed and always had time for his family, because his penetrating intellect and powers of concentration made all work easy. Instead of losing himself in a labyrinth of conflicting evidence which he would rarely bother to read, he would think of the best way interatomic forces could be satisfied to give stable structures.

Nowadays, cynics want us to believe that scientists work only for fame and money, but Bragg slaved away at hard problems when he was a Nobel Laureate of comfortable means. He was driven by curiosity. He was not a public figure and he liked to do his work at home rather than in aeroplanes. So often men of genius were hellish to live with, but Bragg was a genial person whose creativity was sustained by a happy home life; typically one would find him tending his garden, with Lady Bragg, children, and grandchildren somewhere in the background, and before getting down to crystal structures, he would proudly demonstrate his latest roses. He died in 1971.

Life's Energy Cycle[*]

On 14 December 1932, the dean of the medical faculty of the University of Freiburg, Professor E. Rehn, reported to the Ministry of Education,

> As an assistant physician Dr. Krebs has shown not only outstanding scientific ability, but also unusual human qualities. His... paper on the synthesis of urea in the animal body... will be regarded as one of the classics of medical research.

Four months later, shortly after Hitler seized power, this same dean sent him a "Notification of Immediate Removal from Office," obediently implementing the Minister's orders against members of the Jewish race.

Krebs kept several such letters and also various press cuttings from those days and had them reproduced as facsimiles in his book. There is the "Manifesto Against the Un-German Spirit" published by the Freiburg Student Union and posted in the University: "Our most dangerous adversary is the Jew and he who serves him. The Jew can think only as a Jew. If he writes German he lies. The German who writes German, but thinks un-German is a traitor..." The Rector of the University, von Möllendorf, ordered the posters to be removed, whereupon the Minister replaced him by the existentialist philosopher Martin Heidegger who endorsed them and proclaimed, "German students!... Your existence must not be ruled by learned axioms and ideas... Only the Führer himself constitutes today's

*A review of the book *Reminiscences and Reflections* by Hans Krebs (in collaboration with Anne Martin; Oxford University, Press).

and tomorrow's reality and law. Daily and hourly must you fortify your faithful obedience..." Most of Krebs' colleagues acquiesced, but one Dr. Arthur Jores dared send his and Krebs' former Jewish chief, by then in New York, a reprint with a personal dedication. The envelope was opened by a Nazi colleague who denounced him. Jores was dismissed from his post, publicly branded as an enemy of his country and, in due course, imprisoned.

Despite these harrowing experiences and the extermination of 20 of his relatives, Krebs does not follow the example of Einstein who refused ever to set foot in Germany again, but declares in the book that "to be anti-German seems to me just as bad as being anti-semitic." After the War he therefore persuaded the British Biochemical Society to re-establish contact with Germany by inviting several known anti-Nazis to the First International Congress of Biochemistry at Cambridge.

Krebs' discovery of the ornithine cycle of urea synthesis had caught the eye of F. Gowland Hopkins, then Professor of Biochemistry at Cambridge and President of the Royal Society. When he heard of Krebs' dismissal, he immediately invited him to Cambridge. The Rockefeller Foundation which had already supported Krebs' research at Freiburg, provided the money, and continued to support his research for the next 30 years. Krebs was deeply touched by the warmth of his reception at the Biochemistry Department at Cambridge and by the generous hospitality and friendliness of English people generally. His book is permeated by affection for England, which he epitomizes by Carl Zuckmayer's, another refugee's, saying that "home is not where a man is born but where he wants to die."

How Prussian Krebs remained all the same! He writes,

> ...I expected a lot from my associates—hard disciplined work and the ability to accept my criticisms... I have tried to be fair, honest and helpful, and not to demand more of others than I do of myself... . We criticised each other ruthlessly—we know it was done honestly, in good faith and in a spirit of helpfulness.

Few Englishmen would write such solemn stuff about themselves. I cannot imagine Francis Crick declaring in his memoirs that he demolished my cherished 1949 model of haemoglobin, "honestly, in good faith and in a spirit of helpfulness," rather than admitting to a certain mischievous satisfaction at the dastardly deed.

For people who complain that money for research has become hard to come by, Krebs' early career makes salutary reading. His father, a doctor in

Hildesheim, had to support him throughout his medical studies. After graduating, Krebs wanted to do research, but paid research posts were rare, never advertised, and obtainable only by what used to be known in my native Vienna as *Protektion*, which signified a mix of sycophancy, nepotism, and the old boy network. Eventually, Krebs found an unpaid post at the Third Medical Clinic in Berlin where another biochemist who later made his name, Bruno Mendel, became one of his colleagues. One night the Mendels were invited to the Einsteins when Otto Warburg was among the guests. When Warburg told Mendel that he was looking for a collaborator, Mendel recommended Krebs and also raised the money privately to pay him a modest salary. Krebs regards his four years with Warburg as formative and expresses the greatest admiration for his scientific genius, regardless of his despotic, egotistical, and sometimes malicious behaviour, his lack of confidence in Krebs's own talent, and his refusal to help him find a university post where Krebs would have to attach himself "to some old ass of a professor." Only in medicine would Krebs be able to make a living, Warburg told him. Warburg demanded of his staff that they work punctually from eight to six, six days a week, a precept which Krebs himself seems to have followed for much of his life. At 11 o'clock on the Monday morning after reading of Warburg's Draconian regime, I walked into my own biochemistry laboratory to find that only one of my three collaborators had arrived yet. "The other two were still working when I left at midnight," the first reported. This convinced me again that the free coming and going of Cambridge is more conducive to dedicated research than the iron discipline of old Berlin.

Krebs made his first great discovery, the ornithine cycle of urea synthesis, in 1931 and 1932. He used the tissue slice and manometric techniques which Warburg had taught him; another decisive factor was his own invention of the Krebs-Ringer solution. I found it exciting to read of his groping in the dark, methodically testing all conceivable intermediates until he discovered that "one molecule of ornithine could bring about the formation of more than twenty molecules of urea, provided that ammonia was present." From that moment the tracing of the other intermediates followed logical steps. It was the first biological process in which the intermediates were found to play a purely catalytic role. Krebs made this fundamental discovery while in charge of a medical ward with over 40 beds, which makes his feat even more remarkable.

Krebs' unravelling of the citric acid cycle in 1937 was to win him even greater fame, but his letter to *Nature* announcing it was rejected by the editor, Sir Richard Gregory, who at that time took it upon himself to judge the scientific worth of most of the communications sent to him.

Objections were raised at first against both the cycles. Those against the ornithine cycle were later found to be based either on wrong experiments or incorrect interpretation, while those against the precise chemistry of the citric acid cycle seemed fundamental. Biochemists argued that radioactively labelled CO_2 introduced into the cycle should become randomly distributed among the two carboxyl groups of α-ketoglutaric acid, an intermediate two steps after citric acid, because an enzyme would not be able to distinguish between the two symmetry-related carboxyl groups of citric acid. In fact, only the carboxyl nearest to the keto group of α-ketoglutaric acid was found to be labelled; hence, it was concluded, citric acid could not be an intermediate in the Krebs cycle. That was in 1941. I would have been desperate if an apparently valid objection had been made to my most fundamental discovery in which I could detect no flaw, but Krebs writes as though it had never cost him any sleep. Was he really so placid that he did not continuously turn over in his mind all conceivable explanations of the paradox, or did the sunshine of his later glory dissolve the memory of the seven clouded years that were to elapse before A. Ogston, in a brief and classic note to *Nature*, pointed out the fallacy in the objection: a symmetric molecule attaching itself to an enzyme at three points may give rise to only one of two possible asymmetrical reaction products. This was the birth of the concept of prochirality.

The glory was first heralded in October 1952 by eager journalists who told Krebs that he would shortly be awarded the Nobel Prize, but the rumours proved false and S.W. Waksman received it instead. Krebs related proudly that the rumours left him and his wife unruffled. But did they really? Nine years later, similar rumours about John Kendrew and me were floating around our laboratory. We doubted them until my secretary rushed in, flourishing two telegrams, one addressed to Kendrew and the other to me. This was it. When we had eagerly torn them open we found them to be from the Pontifical Academy in Rome, enquiring how many reprints we wanted of the papers we had read there the previous autumn. We also pretended to a stoic calm. Unruffled or not, Krebs did receive the Prize the following year.

After nearly 20 happy years and the founding of a flourishing school of biochemistry at Sheffield, Krebs became Professor of Biochemistry at Oxford and stayed there until his death last November. I was surprised by his statement that Oxford had remained in the forefront of learning for 600 years. Had he never read Edward Gibbons' description of its decline into sloth in the eighteenth century when dons could not even be bothered to teach their students? The picture of Oxford University in Krebs' own time conjured up in the book is of a citadel of unjust privileges jealously guarded by the Party members. Krebs got himself elected to the Central Committee by the underprivileged Solidarity of Active Researchers, but all his brave attempts to reform The System were defeated by the strength and cunning of the Old Guard, who even went so far as to refuse money offered by the University Grants Committee for the creation of additional science professorships, lest this should lead to the appointment of more Dissidents and Enemies of the Party like Krebs. Despite these rebuffs, Krebs describes his life at Oxford as a happy and successful one. Asked once for a guiding motto, Krebs replied "Never put off till tomorrow what you can do today," which sounds like one of those recipes for virtue Victorian children were made to embroider and hang over their beds. The King of France says it less prosaically in *All's Well that Ends Well:*

> Let's take the instant by the forward top;
> For we are old, and on our quick'st decrees,
> The inaudible and noiseless foot of time
> Steals ere we can effect them.

Krebs emerges from this book as a dynamic scientist and an engaging, warm-hearted individual utterly devoted to his research and teaching. He quotes Noel Coward's saying that "work is fun, there is no fun like work." I agree.

*The Hormone That Makes Nerves Grow**

Rita Levi-Montalcini received the Nobel Prize for Physiology or Medicine in 1986, together with Stanley Cohen, for their discovery of growth factors made thirty years earlier while they were working together at St. Louis. Her book recalls her life from her childhood in Turin to her return to Italy from America in 1963. The first part gives an interesting account of life in a Jewish middle-class family in fascist Italy before and during the Second World War, especially during the Nazi terror, when the Levis were hidden in Florence under false names by good-natured, courageous gentiles who pretended not to know that they were Jews. A far greater proportion of Jews survived in Italy than in most other countries of continental Europe because compassionate gentiles helped and hid them, often at the risk of their lives.

Levi-Montalcini derives the title of her book from a poem by Yeats:

The intellect of man is forced to choose
Perfection of the life, or of the work,
And if it take the second must refuse
A heavenly mansion, raging in the dark

She chose the work. The French biologist André Lwoff once wrote that "the scientist's art is first of all to find himself a good master," to which I should

*A review of the book *In Praise of Imperfection: My Life and Work* by Rita Levi-Montalcini (translated by Luigi Attardi; Basic Books).

add "and next, to find himself a good problem." Levi-Montalcini found her problem in a laboratory rigged up in her bedroom during World War II, when Mussolini's anti-Semitic legislation, a cowardly copy of German racial laws, forced Turin University to expel all Jews, including her professor Giuseppe Levi, who was one of Europe's leading anatomists. He had aroused her interest in the development of the nervous system while she was an intern; he worked with her in her improvised laboratory, and encouraged her for the rest of his long life. (They were not related.) Together they wondered what determines the beautifully ordered development of the nervous system in early chick embryos whose spinal cords develop before they grow limbs; nerves then grow from the cord into the limbs. When Rita and Giuseppe Levi excised the budding limbs from the embryos before these nerves had started to grow, the nerves never developed further; this suggested that the growing limbs release a chemical factor that attracts the nerves. After the war their publication came to the attention of Viktor Hamburger, a German émigré and pupil of the great embryologist Hans Spemann. Hamburger invited Levi-Montalcini to spend a semester with him at Washington University in St. Louis. That semester was to last sixteen years.

The decisive observations came in an experiment that illustrates the importance of the prepared mind. Elmer Bueker, a former pupil of Hamburger, sent him an article that described how a cancerous mouse tumor, grafted onto a chick embryo, had become invaded by nerve fibers from the embryo. Bueker concluded that the tumor had provided more ample terrain for the growth of the nerve fibers than the nearby embryonic limb, but Levi-Montalcini thought otherwise. In a euphoric mood, she dropped all current work in order to repeat Bueker's work. On being grafted to her embryos, the tumors Bueker had used became invaded by nerves as Bueker had described, but another tumor sent to her (by mistake?) produced a far more dramatic effect. It caused the organs of the embryo that are normally free from nerve fibers, including its gut and blood vessels, to be invaded by large bundles of nerve fibers. She concluded that the tumor had released a chemical compound, a factor that dissolved in the body fluids, accelerated normal growth of nerve fibers to their predetermined destinations, and also caused excessive growth of abnormal nerves.

If a biochemist suspects tumors of producing such a factor, his next step is to make an extract of the tumor and see if it has the same effect as the tumor itself. Rita Levi-Montalcini made such extracts and tested them, not on whole chick embryos but on nerve ganglia excised from their spinal cords. Such ganglia can be pictured as microscopic telephone switchboards. When Rita Levi applied her tumor extracts to the ganglia, haloes of nerve fibers grew around them, but only if the tumors had first been transplanted into chick embryos and then cut out again, as if the embryo induced the synthesis of the compound she was looking for.

Levi-Montalcini and Cohen spent the next year trying to extract enough of the growth factor from such transplanted mouse tumors but reaped so meager a harvest that it was hard to tell if their factor was made up of nucleic acid and protein or of protein alone. Wondering if the nucleic acid was a contaminant, Cohen asked the later Nobel laureate Arthur Kornberg, then at Washington University, for advice. Kornberg suggested treating the extract with snake venom, which contains an enzyme that breaks down nucleic acids. When Cohen tried this the activity of the extract rose spectacularly. It turned out that snake venom contained a several thousand times greater concentration of the growth factor than the mouse tumors. This discovery allowed Cohen to isolate and characterize the factor and provided Levi-Montalcini with pure factor to inject into her embryos. Had they been able to buy the pure enzyme, which is now on sale commercially, they would never have discovered this.

At first the nerve growth factor was regarded as an isolated phenomenon of no general significance, but scientists have now found several other growth factors that are crucial to animal development. If produced in excess, or if the chemical machinery that they normally set in motion is out of gear, they can also give rise to cancer. Levi-Montalcini's and Cohen's discovery had laid the foundations for that work, and this is why they have been awarded the Nobel Prize. Paul Ehrlich has said that success in research needs four G's: *Glück, Geduld, Geschick, and Geld* (luck, patience, skill, and money). Levi-Montalcini's book shows that she had the first three in good measure and needed little of the fourth.

How Nerves Conduct Electricity[*]

In August 1939, Alan Hodgkin and Andrew Huxley joined forces at the Laboratory of the Marine Biological Association in Plymouth for experiments on the transmission of nervous impulses along the giant axons of the squid. Hodgkin was a young research fellow at Trinity College in Cambridge and Huxley had just finished his undergraduate work in physiology there. On 23 August Hodgkin wrote to his mother: "We ran into a good many difficulties, but Andrew is a wizard at apparatus and got over them in an incredibly short space of time."

Huxley had tried to measure the viscosity of the axoplasm by letting mercury droplets fall through it, but had found to his surprise that it was a solid gel. This gave them the idea of introducing an electrode to measure directly the change in potential between the inside and the outside of the nerve during the propagation of an impulse. It was known already that the inside of a resting nerve fibre is electrically negative relative to the outside, but this negative potential was thought to fall to near zero during an impulse. Hodgkin and Huxley discovered instead that the amplitude of the potential change during the impulse was about double the resting potential, so that the internal potential became positive. These results opened the way to the discovery of the underlying physicochemical events, but just then the Second World War broke out.

*A review of the book *Chance and Design: Reminiscences of Science in Peace and War* by Alan Hodgkin (Cambridge University Press).

Hodgkin writes in his autobiography: "Only by undertaking some physically and intellectually demanding work could I forget the frustration engendered by having to abandon the marvellously exciting nerve experiments after 5 or 6 years' hard work, at the moment when they were most rewarding." He buried himself first in aviation medicine and then in the development of airborne centimetric radar for Bomber and Coastal Commands. Reconnaissance had shown that two-thirds of the bombs dropped at night over Germany, at heavy cost to crews and aircraft, fell in open fields. Radar helped the pilots to find their targets. These were not the railway yards and factories at which the Royal Air Force's communiqués pretended the bombs were directed, but cities, as they were all that radar could find. Hodgkin writes of his relief when the bombers were diverted to help with the invasion of France in the spring of 1944—he "had grown utterly sick of the night bombing offensive."

Hodgkin reports that all scientists supported the view of Patrick Blackett and Henry Tizard that the development of airborne radar should be directed instead at the detection of German U-boats by Coastal Command, and he proudly relates that this eventually reduced allied shipping losses from the unsustainable figure of 600,000 tons a month to only 50,000. He quotes Hitler's lament that "the temporary setback to our U-boats is due to a single invention of our enemies."

A temporary posting to Washington in 1944 led Hodgkin to visit the Radiation Laboratory at the Massachusetts Institute of Technology, where he was impressed by "the seemingly inexhaustible supply of people with a college degree in engineering" available to his colleagues working on radar; in Britain, he notes, "you were lucky to find anyone with a school certificate in physics." Ever since the war, this failing has been recognized as a major cause of the uncompetitiveness of much of British industry, but only now has that lesson sunk into technically illiterate governments, parliament, and the civil service.

In June 1947, Hodgkin could at last return to Plymouth to resume his experiments on the giant nerve fibres of the squid. On 8 July, he wrote to his friend Victor Rothschild that he thought he had proved why the nerve membrane potential reverses during activity: "The reason is simply that the active membrane, instead of becoming freely permeable to all ions (as supposed by classic membrane theory) becomes much more permeable to Na than to K." In September 1947, Hodgkin and Bernard Katz proved that in the physiological range the reversed potential V_{Na} varied with the external sodium concentration as predicted by the Nernst equation

$$V_{Na} = \frac{RT}{F} \ln \frac{[Na]\ \text{outside}}{[Na]\ \text{inside}}$$

In the following year, Hodgkin and Huxley demonstrated rigorously and quantitatively that ionic currents across the nerve membrane are responsible for the propagation of nerve impulses. They measured the time courses of the changes in the permeability of the membrane to sodium and potassium ions when the membrane potential is suddenly changed, and showed that these changes accounted quantitatively for the amplitude and time course of the propagated impulse. To clinch their results, Hodgkin's graduate student Richard Keynes measured the ionic flux with radioactive sodium and potassium he had prepared himself at the cyclotron in the Cavendish Laboratory at Cambridge. In 1952, Hodgkin and Huxley, both versatile at mathematics, developed differential equations to describe the propagation of the action potential in terms of the observed ionic currents.

It might be thought that any reasonably open-minded and perceptive scientist would have found this work convincing, but according to Hodgkin it met with widespread scepticism. Even the Nobel committee was unconvinced that nervous impulses are not transmitted by some energy driven process along the axoplasm until 10 years later, when Peter Baker and Trevor Shaw replaced the axoplasm of a giant squid nerve with a salt solution of the same composition and showed that such a perfused nerve conducts nearly a million impulses without the intervention of any biochemical process. Scientists are apt to close their minds to new ideas, especially to those in their own fields.

There is much in this autobiography besides Hodgkin's classic work on neurophysiology. For the physicist, there is a technical account of the development of airborne radar. For the general reader, the book contains vivid descriptions of Hodgkin's happy childhood among the often eccentric members of his many-branched Quaker family and of the Cambridge scene in the 1930s. He tells us, for example, how as an undergraduate he joined a hunger march of the unemployed to London, accompanied by Guy Burgess, who wore a woollen jumper with a zip-fastener concealing an Eton tie, to be exposed in case he was arrested by the police. And when Anthony Blunt was elected a research fellow of Trinity College, the master, J.J. Thomson, famed for his discovery of the electron but apparently not for his tact, introduced him with the words: "This is the first time we have elected an art historian to a fellowship, and I very much hope it will also be the last."

Hodgkin's accounts of his first visit to New York as a Rockefeller Fellow and of a trip to Mexico in the late thirties are enlivened by the delightful letters written to his mother at the time. At the Rockefeller Institute he was agreeably surprised to be given a well-equipped laboratory, in contrast to Cambridge, where he had had to make all the equipment for his experiments himself. Hodgkin tells of his falling in love with Peyton Rous's eldest daughter Marni, and of their happy marriage after years of separation through the war, and finally of his receipt of the Nobel prize together with Huxley and John Eccles, who was awarded it for his fundamental work on synaptic transmission.

Hodgkin comes across in this enjoyable book as a genial and humane scientist passionately devoted to his research for its own sake both before and after being awarded his Nobel prize. For 36 years he did most of his experiments with his own hands and yet found time to have a family life; to develop wide interests in literature, painting, and music; to watch birds; and to cultivate lasting friendships with a great variety of fascinating people. There are no enemies.

In Pursuit of Peace and Protein

Dorothy Hodgkin (née Crowfoot) was the third woman to receive the Nobel prize in Chemistry—the first and second were Curies, mother and daughter respectively—and the second woman to be decorated, with the Order of Merit—preceded only by Florence Nightingale. She received the Nobel prize in 1964 "for her determination by X-ray techniques of the structures of important biochemical substances." These were cholesterol, medically important as one of the causes of atherosclerosis and for its relationship to several hormones; penicillin; and vitamin B12, the anti-pernicious anaemia factor. No other woman has won the chemistry prize since.

Hodgkin was unique in many other ways. Maggi Hambling, the artist who painted her picture in the National Portrait Gallery, summed her up as "the closest person to a living, walking saint that I have ever met." There was magic about her person, and Georgina Ferry's biography brings her to life through a wealth of letters written by her and those around her.

Tidiness was not one of her strengths (there was a crisis when she had to go to a meeting abroad and had forgotten her passport); and she never discarded any letters, perhaps because she could not be bothered to sort and file them. They proved a treasure-trove to her biographer.

Dorothy Crowfoot was born in Cairo in 1910, the eldest of four sisters. Her father worked in the Egyptian Education Service and was later promoted to become director of education in the British-ruled Sudan. When war broke out in 1914, Mrs. Crowfoot took the girls and their English nanny back to England, installed them in a house in Worthing close to her elderly in-laws and left them there, returning to visit them only once during the four years of the war.

Apparently this neglect did Dorothy no harm; nor did it diminish her affection for her parents, but she wrote that it made her independent from an early age.

Mrs. Crowfoot was an outstanding woman who later had a decisive influence on Dorothy's development. She had been educated at home and discouraged from studying medicine. To make up for it, she had herself trained as a midwife, a skill that she put to good use after her marriage to help deliver the babies of women in the Sudan. Her stay in Egypt and visits to historic and prehistoric sites in Palestine roused her interest in archaeology and also made her an authority on ancient weaving techniques, interests that she passed on to her daughter. Two of her brothers were killed in the first world war and the other two died from its after-effects, a tragedy that made her resolve to devote herself to the promotion of international understanding. When Dorothy was a teenager, Mrs. Crowfoot took her to a discussion on disarmament at the League of Nations in Geneva, an event that made a deep impression on Dorothy and may have been the spur for later travels round the world as an angel of peace. It finally led Dorothy to accept the presidency of Pugwash, the organisation founded by Albert Einstein and Bertrand Russell to bring together physicists from East and West to halt the nuclear arms race.

On one occasion, Dorothy's parents took her on an archaeological dig in Palestine, where they had made friends among the Arabs. These friendships lasted, and Dorothy made no secret of them when she visited Israel. When I visited the country recently, some colleagues there were still indignant that she should have shown sympathy for their enemies.

In 1826, Michael Faraday instituted the "Christmas Lectures for a Juvenile Audience" at the Royal Institution in London. They have been given every year since and have inspired many young listeners to embark on a scientific career. Dorothy did not attend them, but as a 15-year-old, her mother bought her Sir William Bragg's lectures "On the nature of things" and "Old trades and new knowledge," where he mentioned the discipline of X-ray crystallography, which he and his son had recently introduced. This had allowed them for the first time to "see" atoms. According to Ferry, Bragg's "elegant introduction...excited the impressionable Dorothy beyond measure." That excitement lasted the rest of her life.

Dorothy's ethereal beauty, soft voice and reticent, gentle manner concealed a razor-sharp mind, a passionate temperament and an iron will to succeed. A fellow undergraduate at Somerville College in Oxford wrote: "She is very quiet and one just has to do all the talking, but anything she says is usually worthwhile. She is an extraordinary person as she is so quiet and yet does all these things. If you can get her to talk she is very interesting too."

This never changed. Dorothy's ambition, enthusiasm and thoroughness often made her work herself sick, but she enjoyed a sound constitution and never broke down. When she was an undergraduate, her parents often left her to look after her younger sisters, but Ferry writes that instead of being distracted from her studies, she learned to cope with having a lot of demands on her time by focusing intensely on each task as it arose and switching her attention rapidly and completely between them. It was a skill that was to prove invaluable in the future. In later years she could switch without effort between mathematical equations and her children's chatter.

After her chemistry studies at Oxford, Dorothy took the first step towards her long-cherished ambition to see atoms by joining the leading X-ray crystallographer J.D. Bernal at Cambridge. He was a flamboyant Irishman of immense charm and learning, which earned him the nickname Sage. Dorothy fell in love with him, and their relationship lasted until she met her future husband, Thomas Hodgkin. Judging by her letters to Bernal, she was as passionate a lover as she was a scientist. She never saw atoms in Cambridge, because all the crystals prepared by various chemists that Bernal passed on to her had structures far too complex for the methods of X-ray analysis then available, but she joined Bernal in one historic experiment. In 1934, an American physiologist returning from Sweden brought Bernal crystals of the digestive enzyme pepsin, which is a protein. By then biochemists had come to realise that proteins are the engines of the living cell. It was known that they are made up of hundreds or even thousands of atoms, but they were black boxes because the spatial arrangement of these atoms was unknown. It was not even known whether these atoms formed well-ordered structures at all, yet understanding their atomic structure seemed to be the key to the understanding of life itself. Several crystallographers had already taken X-ray diffraction pictures of dried protein crystals, but obtained no diffraction patterns. By keeping them wet, as the protein would have been in the living organism, Bernal and Crowfoot got beautiful diffraction patterns, which implied that the pepsin molecule had an ordered atomic structure. X-ray crystallography offered a hope of solving it, but fulfillment of that hope was remote because it was hundreds of times more complex than any crystal structure yet solved.

In 1935 Dorothy returned to Oxford to take up a teaching fellowship at Somerville. For her research she was given a table in the X-ray room of the Oxford Museum, Ruskin's neo-Gothic cathedral of science, whose fossil skeletons and general gloom made it look more like a mausoleum of science. But as soon as one entered Dorothy's room, one was lifted into the clouds by her serenity, brilliance and unbounded enthusiasm. One day she was given crystals of insulin, a protein containing fewer atoms than pepsin, which seemed a more hopeful candidate for X-ray analysis. She put a crystal in front of an X-ray beam and placed a photographic plate behind it. That night, when she developed the film, she saw minute, regularly arranged spots forming a diffraction pattern that held out the prospect of solving insulin's structure. Later that night she wandered around the streets of Oxford, madly excited that she might be the first to determine the structure of a protein, but next morning she woke with a start: could she be sure that her crystals really were insulin rather than some trivial salt? She rushed back to the lab before breakfast...a simple spot test on a microscope showed that her crystals took up a stain characteristic of protein, which revived her hopes. She never imagined that it would take her 34 years to solve that complex structure, nor that once solved it really would have medical applications.

Before I read Ferry's biography, I never realised how passionately Dorothy hoped to be the first to solve the structure of a protein, yet in 1959, when John Kendrew and I succeeded, she seemed delighted and never gave us the slightest hint of the disappointment she must have felt.

Much of Ferry's book is devoted to Dorothy's work for disarmament and conciliation between East and West, but this was biased by her unrealistically rosy pictures of Communist dictatorships in North Vietnam, China and the Soviet bloc. After insulin was solved, she devoted much of her time to these aims; frail though she was and increasingly crippled by rheumatoid arthritis, she became an indefatigable traveller in search of world peace.

Ferry never knew Dorothy Hodgkin, but it seems that the more she learned about her from letters and interviews with her friends and colleagues all over the world, the more she came to love and admire her. Her book is a delight to read, very well-documented and refreshingly untainted with feminism (with which Dorothy had little sympathy), relativism or deconstructivism. I learned a lot about Dorothy that I never knew, especially about her youth. Strangers may enjoy it equally.

Keilin and the Molteno

Introduction

David Keilin, who was born on 21 March 1887, was Lecturer in Parasitology at Cambridge University from 1925 to 1931 and Quick Professor of Biology from then until his retirement in 1954. He first distinguished himself with the discovery of the life cycles of flies whose larvae develop parasitically in animals and plants, and are themselves parasitized by micro-organisms. Keilin then made his fame with the discovery and characterization of a system of coloured enzymes, the cytochromes, which turn the chemical energy gained by the combustion of foodstuffs into a form that organisms can use for growth, movement, reproduction and even for thought. Keilin did most of his life's work at the Molteno Institute, a pleasant small building on the Downing site next to the Anatomy School, built in 1920 from a donation by Mr. and Mrs. Percy A. Molteno, who lived in South Africa and took a keen interest in the parasitological research of Keilin's predecessor, Professor G.H.E. Nuttall.

We used to attend Keilin's lively lectures on haem proteins every year, I mean we, the permanents at the Molteno Institute, together with some undergraduates and the ever-changing visitors who flocked there from abroad, attracted by Keilin's fame as a biochemist and parasitologist. He showed lecture demonstrations, a practice long since abandoned, and always took a special delight in projecting the porphyrin spectrum from a feather of the turaco bird. In fact, the spectroscope was his favourite instrument, not one of those hefty ones that chemists used, but a delicate little Zeiss hand spectroscope replacing the eyepiece of a microscope, with which he first discovered the absorption bands of the cytochrome system in the thoracic muscle of the fly *Gastrophilus intestinalis*. He used it to observe the absorption spectra of living matter. He recalled:

> One day, while I was examining a suspension of yeast freshly prepared from a few bits of yeast shaken vigorously with a little water in a test tube, I failed to find the four-banded spectrum, but before I had time to remove the suspension from the field of vision of the microscope, the four bands suddenly re-appeared. The experiment was repeated time after time and always with the same result. This first visual perception of an intra-cellular respiratory process was one of the most impressive spectacles I have witnessed in the course of my work.

Keilin used to tell us always to work with coloured proteins because their spectra were so revealing.

After his lectures Keilin sometimes asked me: 'Perutz, why can the same haem fulfil such different functions in haemoglobin, catalase and cytochrome c?' Implied in the question was: 'I am a parasitologist, but you as a chemist ought to understand this.' I was clueless at a time when the structures of all these proteins were still unknown, but I doubt that I could provide complete answers even today when they have been determined in detail.

Keilin's Favourite

Cytochrome c was Keilin's favourite, but he never succeeded in crystallizing it. In 1955, Gerhard Bodo, a young Austrian biochemist working with John Kendrew, crystallized it from the muscle of the King Penguin and tried to determine its structure. Keilin's birthday was on the first day of spring, symbolic of his youthful exuberance and easy to remember, so that I always called on him to wish him many happy returns. That year I brought him the first X-ray diffraction pictures of cytochrome c. Keilin's delighted response was: 'Next year you will bring me model.'

He was born in Moscow, brought up in Warsaw, studied in Liège and Paris, arrived in Cambridge in 1915 and still spoke English with an endearing Polish accent and syntax. When doing his round of the laboratory, he would ask: 'How you are getting on Perutz?' That was not a perfunctory question, but sprang from a deep interest in my work and everyone else's at his laboratory.

The physiologist Gilbert Adair made the first haemoglobin crystals for me and later taught me how to make my own in his room at the Low Temperature Station, but his working habits were eccentric. He never washed any of his glassware nor let anyone else touch it until he actually needed it so that all his benches were cluttered as in an alchemist's shop. Sometime in 1938, Keilin offered me bench space in his laboratory which was always clean, tidy and warm. Keilin was no pedant, but he suffered from allergic asthma which was provoked by house dust and aggravated by cold. His asthma caused him much discomfort and he kept it in check with an adrenaline spray that may have brought on the heart attack that finally killed him.

Charles Darwin wrote in one of his letters from South America: 'In short, I am convinced it is a most ridiculous thing to go round the world when by staying quietly at home, the world will go round with you.' Some science professors are in their departments only when they are not attending committees in London or touring abroad, but Keilin was always at his post and accessible to his staff. Scientists from all over the world sought him, like an oracle, for his knowledge, wisdom and humanity. He would always *listen*, not only to the president of some foreign academy, but to anyone of us, his students, watching us benignly from behind his mahogany desk in his small office on the first floor, an office without a secretary, because she was downstairs and also answered the lab telephone which had only two or three extensions. So there was not that barrier that other professors erected between them and their staff, the secretary whom you had to ask to be admitted into their presence: you could just knock at the door and walk in if there was no-one else in already. Keilin was never too busy to see you when he was in his office, but we did feel shy interrupting him when he was doing experiments. In 1957, I called at the Molteno to give him my good wishes on his 70th birthday. He arrived a few minutes after me and found a pile of mail waiting for him, but put it aside saying proudly: 'I am not going to read it, I am going to do experiments.' He continued them right to the end of his life. He was averse to theorizing and disapproved of young Crick's flights of fancy, telling me to 'keep him to the bench.' I ignored that advice, because Crick was irrepressible, but had he followed it, DNA would not have been solved. There are many ways of doing science.

Outside Keilin's office and all along the passage between the laboratories hung photographs of the world's leading parasitologists, collected by

Keilin's predecessor G.H.E. Nuttall, and himself: dignified elderly gentlemen, mostly, with magisterial beards and studious pincenez. As leader of the field and editor of the journal *Parasitology*, Keilin had known most of them personally. Besides his native Polish, he spoke fluent French, Russian and also German, so that he could converse and correspond in several languages at a time when English was not yet the scientific lingua franca. At one end of the passage was the tea room, where the monotony of dry biscuits was broken by an occasional cake to celebrate the publication of a paper. One day we had a real celebration.

Warburton and the Siberian Tick

In 1931, when Keilin had succeeded Nuttall as Quick Professor of Biology, an elderly lecturer called Warburton complained that he had been appointed before the University Superannuation Scheme had been instituted, so that he had no pension and would have to die in harness. When Keilin told the University Treasurer that Warburton was in his late seventies and had no pension, the Treasurer agreed that in view of Warburton's advanced age the University could afford to be generous. He failed to foresee that 24 years later we would celebrate Warburton's 100th birthday! On that occasion he told us a wonderful story. In his prime, Warburton had been the world's authority on ticks. One day in the twenties, some of his students were eating their lunch of bread and cheese when they found a tick in their butter. They brought it to Warburton who identified it as a Siberian tick. The discovery was to provoke a diplomatic crisis. The students had bought their butter at Sainsbury's, not knowing where it came from. Impressed by entomology's detective powers that could trace the butter's origin to Russia, they told their story to a don who mentioned it to a visiting MP and he in turn related it to a journalist. The outcome was a headline in one of the London evening papers: 'Disease-Carrying Tick Imported with Russian Butter.' Questions were asked in Parliament, the horse-drawn milkcarts which in those days also distributed butter in London bore placards reassuring housewives that they carried no Russian butter, the Soviet Ambassador called on the Foreign Secretary to protest against the campaign of slander against his country's agricultural exports and Pravda condemned Warburton's deliberate lies. Years later, Russian parasitologists visiting the Molteno Institute reproached Keilin for allowing it to become a tool of anti-Soviet propaganda and refused to believe that Warburton was just an unworldly scholar who had happened to come across a curiosity. Secure in his generous pension, Warburton continued to live in good health in Grantchester to the ripe old age of 103.

Frail Health

Keilin did not ride a bicycle, nor drive a car. In good weather he walked to the lab from the Keilins' comfortable home in Barton Road and on bad days Mrs. Keilin, also from Poland and a G.P., drove him to the lab before she started on her round of visits. His frail health prevented his engaging in any vigorous physical activity, and his asthma kept him from travelling. When he did not work in the lab, he edited the journal *Parasitology* or studied at home. To me, skiing and mountaineering holidays in the Alps and all kinds of physical activity were an essential part of well-being. I learnt to appreciate Keilin's cheerful endurance of his handicaps when an accident condemned me to lead a similar life of all work and no play for several years.

Keilin measured the respiration of tissues with Barcroft manometers manufactured in the Molteno's poky basement workshop by his general factotum, the rotund, easy-going laboratory steward and mechanic Charles, whose instruments had a habit of falling to pieces at awkward moments. There was no money to buy instruments except when Gerald Pomerat, the lean, dapper, meticulous representative of the Rockefeller Foundation, appeared like the rich uncle from New York and asked Keilin whether he needed anything for his research. I remember our joy when the Foundation bought Keilin the first Beckman spectrophotometer that measured optical densities directly, and we no longer had to work them out elaborately by comparison with known absorbents.

In 1944 or 1945, Keilin persuaded the Faculty Board of Biology B to let me give an annual course of eight lectures on what later came to be called Molecular Biology. They were held in the museum on the 2nd floor that housed disgusting tropical parasites pickled in glass jars and which also served as a lecture room. They were extracurricular, advertised on notices to the various science departments and attended mainly by research students and foreign visitors, rather than undergraduates. In 1954, the University appointed me a University Lecturer in Biophysics, and my lectures were advertised in the University Reporter as part of the biochemistry course, which made me think the posting of notices unnecessary. I arrived in the museum full of expectation, soon joined by James, the scraggy, ginger-haired little technician who regularly projected my slides, to await the students. Five o'clock came but no students. James and I exchanged little jokes which became feebler as time passed, but still not a single student appeared, to the joy, possibly, of the plant virologists whose rooms opened onto the museum and who could not run their roaring centrifuges while I lectured. But here I was, a newly appointed University Lecturer without a single student to listen to his lecture. I nearly wept with humiliation. After a while, I did find out what happened. The Professor of Biochemistry who had supported my appointment had also advised his students that my lectures were too specialized for them.

Lack of Tenure

Although Keilin was one of the most distinguished Cambridge scientists, he did not have tenure because the statutes of the Quick Professorship, the first research Chair to be established in Britain, required his appointment to be renewed every 3 years. Apparently, the Quick Trustees had made that provision because it seemed inconceivable to them that anyone could successfully concentrate on research for a period longer than 3 years at a time. Keilin once asserted to me that this never worried him one little bit, but I have wondered whether he really possessed such complete self-confidence or said it only to console me for the insecurity of my own position. Going backwards and forwards on my bicycle between the Cavendish Laboratory in Free School Lane and the Molteno Institute in Downing Street, I was the living link between Biology and Physics, but I was a chemist, incapable of teaching either subject, and neither the University nor any of the Colleges had a place for me. I was officially beyond the pale, but Keilin gave me confidence that my crazy attempt to solve the structure of haemoglobin would succeed, and he suggested W.L. Bragg's approach to the Medical Research Council which led the foundation of the Research Unit for the Study of the Molecular Structure of Biological Systems. This was the forerunner of today's Laboratory of Molecular Biology. Fifteen years later, when I proudly showed Keilin our new laboratory, he said to me: 'Now you merely have to win Nobel Prize!' I protested that he deserved it first, but he waved me off with the resigned remark that all this was long past. Clearly, he should have shared either the 1931 Prize for Physiology or Medicine that was given to Otto Warburg 'for the discovery of the nature and mode of action of the respiratory enzyme' or that given to Hugo Theorell in 1955 'for his discoveries concerning the nature and mode of action of oxidation enzymes.' When Theorell's prize was announced, Keilin's long-term collaborator, compatriot and friend, Thaddeus Mann, went to see him to express his disappointment at his exclusion. Keilin took him affectionately by the elbow and said: 'When I stand at the Gates of Heaven, St. Peter is not likely to ask me whether I have brought with me Nobel Prize.' He was thrilled all the same when the Royal Society awarded him their highest honour, the Copley Medal, and he proudly showed it to all of us in the lab.

Peter Mitchell, who received the Nobel Prize for Chemistry in 1978 for his chemiosmotic theory, had this to say about Keilin in his Nobel lecture:

> But let me first say that my immediate and deepest impulse is to celebrate the fruition of the late David Keilin, one of the greatest of biochemists and—to me, at least—the kindest of men whose marvellously simple studies of the cytochrome system, in animals, plants and microorganisms, led to the original fundamental idea of aerobic energy metabolism: the concept of the respiratory chain. Perhaps the most fruitful (and surprising) outcome of the development of the notion of chemiosmostic reactions is the experimental stimulus and guidance it has provided in work designed to answer the following three elementary questions about respiratory chain systems and analogue photoredox chain systems: what is it? what does it do? how does it do it? The genius of David Keilin led to the revelation of the importance of these questions.

Keilin personified the saying of Montesquieu that knowledge makes men gentle, and also Bertrand Russell's remark that 'Einstein's is a kind of simplicity which comes of thinking only about the subject concerned, and forgetting its relation to one's own ego.' John Kendrew and I owe Keilin a tremendous debt, for he was one of the first to see the potentialities of our physical approach to biochemistry. Until the 1950s, we had no facilities for biochemical work in the Cavendish Laboratory; Keilin gave us bench space in his institute even though he was short of space himself, and he helped us to grow protein crystals. When Kendrew's and my research was in danger of closing down for lack of support by the University, Keilin suggested, and supported, Sir Lawrence Bragg's approach to the Medical Research Council which saved us. But to me his most important gifts were his confidence in me and the warmth of his friendship, which helped me to gain confidence in myself.

After his death, Francis Turner wrote of him in the Magdalene College Magazine and Record (I, 13, 1963):

David Keilin wore the mantle of his world-wide distinction as an invisible garment: what was visible was disarming modesty, true friendliness, and a mind so quick that he immediately understood all that his neighbour meant, however imperfectly expressed. But this understanding was not only quickness of wit: it arose from the deeply charitable view he had of human nature, which, together with his wide experience, gave him an unusual insight into those with whom he came in contact. He was a most remarkable and loveable person, patient and wise, with so great a spirit in so small a frame.

Growing Up among the Elements[*]

London's Science Museum in South Kensington was closed during the Second World War. When it reopened in 1945, the twelve-year-old Oliver Sacks discovered there the periodic table of the chemical elements. The names were written in large letters on a wall, with samples of each element or one of its compounds attached to each. That night Oliver could hardly sleep for excitement. To a boy who was already a keen amateur chemist, the revelation that the apparently disconnected properties of the elements could be fitted into a logical system gave the first sense of the power of the human mind. Sacks writes:

> In that first, long, rapt encounter in the Science Museum, I was convinced that the periodic table was neither arbitrary nor superficial, but a representation of truths which would never be overturned, but would, on the contrary, continually be confirmed, show new depths with new knowledge, because it was as deep and simple as nature itself. And the perception of this produced in my twelve-year-old self a sort of ecstasy, the sense (in Einstein's words) that "a corner of the great veil had been lifted."

This sounds a little precocious for a twelve-year-old, but it reminded me of my own excitement when, as a student, I read Linus Pauling's just-published book *The Nature of the Chemical Bond*. It transformed the

*A review of the book: *Uncle Tungsten: Memories of a Chemical Boyhood*, by Oliver Sacks (Knopf).

empirical edifice of chemistry that I had been required to memorize into a science that could be understood, because it was based on fundamental properties of atoms, their sizes and charges, and the configurations of their electron shells, based on the new quantum mechanics.

The discoverer of the periodic system was the Russian chemist Dmitry Ivanovich Mendeleev. The romantic story of Mendeleev's early career made him one of Sacks's heroes. He was born in a small town in Siberia in 1834, the youngest of fourteen children. His father, the head of the local high school, went blind shortly after his birth and died not many years afterward. Dmitry's mother recognized the boy's outstanding talents and walked with the fourteen-year-old thousands of miles to Moscow to enroll him at the university there, only to learn that as a Siberian he was ineligible for admission. The same happened in St. Petersburg, but there she finally found him a place in the Pedagogical Institute to train as a teacher.

Despite those setbacks, Mendeleev became professor of chemistry at the Institute of Technology in St. Petersburg at thirty, and four years later professor at the university there. In the same year, aged thirty-five, he published the periodic table as part of his monumental *Principles of Chemistry*. Mendeleev had begun by ordering the eighty-one elements known at his time in sequence according to their atomic weights. He noticed that many of their chemical properties were repeated at regular intervals, which led him to order them into horizontal rows and vertical columns. He placed the two lightest elements, hydrogen and helium, at either end of the first row, followed by two rows of eight elements, like two octaves of musical chords, followed by three rows of eighteen. This left some vacant places where Mendeleev predicted new elements, which were later found.

There are profound physical reasons underlying the order, the periodicities, of Mendeleev's table, but it it took another fifty years before Niels Bohr, one of the greatest physicists of all time, discovered them. Oliver Sacks must have already absorbed an astonishing amount of chemical knowledge to have become so excited on seeing the table for the first time. Sacks calls the *Principles* the most delightful and vivid chemistry text ever published. Mendeleev's romantic, magisterial, Victorian introduction sets the tone:

> In comparing the science of the past, the present, and the future, in placing the particulars of its restricted experiments side by side with its aspirations after unbounded and infinite truth, and in restraining myself from yielding to a bias towards the most attractive path, I have endeav-

> oured to incite in the reader a spirit of inquiry which, dissatisfied with speculative reasonings alone, should subject every idea to experiment, to encourage the habit of stubborn work, and to excite a search for fresh chains of evidence to complete the bridge over the bottomless unknown.[1]

Sacks writes that he devoured Mendeleev's classic, but I wonder if he really absorbed its 1,168 pages.

Oliver Sacks' parents were both doctors living in a large, comfortable brick house in northwest London. His father was an expansive, outgoing man. Like the young Lewis Thomas in *The Youngest Science*,[2] Oliver sometimes went along on his father's Sunday morning house calls. Sacks writes:

> He loved doing housecalls more than anything else, for they were social and sociable as well as medical, would allow him to enter a family and home, get to know everybody and their circumstances, see the whole complexion and context of a condition. Medicine, for him, was never just diagnosing a disease, but had to be seen and understood in the context of patients' lives, the particularities of their personalities, their feelings, their reactions.
>
> ...I loved to see him percuss the chest, tapping it delicately but powerfully with his strong stubby fingers, feeling, sensing, the organs and their state beneath. Later, when I became a medical student myself, I realized what a master of percussion he was, and how he could tell more by palpating and percussing and listening to a chest than most doctors could from an X-ray.

Unlike Lewis Thomas' father, he did not tell his son how little he was able to do for most of his patients at a time before antibiotics became generally available. At home his father spent his leisure with ancient Hebrew texts rather than *The Lancet* or *The New England Journal of Medicine*, which might have kept his medical knowledge up-to-date.

Oliver's mother, a teacher of medicine, was the sixteenth of eighteen children, which provided him with about a hundred cousins, mostly living in London. The family was exceptionally gifted, and Oliver writes that his uncles and aunts were as good as a reference library. Uncle Yitzak was a radiologist, Uncle Abe a physicist, Uncle Dave a chemist, mineralogist, and metallurgist, Auntie Len an amateur botanist who told Oliver that God thinks in numbers, and this may have led to his early fascination with prime numbers. Uncle Dave ran a factory making tungsten filaments for electric light bulbs, hence the nickname that gave Sacks' book its title. Oliver never knew his youngest uncle, because he had become an outcast, a nonperson whose name was never spoken again after he married a gentile, a shiksa. Each year, Oliver's father told his family that he was off to Wales on a slimming cure; he would return tanned, invigorated, but not much slimmer. It was only after his death that Sacks found the ticket stubs showing that he had in fact secretly visited his excommunicated brother in Portugal. Home was also a venue for meetings of Zionists, but their bullying manners turned Oliver into an enemy of Zionism, evangelism, and politicking of every kind for life.

Oliver describes his mother as intensely shy, a woman who retreated into silence or her own thoughts on social occasions, but an exuberant performer with her students (of surgery, anatomy, or histology?—Sacks does not say). He believes that she was drawn into medicine because it was part of natural history or biology and writes that she had a love of structure extending from plants to human anatomy. When Oliver was only eleven, she had him dissect malformed, stillborn babies. At fourteen she took him to the dissecting hall of the anatomy school of the Royal Free Hospital for Women and told him to dissect the leg of a fourteen-year-old girl. It was his first encounter with a corpse, and he spent a month on the task. He writes:

> I lacked my mother's powers of visualization, her strong mechanical and engineering sense, but I loved it when she talked of the foot and drew, in rapid succession, the feet of lizards and birds, horses' hooves, lions' paws, and a series of primate feet. But this delight in understanding and appreciating anatomy was lost, for the most part, in the horror of the dissection, and the feeling of the dissecting room spread to life outside—I did not know if I would ever be able to love the warm, quick bodies of the living after facing, smelling, cutting the formalin-reeking corpse of a girl my own age.

Oliver's fascination with chemistry began as an escape from the aftermath of his traumatic school experiences during and after the war. His life at home had been happy until the war broke out in September 1939, when he was six years old; but then the government, fearful of immediate, devastating German air raids, decreed the evacuation of women and children from London. As a doctor, Oliver's mother had to stay, so she sent Oliver and his elder brother Michael to a small boarding school in the Midlands that had just been set up. It was headed by a master who, faced with his new responsibilities, soon turned into a sadistic, avaricious monster. He beat many of the boys, Oliver included, almost daily. When he beat Oliver so hard that his cane broke, he sent the bill for a new one to his parents. He fed the boys on a diet of turnips and coarse beets grown for cattle. Sacks writes:

> The horribleness of the school was made worse for most of us by the sense that we had been abandoned by our families, left to rot in this awful place as an inexplicable punishment for something we had done.

He felt trapped without hope, without recourse, forever. Yet on their rare visits, his parents noticed nothing. At home his mother had said a prayer with him before she kissed him goodnight, but here Sacks replaced his childhood religion with a raging atheism, a fury with God for not existing, not taking care, for allowing the war with all its horrors to occur.

When the school finally closed because most other parents had withdrawn their children, Oliver and Michael's parents sent them to another boarding school where the bullying drove Michael insane. Sacks does not say whether he ever recovered. Oliver withdrew into a world of fantasy, telling the other nine-year-olds, not unreasonably, that his parents had thrown him out as a small child and that he had been brought up by a she-wolf. The parents finally realized that he was on the brink and took him home to London, where he recovered.

Despite those haunting experiences, I found no hint of resentment of his parents in Sacks' book. On the contrary, he writes affectionately about them both, and the only mild criticism he allows himself is a remark that "they were intensely sensitive to the suffering of their patients, more so, I sometimes thought, than to those of their children."

To make Oliver more sociable and teach him practical skills, they enrolled him in the Cub Scouts, but it was a failure. The fires he laid never started to burn and the tents he pitched invariably collapsed. One day the scoutmaster told the boys to bake disks of unleavened bread, and bring them to the next outing. When he found the flour tin empty, Oliver, ever resourceful, discovered some cement outside, made it into a paste, flavored it with garlic, shaped it into an oval disk, and baked it in the oven. When he tempted the unsuspecting scoutmaster with that delicacy he cracked a tooth, and Oliver was expelled.

He turned to chemistry instead, and Uncle Tungsten became his teacher. He showed Oliver how to make tungsten metal by smelting tungsten ore with charcoal. The ore was named scheelite after its eighteenth-century Swedish discoverer, Karl Wilhelm Scheele. His uncle told Oliver that Scheele had been an apothecary who worked all on his own, cared nothing for money, and just explored chemistry for its own sake. This made Oliver want to become a chemist and to discover a new mineral that would be named Sacksite.

His uncle told Oliver that the tungsten filaments made in his factory had been invented after many years of experiment that began with the observation that lime shone brightly in a gas flame. Lime was soon used to light theater stages, hence the figurative term "limelight" that has survived after the real lime lights have long been forgotten. The Austrian inventor Carl Auer von Welsbach (not a German, as Sacks writes) replaced them with fabrics impregnated with a mixture of cerium and thorium oxides. They produced a brilliant glow and were soon used for domestic and street lighting. Oliver used to watch the lamplighters go round the streets with their long poles lighting one gas light after another, as my own children did.

Oliver's first London school concentrated on Greek and Latin. He writes that

> this did not matter, for it was my own reading in the library that provided my real education, and I divided my spare time, when I was not with Uncle Dave, between the library and the wonders of the South Kensington museums, which were crucial for me throughout my boyhood and adolescence.
>
> The museums, especially, allowed me to wander in my own way, at leisure, going from one cabinet to another, one exhibit to another, without being forced to follow any curriculum, to attend to lessons, to take exams or compete. There was something passive, and forced upon one, about sitting in school, whereas in museums one could be active, explore, as in the world. The museums—and the zoo, and the botanical garden at Kew—made me want to go out into the world and explore for myself, be a rock hound, a plant collector, a zoologist or paleontologist....
>
> One gained entrance to the Geological Museum, as to a temple, through a great arch of marble flanked by enormous vases of Derbyshire blue-john, a form of fluorspar. The ground floor was devoted to densely filled cabinets and cases of minerals and gems. There were dioramas of volcanoes, bubbling mudholes, lava cooling, minerals crystallizing, the slow processes of oxidation and reduction, rising and sinking, mixing, metamorphosis; so one could get not only a sense of the products of the earth's activities—its rocks, its minerals—but of the processes, physical and chemical, that continually produced them.

Oliver liked the minerals' personal names like wollastonite, montmorillonite. Entrance to all the museums was free, and in the 1940s it was still safe for a young boy to travel across London by underground on his own.

At home Oliver repeated the experiments that led the French eighteenth-century chemist Antoine Lavoisier to disprove phlogiston, the substance that was supposed to be given off by inflammable substances on burning, and to discover that it is the combination with the oxygen in the air that makes them burn. One of Oliver's favorite books was Mary Elvira Weeks's *Discovery of the Elements*,[3] which ranged from the prehistoric discoveries of copper and iron to Otto Hahn, Fritz Strassmann, and Lise Meitner's discovery of nuclear fission in 1939. It gave details of the methods used for each element and short biographies of their discoverers. Oliver read how on October 6, 1807, the twenty-eight-year-old English chemist Humphry Davy had the idea of using a battery to pass an electric current through crystals of potash and saw for the first time tiny lumps of metallic potassium forming at one of the battery's terminals. A few days later, passage of a current through crystals of soda led Davy also to the discovery of metallic sodium. Oliver could repeat these experiments. At the Science Museum his mother showed him the miner's lamp named after Davy, designed to prevent the flame from igniting explosive mixtures of gases in underground mines, and she also showed him the improved version of the lamp named after her father, Marcus Landau. Sacks writes:

> It was Davy's personality that appealed to me—not modest, like Scheele, not systematic, like Lavoisier, but filled with the exuberance and enthusiasm of a boy, with a wonderful adventurousness and sometimes dangerous impulsiveness—he was always at the point of going too far—and it was this which captured my imagination above all.

Another of Oliver's favorites was a short volume that looked like a prayer book: *Chemical Recreations* by John Griffin, published in Glasgow in 1825. Its introduction is even more uplifting than Mendeleev's:

> If we consider Chemistry purely as a science, we shall find no study better calculated to encourage that generous love of truth which confers dignity and superiority on those who successfully pursue it. No science holds out more interesting subjects of research, and none affords more striking proofs of the wisdom and beneficence of the Creator of the universe. Chemistry is a science that is founded entirely upon experiment; and no person can understand it unless he performs such experiments as verify its fundamental truths. The hearing of lectures, and the reading of books, will never benefit him who attends to nothing else; for Chemistry can only be studied to advantage practically; chemical operations are in general, the most interesting that could possibly be devised. Reader! What more is requisite to induce you to MAKE EXPERIMENTS?

The book taught Oliver how to make a battery by putting down alternate plates of zinc and copper interleaved with moist paper, and how to make a balloon out of a turkey's stomach filled with hydrogen gas, as well as a chemical chameleon from salts of manganese. He learned several ways of preparing invisible ink and found the chapter on combustion and detonation especially appealing.

Oliver succeeded in making his own photographic emulsions by suspending silver chloride in gelatin and spreading it on a glass plate, and his own batteries by sticking rods of copper and zinc into raw potatoes. After reading a chapter on "The Smells We Dislike," Oliver prepared hydrogen sulfide and selenide, two evil-smelling and toxic gases that made his parents' house barely habitable, but instead of forbidding further experiments, they equipped his room with a glass-fronted, ventilated cupboard that extracted the smells into the open air. He also loved setting off explosions in the garden, and showing the wonders of chemistry to others helped him to overcome his shyness.

Uncle Dave had specimens of many of the rare heavy metals and the ores from which they are mined: platinum, palladium, osmium, tantalum, "given its name because its oxide was unable to drink water, i.e., to dissolve in acid." Tantalus suffered agonies of thirst in hell because water retreated from him whenever he tried to drink it. Uncle Dave also told him that some of these heavy metals acted as catalysts of chemical reactions, meaning that they accelerate them many thousandfold without undergoing any perceptible change themselves. For instance, a platinum wire instantly ignited an otherwise stable mixture of oxygen and hydrogen and made it explode.

Uncle Tungsten apparently failed to tell him that catalysis is the chemical secret of life. Nearly all chemical reactions in living cells need catalysts to make them work, and nearly all catalysts are proteins, complex molecules made of thousands of atoms. Each of the thousands of reactions is catalyzed by a protein that exists specifically for that single purpose. Coding for these proteins is the main function of genes. That much was known when Oliver became an amateur chemist; but proteins were black boxes, and their amazing catalytic powers were a mystery. Solving that mystery was the challenge that inspired my own research.

Uncle Dave had a cathode ray tube in his attic. When he pumped the air out of it, wired its two metal plates to an induction coil, and put them under electric tension, a miniature aurora borealis lit up between them. He told Oliver about Wilhelm Konrad Röntgen, the German physicist who one day covered such a cathode ray tube with black cardboard and saw crystals nearby lighting up brightly in the darkened room. This was his first glimpse of the X-rays whose discovery made him famous. Sacks writes that Röntgen told only his wife about his observation, but I could find no evidence for this. According to her he would arrive late and ill-tempered for meals, speak not a word, and hurry back to his lab immediately afterward. He told no one, because he realized that any one of the many physicists who experimented with cathode rays at the time could have repeated his observation in an afternoon. If his secret had leaked out before publication his priority might not have been recognized. Röntgen was the first recipient of the Nobel Prize for Physics just one hundred years ago.

Sometime after Oliver had studied Mendeleev's *Principles*, he asked his uncle for the reason underlying the strange periodic arrangement of Mendeleev's table. Uncle Dave told him about Niels Bohr, the Danish genius who brought together the periodic table, Max Planck's discovery of the quantum, Rutherford's discovery of the atomic nucleus, and the already well-known characteristic spectral lines of hydrogen, and from them was able to form a satisfying, unified theory of the atom. Bohr proposed that the rows of the table represented successive shells of negatively charged electrons spinning around the positively charged atomic nucleus like planets round the sun. These shells become filled with electrons one by one as the nuclear charge rises: up to two electrons in the first shell, then up to eight in each of the next two shells, and up to eighteen in the remaining ones, matching the two atoms in the first period, eight in each of the next two periods, and eighteen atoms in the remaining ones in Mendeleev's table. That revelation sent Oliver back to the Science Museum once more, thrilled that the table now made physical as well as chemical sense, but it also made Oliver wonder what need there was for experiments now that theory had become so powerful. He writes:

> I had dreamed of becoming a chemist, but the chemistry that really stirred me was the lovingly detailed, naturalistic, descriptive chemistry of the nineteenth century, not the new chemistry of the quantum age, which, so far as I understood it, was highly abstract and, in a sense, closer to physics than to chemistry. Chemistry, as I knew it, the chemistry I loved, was either finished or changing its character, advancing beyond me.
>
> From this point, chemistry seemed to recede from my mind—my love affair, my passion for it, came to an end.

Parental expectations made him turn to medicine and he became a successful neurologist, but he must now feel that his world has turned full circle, since mental illnesses are increasingly recognized as chemical disorders of the brain and modern diagnostic tools rest on quantum physics.

According to Edmond de Goncourt, originality in literature consists in making something extraordinary out of something ordinary. What could have been more ordinary than the Sacks's huge, rambling, Edwardian house, the sitting room with its dilapidated, comfy chairs for general use (dilapidated not because his parents were poor, but because they were too busy to notice or didn't care); the drawing room with its elegant uncomfortable Chinese chairs for sabbath gatherings of uncles, aunts, and cousins; the library, sacred to the children for their father's Hebrew texts; and the surgery which they were forbidden to enter, with its shelves of medicines? Yet Sacks's vivid description delights the imagination like a lovingly painted Dutch interior, and it sets the stage for his actors: Father and Mother; the affectionate and somewhat limited Auntie Birdie, who lived with them; Aunt Lina, who blows her nose in the tablecloth at meals and whose sharp-eyed judgment of people he likens to Keynes's description of Lloyd George; the formidable Aunt Anna, more English than the English and more Jewish than the Zionists, whose orthodoxy is offended by little Oliver riding his bicycle on a sabbath; Uncle Dave, whose hands are stained black with tungsten; and the whole family whose tongues are black from munching charcoal biscuits against wind. I know them all now and feel as though I had lived in their house.

Sacks transforms himself into the wide-eyed, playful, eager, mischievous, and indefatigably questioning boy to whom chemistry has become at once a toy and a window to the miracles of the natural world. Such is his precocity, performing sophisticated experiments and developing an understanding far beyond his age, and so numerous and minute are the details of Sacks's reconstruction that I wondered if this impressionable boy, shaped by his uncles and aunts, might not also have been reshaped somewhat by Uncle Oliver, his present self. Does Sacks really recognize himself in him or would he be, as François Jacob writes of himself as a schoolboy, almost "a stranger," hard to recognize if he met him in the street? "Like a bird contemplating the shell he has just broken out of, saying, 'Me? In there? Never!'"[4] To the reader the answer does not really matter, because Sacks has treated us to an enjoyable and very human story and to a painless and readable introduction to the elements of chemistry and atomic physics, enhanced by his long, erudite footnotes.

Friendly Way to Science

Lancelot Hogben, who died in 1975, was a leading biologist, a brilliant expositor of science and a dedicated Quaker, socialist and scientific humanist. An inborn dynamism and voracious thirst for knowledge made him rise to these heights from a home that tried to stifle all independent thought. Hogben's "unauthorised" biography (unauthorised, because it has been compiled by his elder son and daughter-in-law from several unedited versions of a projected autobiography by Hogben himself) begins with the words, "I come of poor but intellectually dishonest parents and was spared material poverty because they, with their offspring, lived as pensioners of my maternal grandparents." His father was a fiery fundamentalist preacher intent on spreading the gospel and saving souls from hell-fire. When father and son went on a tram, the father would distribute Methodist pamphlets to all the passengers, much to the boy's embarrassment.

Lancelot was a premature baby brought into the world in 1895 with such difficulty that his pious mother vowed to devote him to her godly mission if he arrived safely, but the adult later devoted his life to missions very different from the one to which his mother had consecrated him. Hogben's austere childhood in a home from which books other than religious tracts were banned and where Sundays were all prayer and no play reminded me of Edmund Gosse's classic *Father and Son*. Gosse's parents belonged to the Plymouth Brethren, a fundamentalist sect founded about 1830 in the west of England. After a difficult birth, they dedicated their child to the Lord, to be kept "unspotted from the world." Once, when young Edmund was invited to a children's party his father ordered him to pray to God to save him from such sinful temptation. To the father's fury, the boy announced after his prayer that God had told him to go to the party.

Lancelot's mother's genteel aspirations led her to send him first to a private school for the sons of gentlemen where he learnt little. When he was ten, his parents moved from Southsea near Portsmouth to Stoke Newington in London and sent Lancelot to a Middlesex county secondary school for which he was ill prepared. Aged 14, scarlet fever kept him at home for a term, which changed his life, because school books left there by a cousin made him discover his ability to teach himself. That, and his discovery in the public library of several volumes of the *Cambridge Natural History*, awakened his interest in zoology, stimulated his ambition, emancipated him from his pious surroundings, and set him on a successful academic career.

Aged 17, he won a scholarship to Trinity College in Cambridge, and aged 19 a first in Part I of the natural sciences tripos, coupled with a prize for the best candidate in zoology. Hogben writes: "As an educational institution, Trinity was ideally fitted to foster my intellectual development. Perhaps because of an overdose of family prayers, I had acquired a lifelong resistance to information transmitted by the spoken word, especially to monologues. There was very little pressure on a scholar of Trinity to attend lectures or, if present, to attend to what a lecturer was saying. Few of the lectures on biological topics were inspiring, but the organisation of laboratory work and the equipment provided could not have been much better."

Once a fellow student took him to a Quaker Adult School. He writes: "I already knew of the Quakers taking an active part in the emancipation of the slaves, that they regarded military service as contrary to the profession of a Christian and that they proclaimed no dogma to which a modernist could not subscribe." He joined the Society of Friends and remained a member for most of his life.

In August 1914, Hogben volunteered for service with the Quakers' War Victims Contingent, building bungalows for French families made homeless by the war, and later for the Friends' Ambulance Unit. When conscription came in 1916, he regarded these activities as no longer voluntary and returned to Cambridge to await his call-up. When that came, he refused to serve and was sentenced to three months' imprisonment at

Wormwood Scrubs. He spent them mostly in solitary confinement sewing mail bags. It was a stark contrast to the fleshpots of Trinity College, with dinner menus of six courses and a choice of at least three items for each. "The Tudor oak paneling, the retinue of waiters with white ties and the massive array of solid silver implements gave dinner in my day a flavour of baronial pageantry appropriate to its gastronomic profusion."

The luxury grated on Hogben's frugal, puritanical upbringing and may have contributed to his conversion to socialism. Hogben's conversion reminded me of my former Cambridge teacher, the Irish crystallographer John Desmond Bernal, a great scientist whom we called "The Sage" because he knew everything. He was converted from Catholicism to communism in a single night and clung to his new faith through Stalin's purges, the gulags and Lysenko's persecution of Mendelian geneticists. I wonder if some people's religious upbringing creates in them a need for an absolute faith to guide their lives.

Hogben's philanthropic socialism chimed with his Quaker ethics, which he practised throughout his life. In the words of Frank Landgrebe, who worked with him for many years: "His training of postgraduates in biological research was an inspiration. Full of ideas himself, he was also generous in encouraging the ideas of others; intensely critical, yet his criticism always carried an affirmation of hope. Even when the ideas were his, he scrupulously refrained from putting his name to a paper unless he had done a fair share of the work with his own hands." Today, some heads of laboratories put their names on all the papers published there, even on those that they have hardly read.

Hogben believed that "scientific knowledge rationally applied in a socially reconstructed world could bring about The Age of Plenty— not an age of luxurious extravagance, but an age in which war and poverty have been abolished and the fundamental needs of all humanity are satisfied. However, as an essential preparation for the necessary reorganization, the full potential of scientific discovery must be known and understood by all responsible people. The current education system was designed to produce docile experts rather than citizens prepared for the drastic social changes." Bernal tried to achieve this with *The Social Function of Science*, in which he argued for a rational organisation of science from above, somewhat on the lines practised by the former state-controlled Soviet Academy. Hogben, to his credit, endeavoured to do it from below, by bringing mathematics and science to the people.

While confined to hospital with a persistent throat infection, he turned a series of lectures on mathematics into a 650-page introduction to basic mathematics from Euclid to integral calculus, with a strong emphasis on practical applications. He called it *Mathematics for the Million*. It was written in accordance with his humanistic creed that "without a knowledge of mathematics, the grammar of science and order, we cannot plan a rational society in which there is leisure for all and poverty for none." The book became a bestseller, and so did his *Science for the Citizen*, an 1,100-page, richly illustrated volume that takes the reader from "The conquest of time reckoning and space measurement" by the Babylonians to the "Conquest of power, hunger and disease" and the "Conquest of behaviour" in modern times.

Both books are self-educators, taking readers from basic principles to practical applications; each chapter is followed by a list of problems. Both are still in print, more than 60 years after they were written. Hogben's utopian faith was characteristic of many idealistic scientists and other intellectuals earlier in this century who failed to foresee or recognise the disastrous consequences of centralised planning in eastern Europe and China.

While Hogben campaigned for a rationally ordered society, he debunked the eugenicists' ideas of planning the future composition of the human race. In *Genetic Principles in Medicine and Social Science*, published in 1931, he recounts the obstacles encountered in studying the gentic basis of physical differences among human beings and argues that these obstacles "are as nothing to the pitfalls which beset the study of hereditary factors contributing to man's social behaviour. The extreme complexity of man's social behaviour is evident to every intelligent man or woman whose outlook has not been biased by a prolonged occupation with the varieties of sweet-peas and mice or the patterns of feathers of poultry" (to which I would add the social life of ants). Hogben counters contemporary talk about the superiority of the Nordic race by quoting a 15th-century Moorish savant, Said of Toledo, who describes the people beyond the Pyrenees as "of cold temperament, never reaching maturity; they are of great stature and of a white colour, but they lack sharpness of wit and penetration of intellect."

Hogben's academic career began in 1917 with a lectureship at Birkbeck College. He next found a post at the Animal Breeding Research Laboratory in Edinburgh, and then as assistant professor of medical zoology at McGill University of Montreal. He was overjoyed when he was offered the post of professor and head of department of zoology at the University of Cape Town, where he found a zoologist's paradise. The beginnings of apartheid, however, drove him away after only three years to a new chair of social biology at the London School of Economics. He was befriended by its director, William Beveridge, whom he admired, and by the school's founders, the socialists Sidney and Beatrice Webb, whom he looked upon as foster parents because they had made higher education possible for him by introducing London County Council maintenance grants for winners of open scholarships to Cambridge or Oxford.

According to Hogben, "the Webbs devoutly believed that lads and lasses who studied economics deeply would eventually become good collectivists. From the start both believed that a mariage de convenance between economic theory and factual social studies, if solemnised with a sufficient dowry to the latter, would advance the Fabian cause on a wide front. They continued to believe this despite the fact that at the time of my advent the school was the last stronghold of the most ultra-individualist metaphysical nonsense masquerading as economic science west of Vienna." Hogben enjoyed his contacts with M. M. Postan, the economic historian, because he had a lively appreciation of the scientific basis of economic advance for which Nathan Rosenberg and L. E. Birdzell argued so brilliantly in their recent book, *How the West Grew Rich*. On the other hand, Hogben regarded Hayek's free-market economics, which later influenced Margaret Thatcher, "as a mental exercise comparable with astrology." Hogben did not take kindly to the students, "a sizeable and truculent minority of whom had Communist Party Cards," and whose outlook and demeanour he describes as a by-product of the disputatious attitude to learning encouraged by the Hayek-Robbins circus and by Harold Laski. After seven years at the LSE, Hogben moved to Aberdeen and finally to Birmingham, where he occupied a chair in medical statistics specially created for him.

My interest in Hogben's autobiography waned after the first few chapters describing his childhood and youth, because, apart from the occasional vivid vignettes about well-known people and institutions, it is filled with a humdrum account of his many activities and moves. His research, which earned him his fellowship of the Royal Society, was concerned with endocrinology, mainly of amphibians, but he fails to explain what was interesting about it, or what he achieved as professor of social biology at the LSE or of medical statistics at Birmingham; one can get some idea from the long list of his publications appended to the text.

The article by G. P. Wells, the biologist son of H. G. Wells, in the *Biographical Memoirs* of the Royal Society (1978), gave me a much better picture of the man than his autobiography. He extols Hogben's talent, tremendously broad erudition and amazing productivity, but he also describes him as "a difficult colleague, incapable of concealing personal dislike; ambitious, intent on reforming university education according to his humanist faith; frustrated and unforgiving in defeat; a tremendous rocker of other people's boats, and never very calm in his own." Wells' article and Hogben's other books convinced me that he was a very remarkable man, something that his autobiography would not have made me guess, perhaps because the later parts were clearly compiled from notes he left after his death.

*The Scientific and Humane Legacy of Max Perutz (1914–2002)**

By John Meurig Thomas

The son of a prosperous Viennese textile manufacturer, educated first at the Theresianum (a grammar school derived from an officers' academy of the days of the Empress Maria Theresa) and then at the University of Vienna, Max Perutz was to have studied law in preparation for entering the family business. At the Theresianum, he was a small, dreamy and sleepy boy; and many years later he said: *I owe my first step to popularity to scarlet fever which I caught when I was fourteen. To disinfect the classroom, my schoolmates got three days off, for which they thanked me solemnly in a letter signed by the entire class.* At sixteen, he, plus two of his colleagues, won the cup for their school in the skiing competition of the Vienna High Schools, and as a result of that victory his standing at school was transformed. *For the first time in my life I was treated with a certain degree of respect. From then on our gym teacher always gave me top marks, but they happened to be the only ones in my otherwise mediocre school reports.*

A good schoolmaster kindled his interest in chemistry, and he made this the subject of his studies at the University. Although largely disap-

*Reprinted from *Angewandte Chemie Int. Ed.* (2002), vol. 41, p. 3255.

pointed with the way in which the subject (especially inorganic analysis) was taught, he acquired a special interest in organic biochemistry and heard about the work of the Nobel Prizewinner (and discoverer of vitamins) Sir Gowland Hopkins at Cambridge. His teacher, Herman Mark, the polymer specialist, visited Cambridge and had planned to pave the way for Perutz to join Hopkins' group. But Mark met J D Bernal, who said that he would take Perutz as his student.

Writing in the *Scientific American* in 1978, Max Perutz said:

> When I was a student, I wanted to solve a great problem in biochemistry. One day I set out from Vienna, my home town, to find the Great Sage at Cambridge. He taught me that the riddle of life was hidden in the structure of proteins, and that X-ray crystallography was the only method capable of solving it. The sage was John Desmond Bernal, who had just discovered the rich X-ray diffraction patterns given by crystalline proteins. We really did call him Sage,[1] because he knew everything, and I became his disciple.

Perutz was admitted a graduate student[2] at the oldest college in Cambridge, Peterhouse, on 1 October 1936 and had begun as a researcher in the Cavendish Laboratory (where Bernal taught and researched in physics) some days earlier. It is hardly necessary to recall that Max Perutz found out how to solve the structure of proteins and used his method for the solution of the structure and mode of action of haemoglobin. But it is instructive and inspirational for scientists and non-scientists alike to trace the trajectory of his life and to ponder the ingredients of his greatness as a researcher, as a communicator and as a human being.

A Summary of His Achievements

At the memorial meeting in honour of Sir John Kendrew held in Cambridge, 5 November 1997, Max Perutz said:

> John and I shared three great scientific adventures: founding the MRC Unit for Molecular Biology,[3] solving the first protein structure and founding the European Organization for Molecular Biology (EMBO).

Max Perutz and John Kendrew, who shared the Chemistry Nobel Prize in 1962, for their pioneering work on the elucidation of the structure of the biological macromolecules haemoglobin and myoglobin, the respiratory proteins of red blood cells and of muscles, respectively, were very different from one another. They were phenomenally able and perspicacious individuals, among the greatest scientists ever to be members of their

Figure 1. Max Perutz (right) constructing a model of the first ever protein structure determined by X-ray crystallography, with John Kendrew looking on (1959).

small Cambridge college, Peterhouse,[4] which boasts as its former students Henry Cavendish, William Thomson (Lord Kelvin), Charles Babbage,[5] Clerk Maxwell, James Dewar and Frank Whittle.[6] Kendrew was a precise organiser and a gifted computer programmer, at a time when that breed of scientist was very rare. He was a man who knew exactly where he was going and how to get there. Although possessed of urbane charm, linguistic fluency, a love of music and a great sense of humour, Kendrew was always a more distant, detached character than Perutz. Perutz was a gentle, kindly and tolerant lover of people (particularly the young), passionately committed to social justice and intellectual honesty; and the warmth of his personality radiated a sense of human goodness and decency which induced others to behave sanely, especially because he exuded an inner excitement that stems from a love of knowledge for its own sake. A distinguished American molecular biologist who worked in the Perutz-Kendrew team as Kendrew's postdoctoral assistant, 1957–58, in a reflective article published in *Protein Science*,[7] wrote:

> Max and John were utterly different in personality. Kendrew came in two or three mornings a week to discuss the progress of the research, and to give help where help was needed. He was a great mentor for someone who wanted to learn how to be an independent investigator. At other times he was busy as a science advisor to the British government (on the Polaris missile system as I recall), as an administrator of Peterhouse (college), and on other affairs. In contrast, Max was never so happy as when in the laboratory at the bench, doing science. One learned by talking with John, but by watching Max.

Perutz's inspirational scientific leadership, more than that of any other, made it possible to unite structural crystallography and molecular pathology and helped to unfold the vast ramifying fields of present day molecular biology. In communicating knowledge to other scientists and the general public—and as an ambassador for science, its methods and philosophy —few of his contemporaries rivalled him. So central were his interests in the vast corpus of scientific endeavour that he was regularly called upon to address biochemists, physicists, physiologists, pathologists, geneticists and medical scientists of several other kinds. For the last forty years of his life he maintained a punishing round of lecture commitments: one week it was to school sixth formers or to a University of the Third Age[8] audience; another it might be to the assembled researchers at places like Rockefeller University, New York; the next to frontier scientists in the California Institute of Technology—or in Rome (lecturing in Italian), in Berlin (in German) or Paris (in French).

The Road to Haemoglobin

Perutz, on Bernal's advice, first learned of X-ray crystal structure analysis in the Department of Mineralogy and Petrology, Cambridge, where he was handed—as he later put it— "a nasty crystalline flake of a sil-

Figure 2. Max Perutz lecturing in Berlin, 1999. An enlarged photo of "Sage" (J D Bernal) is seen on the screen behind him.

icate mineral picked off a slag heap" to investigate.[9] This training proved crucial, for he became adept at X-ray crystallography and he set out to pursue his "great problem in biochemistry". Guided a little later by a cousin, who lived in Prague, he soon convinced himself that an appropriate target for his ambitions was the structure of haemoglobin, especially since it was the protein that was most abundant and easiest to crystallize. It was not until the late 1950s that he finally reached his target of elucidating the structure of haemoglobin; and when he did, it made him famous. But his path to success was frequently thwarted by scientific, personal and political obstacles.

Perutz had obtained tantalizing X-ray diffraction patterns, published in 1938—just as Dorothy Crowfoot (later Hodgkin) and JD Bernal had earlier with crystals of the enzyme pepsin—and it could then be seen that the thousands of reflections he had obtained could, in principle, lead to the atomically-resolved structure of haemoglobin. But there was no way of interpreting these data, because the so-called phase angles of the reflections could not be evaluated. Previous X-ray structural determinations had been limited to molecular entities possessing a small number of atoms, and often with a known chemical composition. The analytical techniques then available in the solution of the structure of, for example, polynuclear aromatic hydrocarbons, carbohydrates and other "small" molecules, simply could not be applied to proteins with thousands of atoms and an unknown chemical composition. Perutz, however, felt that proteins had to take up well-defined structures within their crystals, as, otherwise, those sharp spots that he, Crowfoot and Bernal had obtained were altogether inexplicable.

As Bernal's research student, Perutz was disappointed that Lord Rutherford, then the Head of the Cavendish Laboratory, never looked in to find out what the crystallographers were doing. Perutz later discovered that it was not because of any antipathy towards the X-ray crystallography that Rutherford kept his distance. According to Max Perutz, "The conservative and puritanical Rutherford detested the undisciplined Bernal, who was a communist and a woman chaser and let his scientific imagination run wild."[10] Perutz was obviously disappointed when Rutherford died before he had a chance of attending his lectures or getting to know him. After his death, graduate students could help themselves to left-over reprints of Rutherford's classic scientific papers. Their clarity and rigour, Rutherford's imaginatively conceived experiments and his determination to prove every one of his results experimentally beyond any possible doubt convinced Perutz that this was the way to do science.

Perutz finished his Ph.D in 1940, working with Sir Lawrence Bragg, who succeeded Rutherford as the Cavendish Professor, and who had obtained a small grant for him from the Rockefeller Foundation to enable his research to continue. However, the project was interrupted by Perutz's internment in 1940, along with several hundred German and Austrian refugee scholars, mostly Jewish and all anti-nazi.[11] They were rounded up and sent to an internment camp in Quebec, Canada. Fortunately, Perutz's Cambridge friends rallied round and tried to secure his release. He returned to his studies and to work of national importance, which involved him in an unsuccessful scheme to make ships of ice for refuelling aircraft in the North Atlantic.

Perutz resumed work in Cambridge in 1944, and he was joined in 1945 by John Kendrew (whose Ph.D had started with the physical chemist, E A Moelwyn Hughes, and whose interest in protein structure had been kindled first by meeting J D Bernal during war service in operational research and later reinforced by discussions with Linus Pauling in Pasadena on his return journey to the UK). Bragg continued his enthusiastic support, and in 1947 he approached the secretary of the MRC,[3] who made the crucial decision to back this small group despite the continuing lack of tangible results—a far cry from present-day attitudes! Francis Crick, a physicist, joined the group as Max Perutz's Ph.D student in 1948. Jim Watson, a geneticist, came later, in 1951, and was soon working with Crick on DNA.

As Hugh Huxley,[12] the distinguished muscle biophysicist and molecular biologist, has recently written[13]:

> Because there was no direct way of calculating protein structure, Perutz and Kendrew used the huge amounts of data (the intensities of all the X-ray reflections) to produce contour maps, known as Pattersons, to display the most frequently occurring distances between high-density regions in a structure. It was hoped that these might enable prominent features of the structure to be identified. To this end, Perutz and Kendrew, aided by Bragg, enumerated various helical configurations into which polypeptide chains might fold, based on stereochemical data.

However, they made what seemed then the plausible — later found to be erroneous — assumption that the helices would contain a whole number of residues per turn. They also did not take into consideration that the chemical bonds on either side of the peptide bond are coplanar. (Linus Pauling, working independently, made no such assumptions, and it was he who first arrived, in 1951, at the now well-known non-integral α-helical structure of proteins).

In early 1951, Crick showed that the Patterson maps for haemoglobin and myoglobin, did not provide support for the view that there was some regular packing arrangement of polypeptide chains. Indeed he demonstrated that any such arrangement must be extremely irregular. This conclusion, in particular that no useful information was extractable from the Patterson maps, made Crick unpopular, and for a while, depressed Perutz.[14] But his resourcefulness as a scientist knew no limits — and it remained thus to the very end of his life.

Just as his two colleagues Watson and Crick were constructing what was later to become their monumental model for the structure of DNA in early 1953, Perutz ruminated on the potential of the heavy-atom substitution method of determining the phases of diffracted X-ray waves. This had its origins in the work of J M Robertson at the University of Glasgow and that of the Dutch crystallographer J M Bijvoet in Utrecht; Bernal had suggested in 1939 that it might work for proteins; Crick also made this suggestion somewhat later. Perutz argued that if he could attach a heavy atom to a specific site in the haemoglobin molecule, *and* if it did not disrupt the structure of the molecule, *and* if he could make it crystallize in just the same way as ordinary haemoglobin, *and* if it made changes large enough to be measurable — if *all* these conditions were met, Perutz could see a way to use the methods of X-ray crystallography to image the haemoglobin molecule.

What Perutz realized, with increasing clarity and conviction, was that the X-ray diffraction (scattering) intensities from haemoglobin and other proteins were very weak, despite the large number of electrons contributing to them, and that this arose as an inevitable consequence of the fact that the scattering matter was distributed over a relatively large volume. As Huxley recently remarked,[13] "Perutz also realized that scattering from a single heavy atom—mercury, say, with its 80 electrons, or gold with its 79 electrons within a single atomic diameter—would be relatively strong. So if such a marker (heavy atom) were attached at a specific position on each protein molecule in a crystal, it would produce a measurable change in the intensities of the reflections (diffraction spots), and this could be used to obtain (X-ray) phase information necessary for the determination of the structure (the imaging) of the haemoglobin molecule. His peers and contemporaries had wrongly assumed that any such change (to the intensities of X-ray patterns) would be immeasurably small; but Perutz did the experiment carefully, using haemoglobin labelled with mercury, and he found quite measurable changes." This constituted a major breakthrough in the methodological approach to the structure of proteins.[16] And when the world's experts on proteins, assembled by Linus Pauling at the California Institute of Technology in September 1953, heard Perutz adumbrate his method, they were conscious that a major breakthrough in protein science was imminent.

Perutz later wrote:

> As I developed my first X-ray photograph of mercury haemoglobin my mood altered between sanguine hopes of immediate success and desperate forebodings of all possible causes of failure. I was jubilant when the diffraction spots appeared in exactly the same positions as in the mercury-free protein, but with slightly altered intensity, exactly as I had hoped.

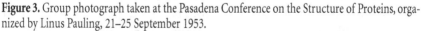

Figure 3. Group photograph taken at the Pasadena Conference on the Structure of Proteins, organized by Linus Pauling, 21–25 September 1953.

For Perutz and Kendrew (who also adopted the heavy-atom approach to the determination of the structure of myoglobin), great difficulties still lay ahead before the new method could lead to an interpretable structure.[13,17] First, it was necessary to have several different heavy-atom derivations before the phases of all the reflections could be reliably assigned. Second, all these derivatives had to crystallize with exactly the same (unit cell) dimensions. Third, appropriate mathematical procedures had to be devised for calculating the phases. Numerous assistants had to make intensity comparisons (using relatively primitive micro-densitometers) on hundreds of thousands of diffraction spots; and, in turn, the electron density distributions had to be calculated on very early — and by present-day standards extremely primitive — computers in the University of Cambridge Mathematics Department.

Kendrew, assisted by two visiting American scientists who came to the Perutz-Kendrew unit in the Cavendish Laboratory, Dickerson and Dintzis, succeeded in obtaining crystals of gold- and palladium-substituted myoglobin. With their resulting X-ray diffraction patterns Kendrew, who possessed formidable quantitative skills and who took advantage of the emergence in Cambridge of the programmable EDSAC-1 and EDSAC-2 digital computers, which he exploited for the Fourier analysis of his diffraction data, solved the structure of myoglobin in 1957.[19] But within two years, Perutz, along with his MRC associates Ann Cullis, Hilary Muirhead, Michael Rossmann and Tony North, had unravelled the architecture of the haemoglobin molecule, which contains four times as many non-hydrogen atoms (10,000) as myoglobin.[20]

Here were two quite independent structural determinations of related proteins, done by pure physics, without any assumptions about the chemical nature of haemoglobin and myoglobin or the relation between them. This exhilarating information revealed that the intrinsic structures of the two proteins, replete with haem groups, numerous folds and (Pauling's) α-helices, were essentially similar. The inescapable conclusion was that each had to be right. This galvanised activity in molecular biology worldwide. Adolf Butenandt, the eminent German Nobel Laureate, set his colleagues in Munich the task of using chemical methods (such as those pioneered by Frederick Sanger) to trace the sequence of amino acids in the proteins studied by Perutz and Kendrew. The chemical results harmonized beautifully with those of X-ray crystallography. (Butenandt, among others, nominated Perutz and Kendrew for the Nobel Prize).[21]

The Nobel Prize and Beyond

By 1962, the number of staff of the Perutz-Kendrew MRC unit devoted to molecular biology had grown to ninety. And it had long been apparent that the Cavendish Laboratory (now led by Sir Nevill Mott) could no longer satisfactorily house this thriving, expanding community. It seemed eminently sensible that the new breed of biologists, spawned by Perutz's and Kendrew's work, should be accommodated within one of the many existing departments of pure or applied biology of the University of Cambridge. However, cogent though his arguments were, there was little or no enthusiasm in these departments, or among the central authorities of the University to welcome Perutz and his colleagues into their fold. Fortunately, the pertinacious and far-sighted Perutz found a ready ally in Sir Harold Himsworth at the MRC Headquarters in London; and, largely through the latter's advocacy, the MRC built the Laboratory of Molecular Biology (LMB) with Perutz as its Chairman, on the site of the new Addenbrooke's hospital on the outskirts of the city of Cambridge. This allowed several different aspects of molecular biology to be brought together in one place, along with all the necessary technology. Fred Sanger, who had already won[22] his first Nobel Prize (in 1958), was invited to join the new laboratory, and he brought with him two bright stars, Brian Hartley and Ieuan Harris (the latter in turn also brought along his outstanding young colleague John (later Sir John) Walker, who was to win the Nobel Prize in Chemistry in 1997).[23] In 1961, Hugh Huxley of University College, London, who had been Kendrew's first research student, came to join the Structural Studies Unit of which Kendrew was head. In 1958, with J D Bernal's retirement looming at Birkbeck College, London, Perutz had invited Aaron Klug and Rosalind Franklin to join the planned new laboratory when their National Institute of Health (USA) grant ran out. Klug, who also joined Kendrew's Division of Structural Studies, came in 1961, bringing Kenneth Holmes (now of the Max-Planck-Institute for Medical Science, Heidelberg) and the expert electron microscopist and all-rounder, John Finch, with him. (Rosalind Franklin, who died in 1958, never came back to Cambridge, from where she had graduated in chemistry many years earlier).

The LMB was duly opened by Her Majesty Queen Elizabeth one fine day in May, 1962 and on that occasion Max Perutz took particular pleasure in explaining the significance of the structure of DNA to his Sovereign. In October of that year, the Swedish Academy announced that Perutz and Kendrew were to share the Nobel Prize in Chemistry, and that their two associates, Watson and Crick, along with Maurice Wilkins of King's College, London, were to share the Prize in Medicine and Physiology.

Figure 4. Official opening of the Laboratory of Molecular Biology, at which Max Perutz (*extreme right*) explains the double helix to H M Queen Elizabeth II, May 1962.

The unique success of the LMB, measured, for example, by the numerous awards gained by members of this one laboratory — nine Nobel Prizes, four Orders of Merit (the highest civil honour that the British Sovereign can bestow), eight Copley medals (the highest honour of the Royal Society) and over a hundred Fellowships of the Royal Society conferred upon its staff — merits closer enquiry. An overwhelmingly important factor in all this is the way that Perutz organized the Laboratory and his own role within it.

Perutz, from the outset, adopted a new style of management. Thus he once wrote:

> I persuaded the MRC to appoint me Chairman of a Governing Board, rather than as Director.... This arrangement reserved major decisions of scientific policy to the Board, and left their execution to me.... The board met only rarely.... This worked smoothly and left me free to pursue my own research. Seeing the Chairman standing at the laboratory bench or

Figure 5. Photograph taken after the Nobel Prize ceremony, Stockholm: M.F.H. Wilkins, M.F. Perutz, F.H.C. Crick, John Steinbeck, J.D. Watson, and J.C. Kendrew.

> the X-ray tube, rather than sitting at his desk, set a good example and raised morale. The Board never directed the laboratory's research but tried to attract, or to keep, talented young people and gave them a free hand.

As his former Ph.D student and distinguished molecular biophysicist, Professor David Blow,[24] who was a member of the LMB in its early days, said recently of Perutz:

> He always recognized the importance of new instrumental developments, and maintained large mechanical and electronic workshops, to which research workers had full access, directly passing their enthusiasm to the technical staff.

But over and above these technical features of running a laboratory Max Perutz, ably assisted by his devoted and shrewd wife Gisela, made the canteen at the LMB a focal point for intellectual stimulus. It was — and still is — visited three times a day by most scientists and associated staff, and it remains an important centre for exchange of ideas and scientific news. One American molecular biologist[25] on sabbatical leave in 1985 at the LMB, commented that, because the canteen was the intellectual centre

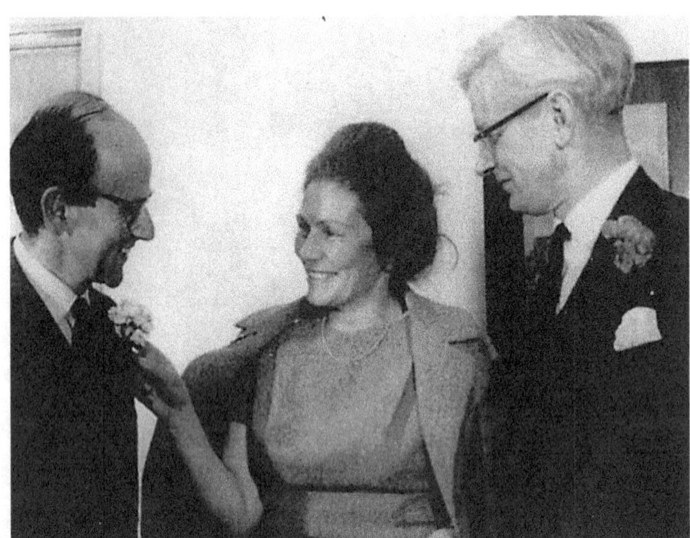

Figure 6. Informal photograph (Dec 1962) of Max Perutz (*left*), Gisela Perutz, and John Kendrew.

of the laboratory, one could not stray off course in one's research work for more than a period of three hours! Max Perutz himself kept abreast of everybody's work, by making a habit of sitting with different groups of people at coffee time, lunch or tea. (He did likewise whenever he came to lunch at his college, Peterhouse.)

Perutz led by example,[26] aiming to spend some 90 per cent or more of his time working at the bench, and he expected others to do likewise. But he recognized and could cope with differences in individual style. In reflective mood, he once wrote:

> When Crick and Watson lounged around, arguing about problems for which there existed as yet no firm experimental data instead of getting down to the bench and doing experiments, I thought they were wasting their time. However, like Leonardo, they sometimes achieved most when they seemed to be working least, and their apparent idleness led them to solve the greatest of all biological problems, the structure of DNA. There is more than one way of doing good science.

"Every now and then," Max Perutz wrote in one of his recent books (*I Wish I'd Made You Angry Earlier: Essays on Science, Scientists, and Humanity*), "I receive visits from earnest men and women armed with questionnaires and tape recorders who want to find out what made the LMB (where I work) so remarkably creative.... I feel tempted to draw their attention to 15th century Florence with a population of less than 50,000, from which emerged Brunelleschi, Donatello, Ghiberti, Masaccio, Botticelli, Leonardo, and Michelangelo and other great artists. Had my questioners investigated whether the rulers of Florence had created an interdisciplinary organisation of painters, sculptors, architects, and poets to bring to life this flowering of great art?... My question is not as absurd as it seems, because creativity in science, as in the arts, cannot be organised. It arises spontaneously from individual talent. Well-run laboratories can foster it, but hierarchical organisation, inflexible, bureaucratic rules, and mountains of futile paperwork can kill it. Discoveries cannot be planned; they pop up, like Puck, in unexpected corners."

In an age when the Paladins of accountability and the Funding Councils persist in preaching the necessity for all academic research centres to have a Mission Statement and Strategic Plans, it is prudent to recall how Perutz set about founding and running the extraordinarily successful LMB. The principles he used were: choose outstanding people and give them intellectual freedom; show genuine interest in everyone's work, and give younger colleagues public credit; enlist skilled support staff who can design and build sophisticated and advanced new apparatus and instruments; facilitate the interchange of ideas, in the canteen as much as in seminars; have no secrecy; be in the laboratory most of the time and accessible to everybody where possible; and engender a happy environment where people's morale is kept high.

But the LMB in Cambridge is not the only major laboratory in which Max Perutz had a hand in creating. Shortly after the Nobel ceremony in Stockholm in 1962, Leo Szilard and Victor Weisskopf of the United States called to see Jim Watson, John Kendrew and Max Perutz to discuss the prospects of establishing a European Molecular Biology Organisation (EMBO) like the Nuclear Science Centre (CERN) in Geneva. All three responded enthusiastically. Supported by other European molecular biologists, notably, François Jacob (France), Friedrich Freska (Germany), Ole Maalo (Denmark), Jeffries Wyman (USA) and, crucially, Ephraim Katchalski-Kazir (Israel), Kendrew led the way, and soon EMBO came into being with Max Perutz as its Chairman from 1963-69. In due course, the European Molecular Biology Laboratory (EMBL) at Heidelberg, of which John Kendrew was the founding Director (for ten years) also came into being.[27]

Haemoglobin and Its Impact on Medicine

In 1959 when Max Perutz and his colleagues first unravelled the architecture of the haemoglobin molecule in outline, it was appreciated by all concerned that it was not the end of their journey. As Perutz put it:

> ...our much-admired model (of haemoglobin) did not reveal its inner workings—it provided no hint about the molecular mechanism of respiratory transport. Why not? Well-intentioned colleagues were quick to suggest that our hard-won structure was merely an artefact of crystallisation and might be quite different from the structure of haemoglobin in its living environment, which is the red blood cell.

For a long time after Perutz solved the structure of haemoglobin, he zealously pursued the secret of its mode of action. How was it that, in effect, haemoglobin functions as a molecular lung? It was not until early in the 1970s that a satisfactory answer to this question came and the fundamentals of the mechanism were elucidated.[28]

The firm experimental evidence on which the now accepted (Perutz) mechanism for the so-called allosteric change that accompanies the reaction of haemoglobin with oxygen — an allosteric protein[29] is one that changes from one conformation to another when it binds another molecule (like oxygen) — came not only from detailed X-ray crystallographic work but also from Perutz's imaginative use of other techniques, including spectroscopy and magnetic measurements. First he recalled the significant earlier work of Faraday (in the 1840's) and Linus Pauling (in 1936) who showed that when haemoglobin binds oxygen it loses its paramagnetism. Then he found that the structural changes in haemoglobin accompanying oxygenation were large. In the "deoxy" form the iron atom of the heme is displaced a little from the plane of the heme group, whereas in the "oxy" structure it lies almost in the plane. Perutz recognized, as Leroy Hood did independently for haem groups, that this is so because of a change in the (electronic) spin state of the iron atom — from so-called "high spin" in the deoxy to "low spin" in the oxy state — and hence to a diminution in the radius of the iron. When the iron moves closer to the plane of the heme in the oxy state it drags with it the α-helix of the protein to which it is connected; and this is the trigger that initiates a sequence of "molecular levers" that loosen and rearrange the subunits in the (tense) deoxy structure into a new (relaxed) oxy structure. This is the basis of the molecular mechanism — "infinitely rewarding in its simple beauty" as Perutz used to say — that governs the oxygen affinity of haemoglobin in response to physiolog-

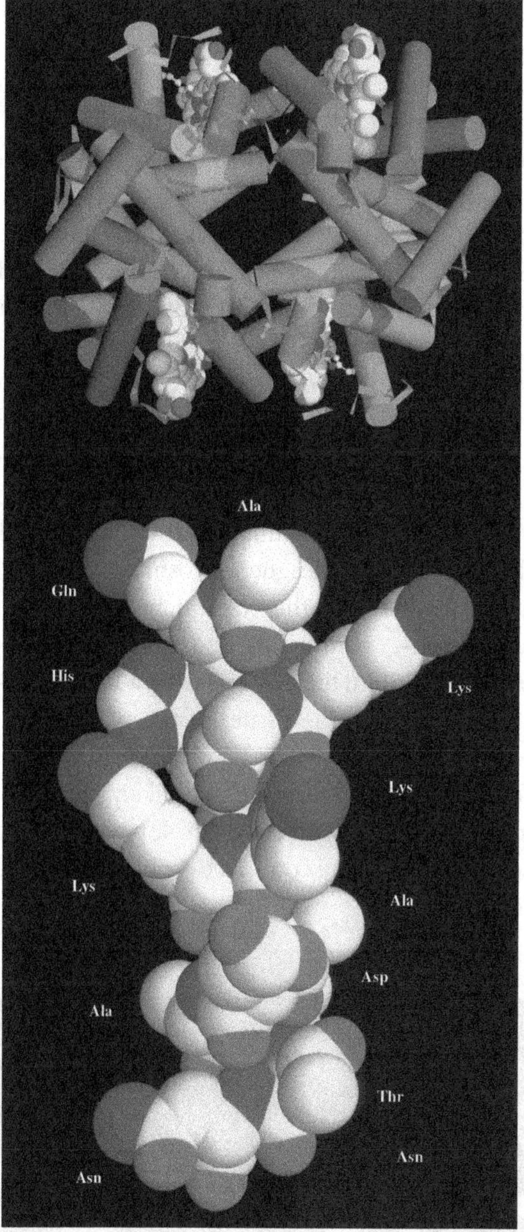

Figure 7. (*Top*) Graphic representation of a molecule of haemoglobin, with the four heme groups shown in atomic detail, and the two pairs (the so-called α and β subunits) of helically arranged aminoacid residues shown as cylinders. (*Bottom*) Atomic representation of the aminoacid residues (labelled: Ala = alanine; His = histidine; etc) that constitute the α-subunits shown as cylinders above.

ical needs, and the release of bound oxygen when it is needed under conditions of oxygen scarcity.

In due course this mechanism was to lead, via Perutz's work on haemoglobin mutants, to a fuller understanding of several inherited diseases and it opened up the new field of molecular pathology, a subject which relates a structural abnormality to a particular disease. Perutz gained new insights into molecular evolution,[31] and into the delicate (sometimes major) differences exhibited by haemoglobin in a wide range of living species. For example, the frogs of Lake Titicaca, high in the mountains of Bolivia, have evolved a form of haemoglobin that can absorb oxygen better than that in the frogs of Lake Michigan, at a lower altitude. This enhanced absorbability of oxygen is also a feature of the haemoglobin of migrating geese that fly at high altitudes. Crocodiles are able to remain under water for more than an hour without surfacing to breathe[32] and often kill their prey by drowning it. How do crocodiles stay under water for so long? Perutz's colleague, Kiyoshi Nagai, and his associates found that when crocodiles hold their breath, bicarbonate ions — the final products of respiration— accumulate and drastically reduce the oxygen affinity of their haemoglobin, thereby releasing a large fraction of the oxygen bound to the globin into the tissues of the mammal.[33] (In collaboration with an American company, Dr Kiyoshi Nagai, at the LMB, laid the groundwork for an engineered haemoglobin likely to be suitable for use as a cell-free blood substitute. The need to develop a blood substitute is now urgent because of the increasing concern over blood-transmitted viral and bacterial pathogens).

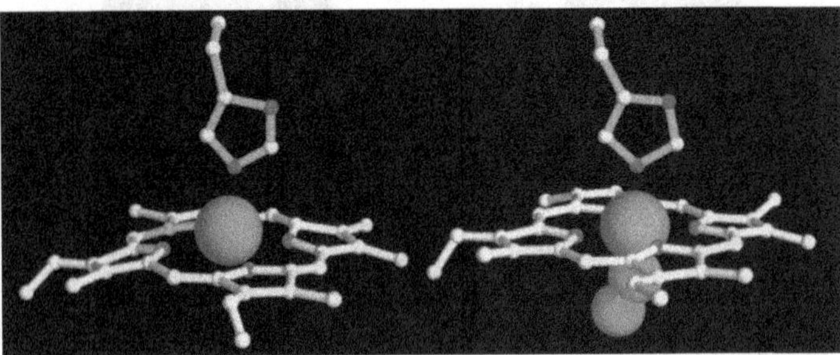

Figure 8. In human deoxy haemoglobin the Fe^{2+} ion, attached to histidine (*above*) lies slightly above the plane of the heme group (*left*). When the Fe^{2+} ion binds oxygen, as in the oxy form, it moves into the plane of the heme (*right*). This is what initiates the sequence of resulting structural changes — see text.

Earlier, Perutz's pioneering work offered a deeper appreciation of such tragic diseases as thalassemia, which arises due to defective synthesis of haemoglobin, and of sickle-cell anaemia. His colleagues at Cambridge, Vernon Ingram and John Hunt, showed that sickle-cell haemoglobin differed from the normal protein merely by a change of a single amino acid residue, in fact the replacement of glutamic acid by valine.

The Exponential Growth of Protein Structure Determination

In 1953, when Perutz discovered that the so-called X-ray phase problem of protein crystallography could be solved by the method of isomorphous replacement with heavy atoms, he expected that the structures, not only of haemoglobin, but also of many other proteins, would presently be solved[34]; but this did not happen. Only three protein structures had been solved by 1965, and only eleven by 1970. The practical difficulties of crystallization, of preparing isomorphous heavy-atom derivatives, and of recording the X-ray diffraction data, were so great that determination of each new structure took many years. Besides, most professional crystallographers were reluctant to enter this risky new field. But, by now, the situation has been fundamentally transformed. Since 1975 there has been an exponential rise in the annual number of protein structures solved. In 1990 over 100 new structures came to light; but by 2000 approximately 16,000 structures were known, and very many of these are of direct practical interest to medicine. If one knows the structure of a protein in atomic detail, then the precise architecture, shape and dimensions of receptor centres and catalytically active sites are also known. This enables the protein engineer, by various modern methods, to improve the performance of existing proteins. This is now one of the major growth areas of enzymology.

It is interesting to note that most, but not all, of the first generation protein crystallographers (1960–1980) came out of British laboratories in Cambridge, Oxford and the Royal Institution, London. (The migration of D C Phillips and his former student Louise Johnson to Oxford in the mid-1960s, after they had solved the structure of lysozyme and explained its mode of action, accounts for a shift of the centre of excellence in protein crystallography outside Cambridge from London to Oxford, although the towering presence and achievements of Dorothy Hodgkin at Oxford had already established it as a world-renowned X-ray structural laboratory). It is also interesting to reflect, with the benefit of hindsight, how wrong-

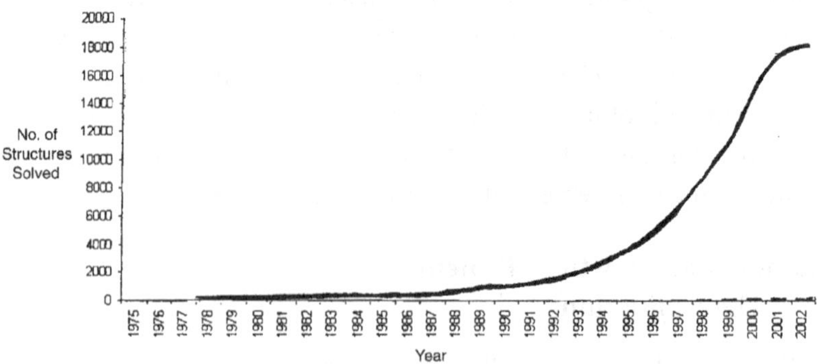

Figure 9. Exponential growth of the number of protein structures determined since 1975. Adapted from the web site of the Research Collaboratory of Structural Bioinformatics (RCSB).

headed the eminent biochemist and Nobel Prizewinner Sir Ernst Chain was when, some thirty years ago, he advised the MRC in London that medical research involving crystal structure analysis was a profligate waste.[35] Such is the unpredictability of science.

From Glaciers to Huntington Disease: Perutz's Earliest and Latest Research

As a teenager Max Perutz gained distinction in his school in Vienna as an expert downhill skier. He was also, at that time, an accomplished mountaineer, and his love of mountains and glaciers never left him. He claims to have barged into Sir Lawrence Bragg's office one day in the Cavendish making the untypical claim: "I've had an honour that you can't match — I've had a glacier named after me!" to which Sir Lawrence retorted "I've had a cuttlefish named after me."[36]

From his early twenties Perutz had a deep interest in glaciers, particularly how they flowed. Did this occur like treacle or honey flowing out of a tilted container, or was there some other mechanism? More or less as a hobby, Perutz, after conducting measurements perched inside his Alpine igloo on the Jungfrau for several months in 1938 and again ten years later in a grotto in Eigergletscher, solved the problem of how a glacier flows. It moves, not like treacle, but more like a ductile metal (such as aluminium when it is rolled into a sheet). The process of slip, in the metallurgical sense of the term, is what occurs here. Perutz gave a Friday Evening Discourse on *The Flow of Glaciers* at the Royal Institution on 5 June, 1953, and a summary of his lecture-demonstration appeared in *Nature* later that year.[37]

Figure 10. Photograph showing Max Perutz with a polarizing microscope examining thin sections of glacier ice on the Jungfraujoch in 1938. (Photograph by G Seligman.)

In the dining Hall at Peterhouse in October 1936, Max Perutz made friends (at the research students' table) with a highly gifted, modest, gentle young man named John Carter whose father, a retired Indian Civil Servant in Budleigh Salterton (a seaside town in Devon), had made a strange discovery. Could Perutz help to explain it? When walking under the red sandstone cliffs to the west of Budleigh Salterton, Carter Senior had noticed grey balls of rock of varying size protruding from the walls of the cliff. They looked like cannon balls and were surrounded by sharply drawn, circular white haloes where the red sandstone had been bleached. Old Carter had himself decided that the bleached haloes must have arisen from intrinsic radioactivity. In a fascinating account entitled *My First Great Discovery*, published in the *Peterhouse Annual Record* eight years ago,[38] Perutz describes how, notwithstanding the obstructionism of the wily old Lincoln (Rutherford's Sergeant Major-like, miserly lab steward), he established that the nodules contained about 0.5 per cent uranium, concentrated at the borders of dark and light material in the sandstone.

J D Bernal, Perutz's supervisor, thought that the nodules were intriguing enough to be exhibited at the Royal Society Conversazione in June 1937.

Approaching his eightieth birthday, Max Perutz took an interest in Huntington disease, a neurodegenerative disorder caused by abnormal expansions of glutamine repeats in the mutant protein (later called huntingtin).[39] In 1994, Perutz and his colleagues showed that polyglutamine polymers could form β sheets in which glutamines of neighbouring strands are linked by hydrogen bonds between the main chain and side-chain amides. Perutz proposed that proteins with long runs of glutamines would form aggregates harmful to the cell. And, indeed, aggregates of huntingtin have by now been observed as intranuclear inclusions in the brain of patients suffering from Huntington disease. It has not yet, however, been incontrovertibly established that they are themselves the cause of neuronal deaths. Perutz, in a promising collaboration (started in 2000) with A H Windle (a materials scientist), drew parallels between rates of nucleation of crystals with the random occurrence of neuron deaths.[40] Their model explains the finding that the age of the patient at the outset of the disease is related exponentially to the length of the glutamine repeals in the pathological protein. Perutz's very last paper[41] was finished just ten minutes before he was admitted to Addenbrooke's Hospital for emergency surgery.

A memorable incident took place in the Henry Cavendish dining room at Max Perutz's College, Peterhouse, on 31st October 2001. At the instigation of the President of the Royal Society of Chemistry, Professor Steven V. Ley, I organized a small lunch party to celebrate the award to Perutz of Honorary Fellowship of the Society. Around the table all of us were chemists of various complexions. Just as coffee was being served, I asked Max, in a rather public manner, to outline his present research. Those who were privileged to listen to his impromptu, ten-minute exposition of the problems that cause neurodegenerative diseases (Huntington and Alzheimer) will never forget it. First he told us that certain clues were known to some scientists in the early 1930s; then he said that it was less than ten years ago that he himself had tumbled (via so-called "polar zippers") to the enigma of the amyloid proteins. Then he proceeded to proffer the hard experimental evidence that had led him to the structure that he was going to propose. We listened in awe and exhilaration. Here was this gentle, patriarchal, sparkling, eighty-seven-year-old, who had won his Nobel Prize forty years earlier, flanked by his octogenarian friend Fred

Sanger (a double Nobel Laureate), telling us not only a scientifically important, but an absolutely fascinating story—a story which, fortunately for posterity, constitutes his last scientific paper.

Some fourteen weeks later he passed away, having been visited by a succession of his friends and colleagues, lovingly supported by Gisela his wife (to whom he had been married for sixty years) and by his devoted children — Vivien, a historian of art, and Robin, Head of the Department of Chemistry, University of York — grandchildren, and neighbours.

The Expositor and Critic

Max Perutz was an excellent lecturer, with a gift of perceiving how to put his points so that they would be readily grasped by an audience which was not expert in his own line. His lectures had a wonderful lightness of touch and a mesmeric charm, partly because they were interspersed with unforgettable human incidents involving his friends and scientific adversaries. In a memorable lecture, entitled *Science is No Quiet Life*, (organised by the Kelvin Club, the students' scientific society at Peterhouse at which he was a frequent and highly popular speaker, to mark the 60th anniversary of his arrival at the College), he related how a pugnacious American physicist "burst into my room at the LMB, like a gladiator entering the arena in Rome, telling me that the Perutz picture of haemoglobin is wrong."

Figure 11. Max Perutz flanked by Frederick Sanger (*right*) and John Meurig Thomas (*left*). Photograph taken at Peterhouse, Cambridge, 31 October 2001, to celebrate Max Perutz's election as Honorary Fellow of the Royal Society of Chemistry (RSC).

Perutz was also a deeply cultured individual, the breadth and depth of whose knowledge was extraordinary. These qualities, along with his magnanimity and generosity to friends and strangers, and his implacable opposition to any unfairness, were particularly evident in the many popular articles that he wrote in limpid prose for the *New York Review of Books* and the *London Review of Books*, with their frequent allusions to characters, incidents and poems in the works of George Eliot, Tolstoy, Dickens, Shakespeare, Bacon, Donne, Rilke, Hugo, Iris Origo, Lampedusa and Manzoni.

William Blake claimed that only through three human pursuits could one be brought to the edge of eternity: music, poetry, painting. With any one of these, said Blake, one could acquire an ineffable sense of ecstasy. Judging by what Max Perutz has written over the years, I am sure that he would have wished to add science as a fourth such pursuit. He used to say that: Scientific research is an exhilarating and imaginative activity depending on qualities of the human mind that are beyond our comprehension.

The pursuit of science frequently entails the fusion of the aesthetic and the intellectual. Like music, poetry and painting its prosecution demands single-minded devotion. In this context two of Max Perutz other favourite sayings are pertinent:

> Haydn rose early each morning to compose; if ideas failed him, he clasped his rosary and prayed until Heaven sent him fresh inspiration;

and

> Renoir painted every day of his life, and, when old age had made his fingers too arthritic to hold a brush, he got someone to tie his brush to his hand.

Max Perutz spoke with the same respect to young students, and to college and laboratory staff, as he did to Prime Ministers and Royalty. He was a citizen of the world, endowed with a great sense of history and of the continuity of existence. It is easy to appreciate why he so much admired the following passage from the *Memoirs* of the ballerina Dame Margot Fonteyn:

> I cannot imagine feeling lackadaisical about a performance. I treat each encounter as a matter of life and death. The important thing I have learned over the years is the difference between taking one's work seriously and taking oneself seriously. The first is imperative and the second disastrous.

Max Perutz was a wonderful human being: the great can also be good.

Acknowledgements

I am grateful for the constructive comments on the first draft of this article made by R N Perutz, H E Huxley, A M Lesk, N R Plevy, and A Klug.

My Commonplace Book

Van het concert des levens
krijgt niemand een programma
(For the concert of life no-one gets a programme)

(Dutch tile)

Wieviel sind Deine Werke Oh Gott.
Wer fasset ihre Zahl?
(How many are your works, Oh God.
Who can grasp their number?)

Haydn in *The Creation*

About Science and Scientists

It cannot be that axioms established by argumentation can suffice for the discovery of new works, since the subtlety of nature is greater many times than the subtlety of argument.

Sir Francis Bacon

Nature and Nature's laws lay hid in night,
God said: "Let Newton be," and all was light.

Alexander Pope

In experimental philosophy particular propositions are inferred from the phenomena and afterwards rendered general by induction.

Newton in *Scholium to Principia*

Newton, when asked how he made discoveries: "By always thinking about them. I keep the subject contstantly before me and wait until the first dawnings open little by little into the full light."

Quoted by Cyril Hinshelwood, *Nature* **207:** 1057 (1965)

Men that look no further than their outsides, think health an appurtenance unto life, and quarrel with their constitutions for being ill; but I, that have examined the parts of man, and know upon what tender filaments that Fabric hangs, do wonder that we are not always ill; and considering the 1000 doors that lead to death do thank my God that we can die but once.

Sir Thomas Browne, in *Religio Medici* (1643)

Of experiments intended to illustrate a preconceived truth and convince people of its validity: a most venomous thing in the making of sciences; for whoever has fixed on his cause, before he has Experimented, can hardly avoid fitting his Experiment to his cause, rather than the cause to the truth of the Experiment itself.

Thomas Spratt, *History of the Royal Society* (1667)

The utmost effort of human reason is to reduce the principles, productive of natural phenomena, to a greater simplicity, and to resolve many particular effects into a few general causes, by means of reasonings from analogy, experience and observation. But as to the cause of these general causes, we should in vain attempt their discovery; nor should we ever be able to satisfy ourselves by any particular exploration of them.

David Hume, *On Human Understanding*

The initiative for the kind of action that is distinctly scientific is held to come, not from the apprehension of "facts," but from an imaginative preconception of what might be true.

Peter Medawar on Karl Popper's *Hypothetico-deductive Method in Induction and Intuition*

Scientific reasoning is a kind of dialogue between the possible and the actual, between what might be and what is in fact the case.

If we accept falsifiability as a line of demarcation, we obviously cannot accept into science any system of thought (for instance psycho-analysis) which contains a built-in antidote to disbelief: to discredit psycho-analysis is an aberration of thought which calls for psycho-analytical treatment.

Peter Medawar in *Induction and Intuition*

Psycho-analysis is itself the disease of which it purports to be the cure.

Karl Kraus

It is high time laymen recognised the misleading belief that scientific enquiry is a cold dispassionate enterprise, bleached of imaginative qualities, and that a scientist is a man who turns the handle of discovery: for at every level of endeavour scientific research is a passionate undertaking, and the Promotion of Natural Knowledge depends above all upon a sortie into what can be imagined, but is not yet known.

Peter Medawar,
Times Literary Supplement (25 October 1963)

There is a real world independent of our senses; the laws of nature were not invented by man, but forced upon him by the natural world.

Max Planck,
Philosophy of Physics

Wenn ein Problem im Prinzip gelöst ist, ist es noch lange nicht im konkreten Fall gelöst. Neue Erkentnisse werden im Allgemeinen nicht in derartig deduktiver Weise gewonnen, wenn es auch hinterher so aussieht. Zunächst muss der Zufall zur Hilfe kommen. (When a problem is solved in principle, it remains a long way from being solved in practice. New insights are generally not gained in such a deductive manner. First of all chance must come to the rescue.)

Manfred Eigen, Nobel Lecture (1967)

If you go on hammering away at a problem, it seems to get tired, lies down and lets you catch it.

W.L. Bragg

We do not suggest that science invented intellectual honesty, but we do suggest that intellectual honesty invented science.

Jim Erikson, quoted in footnote 42 of the *Rationality of Scientific Revolutions in Rom Harre, Ed. Problems of Scientific Revolutions* (1975)

Leonardo explained that men of genius sometimes accomplish most when they work least.

Vasari, *Lives of the Artists*

Was fruchtbar ist allein ist wahr. (What is fruitful alone is true).

Goethe

When anybody contradicted Einstein he thought it over, and if he was found wrong he was delighted, because he felt that he had escaped from an error, and that now he knew better than before.

Otto Robert Frisch on Einstein

The ardour of my mind is so ungovernable that every object that interests me engages my whole attention, and is pursued with a degree of indefatigable zeal which approaches to madness.

Count Rumford to Pictet (1800)

Der Zwang zum Wissen ist wie eine Trunksucht, wie Liebesverlangen, wie Mordlust, indem sie einen Charakter aus dem Gleichgewicht wirft. Es stimmt doch gar nicht, dass der Wissenschaftler hinter der Wahrheit her ist, Sie ist hinter ihm her. Er leidet darunter. (Thirst for knowledge is like an addiction or a yearning for love or a lust to kill, as it throws a character off balance. It is not true that the scientist goes after the truth. It goes after him. It is something he suffers from).

Robert Musil, *Der Mann ohne Eigenschaften*

The real scientist is ready to bear privation, if need be starvation, rather than let anyone dictate to him which direction his research must take.

Albert Szent-Györgyi in *Science Needs Freedom* (1943)

Noch nie hat die Natur, nach Erklärung eines ihrer wunderbaren Vorgänge, als ein entlarvter Jahrmarktscharlatan dagestanden, der den Ruf des Zaubern-können verloren hat; stets waren die natürlichen ursächlichen Zusammenhänge grossartiger und tiefer ehrfuchtsgebietend als selbst die schönste mythische Deutung. (The scientific explanation of one of Nature's wonderful phenomena has never left it standing like an unmasked sorcerer who has lost his reputation for magic. The real causes are always grander, more awe-inspiring even than the most beautiful myths).

Konrad Lorenz

I don't know what's the matter with people; they don't learn by understanding: they learn some other way, by rote, or something. Their knowledge is so fragile.

Richard Feynman, *You Must be Joking Mr. Feynman*

Es gibt eine eigentümliche Faszination der Technik, eine Verzauberung der Gemüter, die uns daran bringt, zu meinen, es sei ein fortschrittliches und ein technisches Verhalten, dass man alles was technisch möglich ist, auch ausführt. Mir scheint das nicht fortschrittlich, sondern kindisch. (Technology has a peculiar fascination; it casts a spell on people which makes them believe it to be progressive if they put into practice everything that is technically possible. To me this seems not progressive, but childish).

Carl Friedrich von Weizsäcker, *Bedingungen des Friedens*

One concrete example is better than a mountain of prose.

Freeman Dyson, *Scientific American*

Be not the first by whom the new is tried,
Nor yet the last to set the old aside.

Alexander Pope

It was a reaction from the old idea of protoplasm, a name which was a mere repository of ignorance.

J.B.S. Haldane, *Perspectives in Biochemistry* (1938)

If we were compelled to suggest a model we would propose Mother's Work Basket—a jumble of beads and buttons of all shapes and sizes, with pins and threads for good measure, all jostling about and held together by "colloidal forces."

Francis Crick on cytoplasm (1949)

What is called myself is, I feel, done by something greater than myself within me.

James Clerk Maxwell on his death bed

Man soll einen ehrlichen Menschen achten, auch wenn der andere Meinungen hat und vertritt als man selbst. (One should respect an honest person even if he expresses opinions differing from one's own).

Albert Einstein

Ich habe nichts dagegen, wenn Sie langsam denken, Herr Doktor, aber ich habe etwas dagegen, wenn Sie rascher publizieren, als Sie denken. (I don't mind if you think slowly, doctor, but I do mind if you publish faster than you think).

Wolfgang Pauli

A first rate laboratory is one where mediocre scientists can produce outstanding work.

P.M.S. Blackett

La science n'a pas de patrie, mais le savant doit en avoir une. (Science has no fatherland, but the scientist ought to have one).

Louis Pasteur

I am just a chap who messes around in the lab.

Fred Sanger to the Author (14 October 1978)

What is known for certain is dull.

I rarely plan my research; it plans me.

Max Perutz

L'art du chercheur, c'est d'abord de se trouver un bon patron. (The art of the scientist is first of all to find himself a good master).

André Lwoff

I was obliged, at last, to come to the conclusion that the contemplation of nature alone is not sufficient to fill the human heart and mind.

H.W. Bates, *The Naturalist on the Amazon*, p. 274

I cannot understand what makes scientists tick. They are always wrong and they always go on.

Fred Hoyle in *The Black Cloud*

Cosmologists are often in error, but never in doubt.

Lev Landau

Science sans conscience n'est que le ruine de l'âme. (Science without conscience is the ruin of the soul).

Montaigne, *Essays* (1580–1590)

It takes many years of training to ignore the obvious.

The Economist on *Theories of Economic Growth*

In Eurem Kopf liegt Wissenschaft und Irrtum, geknetet innig wie ein Teig zusammen; mit jedem Schnitt gebt ihr mir von Beiden. (In your head sense and nonsense are kneaded closely as in a dough. With each slice you give me both).

Heinrich Kleist, *Der Zerbrochene Krug*

Success in science comes from people, not equipment.

The more fundamental a scientific law the more briefly it can be stated.

We should not worry if students don't know everything, but only if they know everything badly.

Secrecy in science robs it of the main element that keeps it healthy: scientific public opinion.

Errors are many, truth is unique.

Peter Kapitsa

François Jacob: La Souris, La Mouche et l'Homme

It is not knowledge but ignorance that is dangerous.

Like art, science does not copy nature. It recreates it.

Has science promised happiness? It has promised truth and the question is whether one can ever make happiness from truth.

It is not enough to tell the truth. One has to tell the whole truth. Nothing should be kept secret. It is there that the scientist carries the whole responsibility.

Mythical stories may be more in accord with common sense than the utterances of biochemists and molecular biologists.

Daedalus symbolises an evil of our age: the high-flying technician who puts his talents at the disposal of ideologies without bothering about their meaning and values.

In poetry the content may shrink to a point where the aesthetic character of a piece ends by residing entirely in its rhythm, in the music of the words. In science, on the contrary, it is almost exclusively the contents which give value to the work. And the contents of an article or a scientific book can often be summarised in a few sentences.

Lewis Thomas suggested that one should measure the interest of a piece of work by the degree of astonishment it raises.

General questions never lead to more than limited answers. On the contrary, limited questions have often led to more and more general answers.

In the course of time, as one develops expertise, one becomes a kind of prisoner of what one does and what one knows.

Some have extolled the use of frozen sperm from judiciously chosen donors. Some have even praised the sperm of Nobel Prize winners. Only if one does not know Nobel laureates would one want to reproduce them like that.

On test tube babies. For millennia, one has tried to have pleasure without children. One will finally have children without pleasure.

But the extraordinary about the birth of an infant, the wonderful about the appearance of a new human being, is not the nature of receptacle in which the first stage takes place. It would not even be the success of making the whole development take place in a test tube. The incredible is the process itself. It is that the meeting of the sperm with the egg initiates a gigantic series of reactions, hundreds of thousands which follow each other, overlap and cross each other in a network of unbelievable complexity. All this to result, whatever the conditions, in the appearance of a human baby and never a little canary, or giraffe or butterfly.

The human brain may be incapable of understanding the human brain.

Darwin's Letters

The old saying, *vox populi, vox dei* cannot be trusted in science.

Charles Darwin, *Origin of Species,* 6th Edition, p. 134 (1872)

Whether true or false, others must judge; for the firmest conviction of the truth of a doctrine by its author, seems, alas, not to be the slightest guarantee of truth.

Charles Darwin, letter to Lyell (25 June 1858)

I am just beginning to discover the difficulty of expressing one's ideas on paper. As long as it consists solely of descriptions it's pretty easy; but when reasoning comes into play, to make a proper connection, a clearness and a moderate fluency, is to me, as I have said, a difficulty of which I had no idea.

Charles Darwin near the end of the *Voyage of the Beagle*

Talk of fame, honour, pleasure, wealth, all dirt compared to affection.

Charles Darwin to Hooker (1860)

It is seldom that one individual has the power of giving another such a sense of pleasure as you have this day granted me. I know not whether the conviction of being loved, be more delightful or the corresponding one of loving in return.

Charles Darwin to his sister Caroline (6 April 1832)

In short, I am convinced it is a most ridiculous thing to go around the world when by staying quietly, the world will go around you.

Charles Darwin, Letter from South America (1836)

William Shakespeare

There be some sports are painful, and their labour
delight in them sets off: some kinds of baseness
Are nobly undergone, and most poor matters
Point to rich ends.

Ferdinand, in *The Tempest*

Say, is my kingdom lost? why 'twas my care;
And what loss is it to be rid of care?

Richard II to Scroop

Our remedies oft in ourselves do lie,
Which we ascribe to heaven: the fated sky
Gives us free scope; only doth backward pull
Our slow designs when we ourselves are dull.

Helena, in *All's Well that Ends Well*

............: honours thrive
When rather from our acts we them derive
Than our foregoers.

The King, in *All's Well that Ends Well*

Let's take the instant by the forward top;
For we are old, and on our quick'st decrees
The inaudible and noiseless foot of time
Steals ere we can effect them.

The King, in *All's Well that Ends Well*

Thou art not of the fashion of these times,
Where none will sweat but for promotion.

Orlando, in *As You Like It*

The evil that men do lives after them,
The good is oft interred with their bones.

Mark Anthony, in *Julius Caesar*

Striving to better, oft we marr what's well.

Albany, in *King Lear*

Infirmity doth still neglect all office,
Whereto our health is bound; we are not ourselves
When nature, being oppressed, commands the mind
To suffer with the body.

The King, in *King Lear*

Rich gifts wax poor when givers prove unkind.

Ophelia

The language I have learn'd these forty years,
My native English, now I must forego;...

And now my tongue's use is to me no more
Than an unstringed viol or harp;
Or like a cunning instrument cas'd up:
Or, being open, put into his hands
That knows no touch to tune the harmony:
Within my mouth you have engaol'd my tongue.

Mowbray on banishment, *Richard II*

My lord, wise men ne'er sit and wail their woes,
But presently prevent the ways to wail.

Bishop, to Richard II

George Eliot

If you deliver an opinion at all, it is mere stupidity not to do it with an air of conviction, and well-founded knowledge. You make it your own in uttering it and naturally get fond of it.

The Mill on the Floss

Having early had reason to believe that things were not likely to be arranged for her peculiar satisfaction, she wasted no time in astonishment or annoyance at that fact.

The Mill on the Floss

The Vicar did feel then as if his share of duties would be easy. But duty has a trick of behaving unexpectedly–something like a heavy friend whom we have amiably asked to visit us, and who breaks his leg within our gates.

Middlemarch

He was a creature who entered into everyone's feelings and could take the pressure of their thought instead of urging his own with iron insistance.

Middlemarch

Men outlive their love, but they don't outlive the consequences of their recklessness.

Middlemarch

Boris Pasternak: *Dr. Zhivago*

What the Gospels tell us is that there are no nations, only people.

For us the most important fact is that Christ speaks in parables taken from every-day life, that he explains truth in terms of every-day reality. The idea which underlies this is that communion between mortals is immortal, and that the whole of life is symbolic, becasue the whole of it has meaning.

Consciousness is a beam of light directed outwards, it lights the way ahead of us so that we don't stumble.

In order to do good to others he would have needed, besides the principles which filled his mind, an unprincipled heart,–the kind of heart that knows no general cases, but only of particular ones, and has the greatness of small actions.

He considered art was no more a vocation than innate cheerfulness or melancholy were professions. He was interested in physics and chemistry and believed that a man should do something useful in his practical life.

Why don't the intellectual leaders of the Jewish people ever get beyond facile Weltschmerz and irony? Why don't they say to the Jews: "That's enough, stop now. Don't hold on to your identity, don't all get together in a crowd. Disperse. Be with the rest."

I used to be very revolutionary minded, but now I think that nothing can be gained by violence. People must be drawn to good by goodness.

Bertrand Russell

Uncertainty, in the presence of vivid hopes and fears, is painful, but must be endured if we wish to live without the support of comforting fairy tales.

History of Western Philosophy, p. 8

To teach people to live without certainty, and yet without being paralysed by hesitation, is perhaps the chief thing that philosophy, in our age, can still do for those who study it.

History of Western Philosophy, p.11

Contact with those who have no doubts has intensified a thousandfold my own doubts, not about socialism itself, but as to the wisdom of holding a creed so firmly that for its sake men are willing to inflict widespread misery.

The Theory and Practice of Bolshevism

The success of non-violent resistance depends upon certain virtues in those against whom it is employed. When the Indians lay down on railways, and challenged the authorities to crush them under trains, the British found such cruelty intolerable. But the Nazis had no such scruples.

Autobiography Pt. II, p. 192

One should not demand of anybody everything that adds value to a human being. To have some of them is all that should be demanded.

On the Webbs in Portraits from Memory

He had a great love of mankind combined with a contemptuous hatred for most individual men.

On the Webbs in Portraits from Memory

Einstein's… is a kind of simplicity which comes of thinking only about the subject concerned, and forgetting completely its relation to one's own ego.

The Listener (30 April 1959)

Do not fear to be eccentric in opinion, because every opinion now accepted was once eccentric.

In fact, no opinion should be held with fervour. No-one holds with fervour that 7 x 8 = 56, because it is known that this is the case. Fervour is necessary only in commending an opinion which is doubtful or demonstrably false.

Voltaire, quoted by Bertrand Russell

Miscellaneous Authors

Moral precepts

Non nobis nati sumus. (We are not born for ourselves alone).

Cicero, *De Officiis*

The concept of "if I were you", is the fundamental moral concept.

Philip Toynbee, *The Observer* (23 December 1962)

Example is not the main thing in influencing others; it is the only thing.

Albert Schweitzer (23 October 1955)

Love men, slay errors!

St. Augustine

Without humility rectitude is blind and no amount of competence can save it.

The Times on Nevill Chamberlain

Be not forgetful to entertain strangers: for thereby some have entertained angels unaware.

Hebrews XIII: 2

When Thou givest to Thy Servants to endeavour any great matter, grant us also to know that it is not the beginning but the continuing of the same, until it is thoroughly finished, which yieldeth the true glory.

Sir Francis Drake

There is nothing from without a man that entering him can defile him: but the things that come out of him, those are the ones that defile him.

Jesus according to St. Mark, Chapter 8 on Jewish food taboos

Jamais on ne fait le mal si pleinement et si gaiement que quand on le fait par conscience. (Never is evil done as plainly and gaily as when done from conscience).

Pascal

For Montaigue the conduct of the Spaniards in the New Yorld was the supreme example of the failure of Christianity. Hypocrisy and cruelty go together and are unified in zeal.

Judith Shklar, *Daedalus* (Summer 1982)

Great God, how can we possibly be always right and the others always wrong?

Montesquieu, *Cahiers*

Montesquieu believed that "knowledge makes men gentle" just as ignorance hardens them.

Le pittoresque est la misère des autres. (The picturesque is the wretchedness of others.)

Weill

Ohne Freiheit gibt es kein sittliches Handeln: Unfreiheit hebt die höchsten Wertmasse auf. Gib Dich um keinen Preis in die Macht der Menschen. (Without freedom there can be no moral conduct. Bondage does away with the highest values. Never submit yourself to the power of men).

F.A. Kaufmann, quoted by Herbert Peiser

First they came for the Jews and I did not speak out—because I was not a Jew. Then they came for the communists and I did not speak out—because I was not a communist. Then they came for the trade unionists and I did not speak out—because I was not a trade unionist. Then they came for me—and there was no one left to speak out for me.

Pastor Niemöller

A hundred good turns are forgotten more easily than one bad one.

An injury is never forgiven.

Cosimo de' Medici

Men must consider more than the moment when making judgements with Nature.

R.W. Decker in *Encyclopedia Britannica Science Yearbook* (1971)

And he went on talking about himself, not realizing that this was not as interesting to others as it was to him.

Leo Tolstoy, *The Cossacks*

Never lift a foot until you come to the stile.

Proverb

It is nothing to be proud of that your parents are rich enough to keep your hands clean of joyless, killing toil, at an age when many better men are rich in slavery.

T.H. Huxley

History and Politics

Jede Umkehr der Welt hat jene Enterbte,
Denen das Frühere nicht mehr,
Und noch nicht das Nächste gehört
(Each reversal of the world disinherits those
to whom the last no longer
and the next not yet belongs).

Rainer Maria Rilke

It arose from the common fallacy that attributes too much importance to the part played in human affairs by deliberate intent.

Harold Nicholson, *The Observer* (April 1955)

Professor J.Z. Young urged speakers not to be apocalyptic, but to remember that at each stage of history thoughtful people had maintained that the world was in a very unpromising condition.

Nature 1954; **174:** 817

In all negotiations of difficulty, a man may not look to sow and reap at once, but must prepare business, and so ripen it by degrees.

Sir Francis Bacon, *Essay 47, Of Negotiating*

Nations, like people, are inconsistent in their character, which changes according to mood and circumstance and the way they are treated by others.

Max Perutz

Everything is always decided for reasons other than the real merits of the case.

Maynard Keynes about Lloyd George's government,
quoted by Roy Harrod, *Life of Keynes*

The history of the industrial revolution gives no support to the view that a bleak present is a necessary, or even a plausible, preliminary to a glorious future.

Nathan Rosenberg, *How the West Grew Rich*

The Gods' interference does not acquit men of folly; rather it is men's desire for transferring responsibility for folly.

Barbara Tuchman, *The March of Folly*

Those who clamour the loudest for public economy are those for whom public services do the least. It is evident that tax reduction that affects public services has a double effect in comforting the comfortable and afflicting the poor. The philosophy of modern conservatives will not help erase poverty. The modern conservative is in fact, not especially modern. He is engaged, on the contrary, in one of man's oldest pursuits, best financed and most applauded and, on the whole least successful exercises in moral philosophy. This is the search for a truly superior moral justification for selfishness.

Kenneth Galbraith

Liberalism remains the basis of all essential decencies: scepticism, curiosity, love of the individual and the personal, the concept of an inner conscience.

Malcolm Bradbury in the *Independent* (2 December 1990)

It became clear then, and I believe it is clear today, that military force—especially when wielded by an outside power—cannot bring order to a country that cannot govern itself.

Robert McNamara about the Vietnam War (1995)

People are best judged by their actions.

Max Perutz

Literature

Le secret d'ennuyer est celui de tout dire. (The way to bore is to tell everything).

Voltaire

The writer's task was "to make out of the material of the human spirit something which was not there before."

William Faulkner, Nobel Speech (1950)

Originality in literature and art consists in working something original out of something ordinary.

Edmond de Goncourt

Le génie ressemble au balancier qui imprime l'effigie royale aux pièces de cuivre comme aux écus d'or. (Genius resembles the mint-master who imprints the royal effigy on copper coins and gold crowns alike).

Victor Hugo

Without loss there would be no literature.

Günther Grass

I was so long writing my review that I never got around to reading the book.

Why, a four-year-old child could understand this report.
Run out and find me a four-year-old child. I can't make head or tail of it.

Groucho Marx

The writer must cultivate: "the reader over your shoulder."

Robert Graves

Nobody ever acquired strength by publishing someone else's weakness.

The New Yorker (22 December 1952)

Leichter das Falsche zu geisseln,
Als das Echte zu meisseln.
(Easier to castigate error
Than to chisel the truth).

Rainer Maria Rilke

Ile write, because Ile give,
Your critics means to live:
For sho'd I not supply
The Cause, th' effect wo'd die.

Robert Herrick, *To Criticks*

Gaben, wer hätte sie nicht, Talente, Spielzeug für Kinder,
Erst der Ernst macht den Mann, erst der Fleiss das Genie.

(Gifts, who's without them, talents—mere toys,
Seriousness makes the man, application the genius)

Theodor Fontane

This solitary place formed a wild and deserted retreat, full of the force of beauty which touches sensitive souls and appears horrible to others.

Jean Jacques Rousseau about the Alps in *La Nouvelle Héloise*

I cannot imagine feeling lackadaisical about a performance. I treat each encounter as a matter of life and death. The important thing I have learned over the years is the difference between taking one's work seriously and taking oneself seriously. The first is imperative and the second disastrous.

Margot Fonteyn, *Memoirs*

People who tie themselves to systems cannot encompass the whole truth and try to catch it by the tail; a system is like the tail of the truth, but truth is like a lizard: it leaves its tail in your fingers and runs away, knowing that it will grow a new one quickly.

Turgenev to Tolstoy

Flattery slipped off the Prince like water off leaves in a fountain.

Giuseppe Tomasi di Lampedusa, *The Leopard*

Notes and References

Friend or Foe of Mankind?

1. See his speech later accepting the Nobel Prize, in *Les Prix Nobel, 1918 and 1919*. 1920. Nobel Foundation, Stockholm.
2. See Wilhelm Roggersdorf in cooperation with BASF, *In the realm of chemistry*. 1965. Econ Verlag, Düsseldorf and Vienna.
3. A story Lise Meitner told the author of this review many years ago.
4. Haber L.F. 1986. *The poisonous cloud*. Clarendon Press/Oxford University.
5. Haber L.F. 1986. *The poisonous cloud*. p. 27. Clarendon Press/Oxford University.
6. I asked Dr. von Leitner for the source of this statement. She writes that no written record of that fatal row survives. She heard about it from the late A.H. Frucht, a grandson of the first president of the Kaiser Wilhelm Gesellschaft, Adolf von Harnack. Harnack in turn heard it from F. Schmitt-Ott, a friend of Haber's who was the "Kulturminister" in charge of science and learning at the time, to whom Haber had confessed his feelings of guilt. Dr. von Leitner also kindly informed me of James Franck's recorded statements.
7. Mendelsohn K. 1973. *The world of Walter Nernst: the rise and fall of German science*. Macmillan.

8. Hahn O. 1970. *My life.*, Macdonald, London.
9. 1993. *Operation epsilon: the Farm Hall transcripts.* University of California Press.
10. Zyklon B was not used for killing mental defectives, who were not sent to Auschwitz. They were killed in Germany with coal gas.
11. For documentary evidence of these plans, see Fischer. F. (translated by Jackson M.). 1975. *War of illusion: German policies from 1911 to 1914.* Norton.

Splitting the Atom

1. They found that some beta rays were emitted by radioactive nuclei directly, while others were knocked out of the surrounding shell of electrons by gamma rays.
2. Fermi used a glass vial filled with beryllium powder and radon as his neutron source.
3. Hahn O. and Strassmann F. January 6,1939. Uber den nachweis und das verhalten der bei der bestrahlung des urans mittels neutronen entstehenden erdalkalimetalle. *Die Naturwissenschaften* pp. 11–15.
4. Strassmann F. 1938 (privately reprinted in Mainz, 1978) *Kernspaltung.*
5. Frisch O. 1979. *What little I remember.* pp. 115–116. Cambridge University Press.
6. In this equation E stands for energy, m for mass, and c for the velocity of light.
7. Meitner L. and Frisch O.R. February 11, 1939. Disintegration of uranium by neutrons: a new type of nuclear reaction. *Nature* pp. 239–290.
8. Frisch O.R. February 18, 1939. Physical evidence for the division of heavy nuclei under neutron bombardment. *Nature* p. 276.
9. Bohr N. February 15, 1939. Resonance in uranium and thorium disintegration and the phenomenon of nuclear fission. *Physical Reviews* pp. 418–419.
10. This memorandum was later published in Margaret Gowing's book *Britain and Atomic Energy, 1939–1945* (St. Martin's, 1964).
11. Reed T. and Kramish A. November 1996. Trinity at Dubna. *Physics Today* p. 32.
12. Crawford E., Sime R.L., and Walker M. August 1, 1996. A Nobel

tale of wartime injustice. *Nature* pp. 393–395.

13. von Weizsäcker C.F. and Oelering J.H.J. September 26, 1996. Hahn's Nobel was well deserved. *Nature* p. 294.
14. Hahn O. From the natural transmutations of uranium to its artificial fission. In *Nobel lectures in chemistry 1942–1962*. 1964. Elsevier.
15. Hahn O. 1968. *Mein Leben*. F. Bruckmann, Munich.
16. Medawar P. October 25, 1963. *The Times Literary Supplement* p. 850.

The Man Who Patented the Bomb

1. Entropy is a measure of the degree of disorder in a system.
2. Wigner E. (as told to Andrew Szanton). 1992. *The recollections of Eugene P. Wigner* p. 209. Plenum Press.
3. See the excerpts from the transcripts published in the *New York Review*, August 13, 1992. pp. 47–53.

The Threat of Biological Weapons

1. Figures kindly supplied by Mr. John Odling-Smee at the International Monetary Fund in Washington, D.C.
2. Geneva, 1970, p. 76.
3. Matthew Meselson pointed out to me that their fermentors could not have produced more than ninety pounds of dried anthrax spores a day. He suggests that Alibek's two tons refer to the suspension of spores in water before drying.
4. University of California Press, 1999.
5. Lewis G.N., Postol T.A., and Pike J., Why national missile defense won't work. *Scientific American*, August 1999, pp. 36–41.
6. Abelson P.H. 1999. Biological warfare, *Science*, Vol. 286 (November 26), p. 1677.
7. Schulte P., Chemical and biological weapons: Issues and alternatives. *Comparative Strategy*, Vol. 18, No. 4 (October–December 1999), pp. 329–334.
8. Meselson M. 1999. The problem of biological weapons, *Bulletin of the American Academy of Arts and Sciences*, Vol. 52, No. 5, p. 57.
9. A micron is a thousandth of a millimeter, or a twenty-five thousandth of an inch.

10. Tucker J.B., editor, 2000. *Toxic terror: Assessing terrorist use of chemical and biological weapons.* MIT Press, 2000. p. 253.
11. See Peto R., Lopez A.D., Boreham J., Thun M., and Health C. Jr., 1994. *Mortality from smoking in developed countries, 1950–2000: Indirect estimates from vital statistics.* Oxford University Press.

High on Science

1. Medawar P. 1986. *Memoirs of a thinking radish.* Oxford University Press.
2. Medawar coined this phrase in an essay on the naturalist d'Arcy Thompson, I suspect with himself in mind. Medawar P.B. 1958. *The art of the soluble.* p. 21. Methuen, London.
3. Faraday M. 1857. The Bakerian lecture. *Philosophical transactions of the Royal Society* (London) p. 145.
4. Fermi L. 1954. *Atoms in the family.* University of Chicago Press.
5. Medawar P.B. 1973. On 'the effecting of all things possible.' *The hope of progress.* Doubleday.
6. Medawar P.B. Darwin's illness. *The art of the soluble.*
7. Lewontin R.C. June 14, 1990. Fallen angels. *The New York Review.* p. 6. (Emphasis added).
8. Racker E. 1989. A view of misconduct in science. *Nature* **339:** 91.
9. Medawar P.B. The phenomenon of man. *The art of the soluble.*

Deconstructing Pasteur

1. Dubos R.J. 1950. *Louis Pasteur.* Little, Brown.
2. Dubos R.J. 1950. *Louis Pasteur.* p. 343. Little, Brown.
3. *Bulletin de l'Institut Pasteur.* 1985. No. 83, pp. 301–308.
4. Dubos R.J. 1950. *Louis Pasteur.* p. 352. Little, Brown.
5. *Bulletin de l'Institut Pasteur.* 1985. No. 83, pp. 301–308.

The Battle Over Vitamin C

1. Szent-Györgyi A. 1963. Lost in the twentieth century. *Annual Reviews of Biochemistry.* **32:** 1.
2. Szent-Györgyi A. 1928. Observations on the function of the peroxidase systems and the chemistry of the adrenal cortex. *Biochemical Journal* **22:** 1387.

3. According to Kenneth J. Carpenter's scholarly account of the *History of Scurvy and Vitamin C* (Cambridge University Press, 1986), knowledge about the antiscorbutic effect of cabbage, oranges, and lemons goes back a long way. Captain James Cook reported to the Royal Society in 1776 on "The method taken for preserving the health of the crew of His Majesty's ship the Resolution during her late voyage round the world" (1776. *Philosophical Transactions of the Royal Society* **66:** 402). That method included taking sauerkraut, oranges, and lemons. The name vitamin C for the still unidentified antiscorbutic factor was coined by the British biochemist and nutritionist Jack C. Drummond in 1920 (1920. *Biochemical Journal* **14:** 660).
4. King C.G. and Waugh W.A. April 1, 1932. The chemical nature of vitamin C. *Science* p. 357.
5. Svirbely J. and Szent-Györgyi A. April 16, 1932. Hexuronic acid as the antiscorbutic factor. *Nature* p. 574.
6. Szent-Györgyi suspected King of having drafted his letter to *Science* after he had received Svirbely's news, and Moss seeks to corroborate that suspicion by reproducing King's reply to Svirbely, dated March 15, 1922. The letter states: "The product (vitamin C from lemon juice) appears to be identical with S.-G's product, but further chemical work will have to be done to be sure." Moss takes this to mean that King was not yet ready to publish, but pays no attention to a sentence further on: "In a note that should appear in *Science* in a few weeks, I cite your paper (Svirbely's earlier work on lemon juice published in the *Journal of Biological Chemistry*) ...as leading up close to where we are now." This is the note that appeared on April 1. When a scientist refers to a "a note that should appear," he means that it is about to be printed, as King's paper must have been on March 15 in order to come out on April 1, because the lapse of time between submission and publication of a scientific paper was, and still is, a matter of a few weeks at least. The geneticist Thomas Jukes recently recalled that King submitted an abstract with his news to a scientific meeting several weeks before the arrival of Svirbely's letter, probably before the end of February (March 31, 1988. *Nature* p. 390).
7. Szent-Györgyi A. in *Les Prix Nobel* (1933, Stockholm).
8. Szent-Györgyi A. 1976. *Electronic biology and cancer.* Dekker.

A Mystery of the Tropics

1. Garnham P.C.C. and Shortt H.E. 1988. *Biographical Memoirs of Fellows of the Royal Society* **34:** 715.
2. Laveran A. (translated by Martin J.W.).1893. *Paludism: marsh fever and its organism.* New Sydenham Society Publications 146, London.
3. Goldberg D.E., Slater A.F.G., Cerami A., and Henderson G.B. 1990. Hemoglobin degradation in the malaria parasite *Plasmodium falciparum*: an ordered process in a unique organelle. *Proceedings of the National Academy of Sciences of the U.S.A.* pp. 2931–2935.
4. Manson-Bahr P.H. and Alcock A. 1927. *The life and work of Sir Patrick Manson.* Cassell, London.
5. Ross R. 1923. *Memoirs.* p. 128. John Murray, London.
6. Ross R., Sir. 1931. *In exile.* Harrison and Sons, London.

The Forgotten Plague

1. This had already been shown in 1865 by Jean Antoine Villemin who transferred pus from human patients to rabbits which then developed tuberculosis. *Comptes Rendus de l'Académie des Sciences 61* (1865). p. 1012.
2. Dubos R.J. and Dubos J. 1953. *The white plague: tuberculosis, man and society.* Gollancz, London.
3. d'Arcy Hart P. and Sutherland I. 1977. BCG and vole bacillus vaccines in the prevention of tuberculosis in adolescence and adult life. *British Medical Journal* **2:** 293–295.
4. Citron K.M. January 23, 1993. BCG vaccination against tuberculosis: international perspectives. *British Medical Journal* **306:** 222.
5. Colebrook L. 1964. Gerhardt Domagk. *Biographical Memoirs of Fellows of the Royal Sciety* **10:** 39–50.
6. Colebrook L. 1964. Gerhardt Domagk. *Biographical Memoirs of Fellows of the Royal Society* **10:** 40.
7. Feldman W.H. 1964. Streptomycin: some historical aspects of its development as a chemotherapeutic agent in tuberculosis. *American Review of Tuberculosis* **69:** 850–868.
8. Waksman S. 1958. *My life with microbes.* The Scientific Book Club, London.

9. February 19, 1955. *British Medical Journal* **1:** 435–445. The Medical Research Council scientists who planned and organized the trials were Philip d'Arcy Hart, Wallace Fox, Marc Daniels, Dennis Mitchison, and Ian Sutherland. The trial reported in this paper was carried out with 558 patients in 51 hospitals of the National Health Service.
10. Between 1986 and 1990, 46.4 percent of tuberculosis patients in New York failed to complete the prescribed six months of chemotherapy.
11. April 28, 1994. The effect of directly observed therapy on rates of drug resistance and relapse in tuberculosis. *New England Journal of Medicine.* 1179–1184. The same issue also contains a report on a workshop recently held at Rockefeller University on pathogenic bacteria that are resistant to multiple antibiotics, with recommendations for preventing the emergence of such bacteria.
12. See for a full discussion of resistance to antibiotics *Science* (April 15, 1994) pp. 360–393.
13. Bloom B.R. and Murray C.J.L. 1992. Tuberculosis: commentary on a re-emergent killer. *Science* **257:** 1055–1064. This is an excellent, authoritative review on the present situation in the U.S. and elsewhere.

What Holds Molecules Together?

1. Sidgwick N.V. 1929. The electronic theory of valency. p. 88. Oxford University Press.
2. Sidgwick N.V. The covalent link in chemistry. p. 52. Cornell University Press.
3. Linus Pauling: 28 February 1901 – 19 August 1994; Nobel Prize for Chemistry, 1954; Nobel Peace Prize for 1962,1963. Selected publications: *The nature of the chemical bond*, 1939 (3rd edition, 1960); *College chemistry*, 1950 (3rd edition, 1964); *No more war!*, 1958 (revised, 1962). Married to Ava Helen Miller, 1923 (died 1981).

How the Secret of Life Was Discovered

1. Crick F. 1988. *What mad pursuit.* Basic Books Inc., New York.
2. Bernal J.D. and Crowfoot D. 1934. X-ray photographs of crys-

talline pepsin. *Nature* **133:** 794–795.

3. Haldane J.B.S. 1937. The biochemistry of the individual. In *Perspectives in biochemistry* (eds. Needham J. and Green D.E.), pp. 1–10. Cambridge University Press, Cambridge.
4. Hopkins F.G. 1949. Some aspects of biochemistry: the organising capacities of specific catalysts. Second Purser Memorial Lecture 1932. In *Hopkins and biochemistry* (eds. Needham J. and Baldwin E.), pp. 226–242. W. Heffer, Cambridge.
5. Jordan P. 1938. Zur Frage einer spezifischen Anziehung zwischen den Molekülen. *Phys. Z.* **39:** 711–714.
6. Pauling L. and Delbrück M. 1940. The nature of the intermolecular forces operative in biological systems. *Science* **92:** 77–79.
7. Avery O.T. and Heidelberger M. 1923. Immunological relationships of cell constituents of pneumococcus. *J. Exp. Med.* **38:** 81–85.
8. Griffith F. 1928. The significance of pneumococcal types. *J. Hyg.* (Cambridge, UK) **27:** 113–159.
9. Dubos R. 1976. *The professor, the institute and DNA.* Rockefeller University Press, New York.
10. McCarty M. 1985. *The transforming principle.* Norton, New York.
11. Avery O.T., MacLeod C.M., and McCarty M. 1944. Studies on the chemical nature of the substance inducing transformation of pneumococcal types. Induction of transformation by a deoxyribonucleic acid fraction isolated from pneumococcus type III. *J. Exp. Med.* **79:** 137–158.
12. Olby R. 1974. *The path to the double helix.* Macmillan, London.
13. Dale H. 1946. Presidential address. *Proc. R. Soc. A.* **185:** 127–243.
14. Hotchkiss R.D. 1951. Transfer of penicillin resistance in pneumococci by the deoxyribonucleate derived from resistant cultures. *Cold Spring Harbor Symp. Quant. Biol.* **16:** 457–460.
15. Hershey A.D. and Chase M. 1952. Independent functions of viral protein and nucleic acid in growth of bacteriophage. *J. Gen. Physiol.* **36:** 39–52.
16. Davidson J.N. 1950 and 1953. *The biochemistry of the nucleic acids.* Methuen, London.
17. Davidson J.N. 1960 and 1963. *The biochemistry of the nucleic acids.* Methuen, London.

A Passion for Crystals

1. Dorothy Mary (Crowfoot) Hodgkin: Chemist; Born in Cairo on May 12, 1910 and died in Shipston-on-Stour, Warwickshire on July 29, 1994; Fellow, Somerville College, Oxford 1936–77; FRS 1947; Royal Society Wolfson Research Professor, Oxford University 1960–1977 (Emeritus); Nobel Prize for Chemistry 1964; OM 1965; Chancellor, Bristol University 1970–88; Fellow, Wolfson College, Oxford 1977–82. Married to Thomas Hodgkin, 1937 (died 1982). Mother of two sons and one daughter.

The Top Designer

1. The synthesis of ATP stores the chemical energy that comes from the burning of sugars to carbon dioxide and water. For example, the burning of one molecule of glucose provides energy for the synthesis of about 30 molecules of ATP. In muscle ATP is split. That splitting liberates the energy needed for the muscle's contraction.
2. These and other quotations are my translations from the original French text.

By What Right Do We Invoke Human Rights?

1. Paper read at an International Meeting of Scientific Academies on Human Rights, held at the Royal Netherlands Academy of Arts and Sciences in Amsterdam on 11 and 12 May 1995.
2. Council of Europe. 1990. *Human rights in international law*. Strasbourg.
3. Lauterpacht H. 1945. *An international bill of the rights of man*. New York.
4. Lauterpacht H. 1950. *International law and human rights*. Stevens and Son Ltd., London.
5. Thucydides (translated by Benjamin Jowett). 1960. *The Peloponnesian War*. p. 114. Bantam Books, New York.
6. Aurelius M. *Meditations*. Penguin Classics, Harmondsworth, Middlesex.
7. Stephenson C. and Marcham F.G. 1937. *Sources of English constitutional history*. Harper, New York.

8. By propriety was meant property in its widest sense.
9. Stirk P. and Weigall D. 1995. *An introduction to political ideas.* Pinter, London.
10. Overton R. 1646. *An arrow against all tyrants.* pp. 3–4. London.
11. Paine T. *The rights of man.* (quoted in reference 9).
12. Montesquieu C. Baron de. 1956. *L'Esprit des lois.* Volume 1: pp. 162, 164, 198, 208, 258. Edition Garnier Frères, Paris.
13. Condorcet Marquis de. 1934. *Esquisse d'un tableau historique des progrès de l'esprit humain.* Boivin et Cie, Paris.
14. Mill J.S. 1989. *On liberty, with the subjection of women, and chapters on socialism.* Cambridge University Press, Cambridge.
15. Gewirth A. 1982. *Human rights.* University of Chicago Press, Chicago.
16. Signatories to the U.N. Convention against Torture 1984. Since this article was written, Turkey and Israel have outlawed torture, though according to the London *Independent,* "physical pressure" is still permitted in Israel.

The Right to Choose

1. Baird D. 1965. A fifth freedom. *British Medical Journal* **2:** 1141.

Swords into Ploughshares

1. Gowing M. 1974. *Independence and deterrence: Britain and atomic energy 1945–1952.* Cambridge University Press.
2. *Accident at Windscale no. 1 pile on 10 October 1957.* 1957. HM Stationery Office, Cmnd. 302, London.
3. Hogle J.E. 1983. *Biological effects of radiation.* Taylor and Francis, London.
4. Crick M.J. and others. November 1982; Addendum, September 1983. *An assessment of the radiological impact of the Windscale reactor fire.* HM Stationery Office, National Radiological Protection Board - R135. London.
5. Yalow R.S. 1988. Biological effects of low-level radiation. In *Science, politics and fear* (M.E. Burns, editor). Lewis Publishers.
6. Wilson R. 1986. Chronology of a catastrophe; What really went wrong. *Nature* **223:** 29.
7. In addition, the dose of radiation absorbed is multiplied by weighting factors recommended by the International Commission

for Radiological Protection to take account of different degrees of risk posed by different radiations when they are absorbed by different organs: for example, ultraviolet radiation can cause skin cancer; radon, cancer of the lungs; or gamma rays hitting the gonads can cause genetic damage. Accordingly the weighting factor for radon in houses would be that fraction of the total absorbed radiation that is absorbed by the lungs. One sievert is equivalent in energy to about one quarter of a calorie of heat.

8. Hall E.J. 1988. *Radiobiology for the radiologist.* J.B. Lippincott Company.
9. Webb G.A.M. November 1979. Quantities used in radiological protection—an explanation. Supplement to *Radiological Protection Bulletin.*
10. Eakins J.D. and others. 1981. *Studies of environmental radioactivity in Cumbria part 2: radionuclide deposits in soil in the coastal region of Cumbria.* H.M. Stationery Office, Atomic Energy Research Establishment - R9873. London.
11. Eakins J.D. and others.1982. *Studies of environmental radioactivity in Cumbria part 5: the magnitude and mechanism of enrichment of sea-spray with actinides in West Cumbria.* H.M. Stationery Office, Atomic Energy Research Establishment - R10127. London.
12. One curie is equivalent in radioactivity to one gram of radium which undergoes 3.7×10^{10} atomic disintegrations per second. The international unit for small quantities of radioactivity is the becquerel (after its discoverer, the Frenchman Henri Becquerel), equivalent to one atomic disintegration per second.
13. Hetherington J.A., Jefferies D.J., Mitchell N.T., Pentreath R.J., and Woodhead D.S. 1976. *Environmental and public health consequences of the controlled disposal of transuranic elements to the marine environment.* Proceedings of symposium on transuranic nuclides and the environment. International Atomic Energy Agency. Vienna.
14. Hunt G.J. 1988. Radioactivity in surface and coastal waters of the British Isles, 1987. In *Aquatic Monitoring Report*; Ministry of Agriculture, Fisheries and Food; Lowestoft.
15. 1971. Effect of ionizing radiation on aquatic organisms and ecosystems. International Atomic Energy Agency; *Technical Report Series 172.* Vienna.

16. Independent Advisory Group (Black D. Sir, chairman). 1984. *Investigation of the possible increased incidence of cancer in West Cumbria.* HM Stationery Office. London.
17. Stather J.W. and others. 1984, Addendum 1986. *The risk of leukemia and other cancers in Seascale from radiation exposure.* HM Stationery Office NRPB-R171. London.
18. Committee on Medical Aspects of Radiation in the Environment (COMARE; Bobrow M., chairman). 1986. *The implications of the new data on the releases from Sellafield in the 1950s for the conclusions of the report on the investigation of the possible increased incidence of cancer in West Cumbria.* HM Stationery Office. London.
19. Committee on Medical Aspects of Radiation in the Environment (COMARE; Bobrow M., chairman).1988. *Investigation of the possible increased incidence of leukemia in young people near the Dounreay nuclear establishment, Caithness, Scotland.* HM Stationery Office. London.
20. Gardner M.J., Hall A.J., Downes S., and Terrell J.D. 1987. Follow up study of children born elsewhere but attending schools in Seascale, Cumbria (schools cohort). *British Medical Journal* **295:** 819.
21. Gardner M.J., Hall A.J., Downes D., Terrell J.D. 1987. Follow up study of children born to mothers in Seascale, Cumbria (birth cohort). *British Medical Journal* **295:** 822.
22. Forman D., Cook-Mozaffari P., Darby S., Davey G., Stratton I., Doll R., and Pike M. 1987. Cancer near nuclear installations. *Nature* **329:** 499.
23. Black D. *Investigation of the possible increased incidence of cancer in West Cumbria.*
24. Kinlen L. 1988. Evidence for an infective cause of childhood leukemia: comparison of a Scottish new town with nuclear reprocessing sites in Britain. *Lancet* **11:** 1323.
25. House of Commons. 1986. *First report of the environment committee: session 1985–86.* pp. 226–229. HM Stationery Office. London.
26. Clarke R.H. and Southwood T.R.E. 1989. Risks from ionizing radiations. *Nature* **338:** 197.
27. Hughes J.S., Shaw K.B., and O'Riordan H.C. 1987. *The radiation exposure of the UK population.* Chilton: National Radiological Protection Board - R227.

28. Hughes J.S. and Roberts G.C. 1984. *The radiation exposure of the UK population -1984 review.* Chilton: National Radiological Protection Board - R173.
29. Pentreath R. J. and others. 1984. *Impact on public radiation exposure of transuranium nuclides discharged in liquid wastes from fuel element reprocessing at Sellafield, United Kingdom.* Proceedings of the International Atomic Energy Agency Conference on Radioactive Waste Management - IAEA-CN-43/32. Vienna.
30. Crick M.J. and others. *An assessment of the radiological impact of the Windscale reactor fire.*
31. Sharpe J.A. 1987. *Early modern England: a social history, 1550–1760.* Edward Arnold, London.
32. Laslett P. 1983. *The world we have lost,* 3rd edition. Methuen, London.
33. Perutz M. 1989. *Is science necessary?* E.P. Dutton.
34. U.S. Department of Commerce. 1960. *Historical statistics of the United States: colonial times to 1957.* Washington, D.C.
35. Office of Population Census and Surveys. *Mortality Statistics.* 1975. HM Stationery Office. London.
36. United Nations Scientific Committee on the Effect of Atomic Radiation. 1982. *Ionizing radiation: sources and biological effects.* Table 6, p. 321. United Nations, Report to the General Assembly.

What If?

1. Reynaud P. 1963. *Mémoires: Envers et contre tous,* Vol. 2. Flammarion, Paris.
2. Roberts A. 1991. *The holy fox: A biography of Lord Halifax.* Weidenfeld and Nicolson, London.
3. See Karl-Heinz Frieser. 2000. *Blitzkrieg-Legende: Der West-Feldzug 1940* (Operationen des Zweiten Weltkriegs II, 2nd edition, Munich, 1996). Reviewed by Tobias Jersak, *Historical Reviews,* Vol. 43, No. 2, pp. 565–582.
4. May E.R. 2000. *Strange victory: Hitler's conquest of France,* Hill and Wang, p. 460; reviewed by Tony Judt, *New York Rev. Books,* February 22, 2001.
5. *The diaries of Sir Alexander Cadogan 1938–1945.* 1971. Edited by David Dilts. Cassell, London.
6. Charmley J. 1992. *The end of glory: A political biography.* John Cur-

tis/Hodder and Stoughton, London.

7. *The Goebbels Diaries 1939–1941*. 1983. Translated and edited by Fred Taylor. Putnam. Entry for June 25, 1940, pp. 123–124.
8. Roberts, *The holy fox*, p. 72.
9. *The labyrinth: Memoirs of Walter Schellenberg, Hitler's chief of counterintelligence*, 1956. Translated by Louis Hagen. Harper.
10. *The Shorter Oxford English Dictionary* defines "rhodomontade" as "extravagantly boastful or arrogant saying or speech."
11. *Speaking for themselves: The personal letters of Winston and Clementine Churchill*, 1998. Edited by their daughter Mary Soames. Doubleday, p. 454.
12. Extract from Churchill's *War memoirs*, quoted by Lukacs.
13. Hugh Dalton, *A labour minister's memoirs, the fateful years*. 1957. Friedrich Muller Ltd., London.
14. P.M.H. Bell. *A certain eventuality...: Britain and the fall of France* 1974. Saxon House, Farnborough.

How W.L. Bragg Invented X-ray Analysis

1. Bragg W.L. 1933. *The crystalline state*. G. Bell & Sons, London.
2. Bragg W.L. 1937. *Atomic structure of minerals*. Cornell University Press, Ithaca, New York.
3. Bragg L., Sir and Claringbull G.F. 1965. *Crystal structure of minerals*. G. Bell & Sons, London.
4. Bragg L., Sir. 1975. *The development of X-ray analysis*. G. Bell & Sons, London.
5. Caroe G.M. 1978. *William Henry Bragg 1862–1942: Man and scientist*. Cambridge University Press, Cambridge.
6. Friedrich W., Knipping P., and Laue M. 1912. *Interferenz-Erscheinungen bei Röntgenstrahlen. Sitzungsber. Kgl. Bayrischen Akad. Wiss.*, pp. 303–322.
7. Bragg W.L. 1913. The diffraction of short electromagnetic waves by a crystal. *Proc. Cambr. Philos. Soc.* **17:** 43–57.
8. Bragg W.L. 1913. The structure of some crystals as indicated by their diffraction of X-rays. *Proc. R. Soc. Lond. A* **89:** 248–277.
9. Bragg W.L. 1914. The analysis of crystals by the x-ray spectrometer. *Proc. R. Soc. Lond. A* **90:** 468–489.
10. Bragg W.L. 1914. The structure of copper. *Philos. Mag.* **27:** 355–360.

11. Bragg W.H. and Bragg W.L. 1913. The structure of diamond. *Proc. R. Soc. Lond. A* **89:** 272–291.
12. da C. Andrade E.N.C. 1943. William Henry Bragg: Obituary. *Notices of Fellows of the Royal Society* **4:** 277–300.
13. Bragg W.L. and Warren B. 1928. The structure of diopside, $CaMg(SiO_3)_2$. *Z. Kristall.* **69:** 168–193.
14. Bragg W.L. and West J. 1928. A technique for the x-ray examination of crystal structures with many parameters. *Z. Kristall.* **69:**118–148.
15. Bragg W.L. 1929. The determination of parameters in crystal structures by means of Fourier series. *Proc. R. Soc. Lond. A* **123:** 537–559.
16. Bragg W.H. 1915. X-rays and crystal structure. *Philos. Trans. R. Soc. Lond. A* **215:** 253–275.
17. Havighurst R.J. 1927. Electron distribution in the atoms of crystals. Sodium chloride, and lithium, sodium and calcium fluoride. *Physiol. Rev.* **29:** 1.
18. Cork J.M. 1927. The crystal structure of some of the alums. *Philos. Mag.* **4:** 688–698.
19. Medawar B.P. 1979. *Advice to a young scientist.* Harper & Row, New York.
20. Kuhn T.S. 1970. *The structure of scientific revolutions.* University of Chicago Press, Illinois.
21. Phillips D., Sir. 1979. William Lawrence Bragg, 31 March 1890–1 July 1971. *Biogr. Mem. Fell. R. Soc.* **25:** 75–143.

Growing Up among the Elements

1. Mendeleev D. *Principles of chemistry.* 1905, translated by George Kamensky and Thomas H. Pope. Green and Co., London.
2. Thomas L. 1984. *The youngest science: Notes of a medicine watcher.* 1984. Oxford University Press.
3. Easton, Pa.: *Journal of Chemical Education*, 1945.
4. Jacob F., *The statue within.* 1988. Basic Books. p. 15.

The Legacy of Max Perutz (1914–2002)

1. Sage was J.D. Bernal (1901–1971), see page 241, Anglo-Irish crystallographer, who graduated in the University of Cambridge before proceeding to work as a research student of W.H. Bragg at

the Davy Faraday Research Laboratory of The Royal Institution, London, where he solved the structure of graphite. He was later appointed Assistant Director of Research at the Cavendish Laboratory, Cambridge, where Dorothy Crowfoot (later Hodgkin), and A F Wells of Oxford, Isadore Fankuchen, of Brooklyn Polytechnic, and Max Perutz, joined him. He later became Professor of Physics at Birkbeck College, University of London. During World War II he led a team of operational research and other work of national importance. John Kendrew rubbed shoulders with him during duties in the jungles of Sri Lanka. Max Perutz said that Bernal was the most brilliant conversationalist he had ever met, and that he was a restless genius always searching for something more important to do than the work of the moment. A lifelong member of the British Communist Party, he was much admired by an enormous circle of influential people. Earl Mountbatten of Burma, to whom he reported in the war years, said of him: "Desmond Bernal was one of the most engaging personalities I have ever known... his most pleasant quality was his generosity. He never minded slaving away at other people's ideas, helping to decide what could or could not be done, without himself being the originator of any of the major ideas on which he actually worked."

Bernal gave laboratory space to Aaron (later Sir Aaron) Klug, Nobel Laureate in Chemistry, 1982, and Rosalind Franklin, while at Birkbeck College. Bernal is widely acknowledged as one of the major pioneers of molecular biology. He and Dorothy Crowfoot published in 1934 (*Nature 130: 794*) the X-ray diffraction photographs of crystals of the enzyme pepsin (in its wet state). Up to then, many felt that large biological molecules had no well-defined structure, certainly not in solution, and believed that they resembled spaghetti with intertwined strands of variable length, bent and folded so as to be difficult to disentangle physically and to defy structural description. Bernal and Crowfoot, however, pointed out that "from the intensity of the more distant (X-ray diffraction) spots, it can be inferred that the arrangement of atoms inside the protein molecules is also of a perfectly defined kind."

2. Perutz sought entry as a graduate student to Trinity, King's, Gonville and Caius and St John's Colleges, but he was turned down. In desperation he asked the crystallographer, W A Wooster,

an associate of Bernal, for advice. He told him: "Why not choose the college with the best food, which is the college of which I am a member?" And that is how he came to join Peterhouse (where the food is still among the best in Cambridge).

3. The Medical Research Council (MRC) is one of Britain's most successful scientific seedbeds. It was set up for practical reasons as a consequence of the 1911 National Insurance Act, when David Lloyd George, as Chancellor of the Exchequer, singled out tuberculosis (TB) as a problem needing special attention. In Great Britain and Ireland this dreadful disease was responsible for one in three deaths among males aged between fifteen and forty-four, and half the deaths among females aged fifteen to twenty-four. Germany was making great strides through the building of TB institutions. The Chancellor of the Exchequer felt that something had to be done in Britain. The First World War brought out many other pressing medical problems as well—wound infections (especially tetanus and gangrene), typhoid, cholera, dysentery, civilian malnutrition and even TNT poisoning in munitions factories. The MRC worked successfully on many of these problems, and when the war ended the first Secretary of the MRC, Sir Walter Fletcher, was able to argue that its work should not be restricted to any particular area such as TB, for which the funds were originally committed. Fletcher defined the primary function of the MRC as promoting fundamental scientific research, since this was essential to the development of clinical treatment. He found a ready ally in the Prime Minister of the day, David Lloyd George, who had been the original architect of the National Insurance Act. This is how the MRC was founded: it has become one of the great jewels in the crown of British science.
4. Of which they were each Honorary Fellows. Perutz, although a member of Peterhouse, with dining rights and some other privileges ever since he entered as a graduate student in 1936, was not made a Fellow until he won the Nobel Prize in 1962. Kendrew, on the other hand, from the time he entered the college in 1947 as a College Lecturer, Official Fellow and Director of Studies in Natural Sciences, played a leading part in the academic, social and administrative life of the College for twenty-seven years. He served successively as Librarian, Wine Steward, Steward and Cura-

tor of the College's paintings and portraits.

5. Charles Babbage (1792–1871) is best known as the inventor of the so-called "Difference Engine", a calculating machine that is accepted as the first computer — see *Charles Babbage and his Calculating Engines* by Doron Swade, Science Museum, London, 1991. (ISBN 0901805499). The inventive Babbage once remarked: "All of chemistry, and with it crystallography, would become a branch of mathematical analysis which, like astronomy taking its constants from observation, would enable us to predict the character of any new compound and possibly the source from which its formation may be anticipated."
6. Inventor of the jet engine. Other distinguished living members include the Nobel Laureates A J P Martin (Chemistry 1952) and Aaron Klug (Chemistry 1982).
7. Dickerson R E. 1992. *Protein Science* **1:** 182–186.
8. The University of the Third Age (U3A), an organization that began in France, caters for retired people, and there are now branches in many British (especially university) towns and cities.
9. Rhodonite is a manganese-rich silicate of the pyroxeroid family $(Mn,Ca,Fe)SiO_3$.
10. Perutz subsequently discovered that Rutherford wanted to throw Bernal out of the Cavendish but was restrained from doing so by W L (later Sir Lawrence) Bragg, who became Rutherford's successor at Manchester and at Cambridge. Perutz used to say that "Had Bragg not intervened, Bernal's pioneering work in molecular biology would not have started, John Kendrew and I would not have solved the structure of proteins, and Watson and Crick would never have met."
11. When Hitler invaded Austria, the family business was expropriated, and his parents became refugees. Perutz brought his parents to Britain, but he and his father were interned in 1940. After his release, his father, who had never worked with his hands before, took a manual job (as a lathe operator in Letchworth) to help the war effort.
12. Who also joined the Perutz-Kendrew team in 1948.
13. Huxley H.E. 2002. *Nature.* **415:** 851-52.
14. Perutz, after six years of labour extracting Patterson maps (which consisted of some 25 million lines between the thousands of

atoms in the haemoglobin molecule, felt[15] "elated when they seemed to tell me that the molecule consists simply of bundles of parallel chains of atoms spaced apart at equal intervals. Shortly after my results appeared in print, a new graduate student joined me. As his first job, he performed a calculation which proved that no more than a small fraction of the haemoglobin molecule was made up of the bundles of parallel chains that I had persuaded myself to see, and that my results, the fruits of years of tedious labour, provided no other clue to its structure. It was a heartbreaking instance of patience wasted, an ever-present risk in scientific research. That graduate student was Francis Crick, later famous for his part in the solution of the structure of DNA."

15. Perutz M.F. 1998. *I wish I'd made you angry earlier* (Cold Spring Harbor Laboratory Press, p. xi).
16. It is a fortunate fact that complex molecules such as haemoglobin take no more notice of the isomorphous attachment of a heavy atom than (to use Sir Lawrence Bragg's words) "*a Maharaja's elephant would of the gold star painted on its forehead*".
17. The picture was crude, but two years later, using the linear diffractometer devised and built by Arndt and Phillips[18] at the Davy Faraday Research Laboratory, London (where Perutz and Kendrew were Honorary Readers, 1954–68), a much sharper picture of myoglobin, in all its glorious complexity, was obtained, with the identities of the amino acid residues clearly discernible.
18. Dr U A Arndt later moved from the Davy Faraday Laboratory to the LMB in Cambridge, and Dr D C Phillips (later Sir David, then Lord Phillips of Ellesmere, 1925–1999) was appointed Professor of Molecular Biophysics in the University of Oxford.
19. Kendrew J.C., Bodo G., Dintzis H.M., Parrish R.G., Wyckoff H.W., and Phillips D.C. 1958. *Nature.* **181:** 662–666.
20. Perutz M.F., Rossman M.G., Cullis A.F., Muirhead H., Will G., and North A.C.T. 1960. *Nature.* **185:** 416–422.
21. Perutz loved telling the story of how, on hearing so many people predict that he and Kendrew would be awarded the Nobel Prize, one day an excited secretary brought in to them an important-looking envelope bearing an unusual stamp, and they each thought: "This is it, news of the Prize". But the letter had been sent

from the Pontifical Academy urgently requesting their reprint order forms, duly completed!

22. For his work on the primary structure of proteins, especially that of insulin.
23. For elucidating the enzymatic mechanism underlying the synthesis of adenosine triphosphate (ATP).
24. Blow D.M. *The Independent* (London), 7 February 2002.
25. Eisenberg D. 1994. *Protein Science.* **3:** 1625.
26. In his *Commonplace Book*, Perutz has enumerated his own: "People are best judged by their actions," and that of Albert Schweitzer: "Example is not the main thing in influencing others; it is the only thing."
27. Holmes K.C., *Biographical memoirs of fellows of the Royal Society*, 2001. **47:** 311–332. Kendrew's own interest in fundamental research began to wane in the mid 1960s as he gradually turned his brilliant mind to matters of policy. He had already served as Deputy Chief Scientific Advisor to the Ministry of Defence (1960–1965); and then he became Chairman and Secretary General of the International Council of Scientific Unions (ICSU), centred in Paris, a body of which he became President, 1988–1990. From 1969 to 1975, Kendrew was Secretary-General of the EMBO, and in 1974 he became the (first) Director General of EMBL at Heidelberg, where a splendid new building was opened in May 1978.
28. Perutz M.F. 1971. *New Scientist*, pp. 676–679.
29. In some cases, the binding of a molecule to the protein produces little conformational change. Sperm whale myoglobin[30] is good example: The oxy and deoxy structures may be superimposed almost exactly.
30. Lesk A.M. 1991. *Protein architecture: A practical approach*, Oxford University Press, p. 121.
31. Fermi G. and Perutz M.F. 1981. "*Atlas of molecular structures in biology: Haemoglobin and myoglobin.*" Clarendon Press, Oxford. The aminoacid sequences in the α- and β- subunits of the haemoglobin in the following creatures are enumerated in this Atlas: man, rhesus monkey, orang-utan, slow loris, tupai, savannah monkey, capuchin monkey, hanuman langur, spider monkey, rabbit, dog, horse, cat, pig, camel, llama, Indian elephant, opossum,

rat, chicken, grey lag goose, carp, goldfish, caiman, Nile and Mississippi crocodile, tadpole and shark.

32. Andersen H.T. 1961. *Acta Physiol. Scand.* **53:** 24–45.
33. Hennakas N., Miyazaki G., Jame J., and Nagai K. 1995. *Nature.* **373:** 244–246.
34. Perutz M.F. 1992. *Faraday Discuss.* **93:** 1–11.
35. Private communication from Sir Aaron Klug to M.F. Perutz, July 1993.
36. Neither of these two scientists was given to boasting, idle or otherwise. But the story is undoubtedly true and was told in Max Perutz Royal Institution Discourse in 1990 entitled *How W L Bragg invented X-ray Analysis* published in *The Legacy of Lawrence Bragg* (Ed. J.M. Thomas and D.C. Phillips), 1990.
37. Perutz M.F. 1953. *Nature.* **172:** 929–932.
38. Perutz M.F. *Peterhouse Annual Record, 1994–1995*, pp. 15–17.
39. Klug A. 2002. *Science* **295:** 2382–2383.
40. Perutz M.F .and Windle A.H. 2001. *Nature.* **412:** 143–144.
41. Perutz M.F., Finch J.T., Berriman J., and Lesk A.M. 2002. *Proc. Natl. Acad. Sci.* **99:** 5591–5595.

Subject Index

H

O